Lecture Notes in Artificial Intelligence   5949

Edited by R. Goebel, J. Siekmann, and W. Wahlster

Subseries of Lecture Notes in Computer Science

T0189721

Jacky Baltes   Michail G. Lagoudakis
Tadashi Naruse   Saeed Shiry Ghidary (Eds.)

RoboCup 2009: Robot Soccer World Cup XIII

 Springer

Series Editors

Randy Goebel, University of Alberta, Edmonton, Canada
Jörg Siekmann, University of Saarland, Saarbrücken, Germany
Wolfgang Wahlster, DFKI and University of Saarland, Saarbrücken, Germany

Volume Editors

Jacky Baltes
University of Manitoba, Department of Computer Science
Winnipeg, Manitoba R3T 2N2, Canada
E-mail: jacky@cs.umanitoba.ca

Michail G. Lagoudakis
Technical University of Crete, Intelligent Systems Laboratory
Department of Electronic and Computer Engineering
73100 Chania, Greece
E-mail: lagoudakis@intelligence.tuc.gr

Tadashi Naruse
Aichi Prefectural University, School of Information Science and Technology
Nagakute-cho, Aichi 480-1198, Japan
E-mail: naruse@ist.aichi-pu.ac.jp

Saeed Shiry Ghidary
Amirkabir University of Technolgoy
Computer Engineering and Information Technology Department
Hafez Avenue, Tehran 15914, Iran
E-mail: shiry@ce.aut.ac.ir

The photo for the cover illustration was taken by David Kriesel;
copyright by University of Bonn.

Library of Congress Control Number: 2010920867

CR Subject Classification (1998): I.2, C.2.4, D.2.7, H.5, I.5.4, J.4

LNCS Sublibrary: SL 7 – Artificial Intelligence

ISSN	0302-9743
ISBN-10	3-642-11875-5 Springer Berlin Heidelberg New York
ISBN-13	978-3-642-11875-3 Springer Berlin Heidelberg New York

springer.com

© Springer-Verlag Berlin Heidelberg 2010
Printed in Germany

Typesetting: Camera-ready by author, data conversion by Scientific Publishing Services, Chennai, India
Printed on acid-free paper   SPIN: 12842797   06/3180   5 4 3 2 1 0

Preface

We are really pleased to present the proceedings of the RoboCup International Symposium 2009. The 13th RoboCup Symposium was held in conjunction with the RoboCup 2009 competition in Graz, Austria from June 29 to July 5, 2009.

The symposium highlights the many research contributions and achievements of the RoboCup community. Contributions from all leagues (RoboCupSoccer, RoboCupRescue, RoboCup@Home, and RoboCupJunior) are included in the symposium. The papers published in these proceedings are not limited to practical issues, but also include fundamental research, system evaluation, and robotics education topics.

There were 112 submissions from 25 countries. All papers were carefully reviewed by an international Program Committee of 96 members, who were assisted by 25 additional reviewers. Each paper was reviewed by three Program Committee members and all reviews were carefully considered and discussed by the Symposium Co-chairs, who made the final decisions. The review process was extremely selective and many good papers could not be accepted for the final program. Out of the 112 submissions, 22 papers were selected for oral presentations, whereas 17 papers were selected for poster presentations. Two of these papers were recognized for their outstanding quality. Specifically, Mohsen Malmir and Saeed Shiry received the best paper award for their paper titled "Object Recognition with Statistically Independent Features: A Model Inspired by the Primate Visual Cortex" and Shivaram Kalyanakrishnan and Peter Stone received the best student paper award for their paper titled "Learning Complementary Mutliagent Behaviors: A Case Study".

In addition to the paper and poster presentations, which cover the state of the art in a broad range of topics central to the Robocup community, we were delighted to welcome a number of distinguished invited speakers (Auke Ijspeert of the Ecole Polytechnique Fédérale de Lausanne, Switzerland, Ulises Cortés of the Technical University of Catalonia, Spain, Silvia Coradeschi of Orebro University, Sweden, and Robin Murphy of the Texas A&M University, USA). These talks were complemented by a number of additional talks by leading robotics researchers and representatives of all RoboCup leagues.

We would like to take this opportunity to thank the Program Committee members and the external reviewers for their hard work and, of course, all the authors for their contributions! Furthermore, we would like to thank the authors of the Easychair system, which was used to manage the submission and publication process, for their excellent work.

November 2009

Jacky Baltes
Michail G. Lagoudakis
Tadashi Naruse
Saeed Shiry

Organization

Symposium Chairs

Jacky Baltes	University of Manitoba, Canada
Michail G. Lagoudakis	Technical University of Crete, Greece
Tadashi Naruse	Aichi Prefectural University, Japan
Saeed Shiry	Amirkabir University of Technology, Iran

Program Committee

Carlos Acosta	Singapore Polytechnic, Singapore
H. Levent Akın	Bogaziçi University, Turkey
Luis Almeida	University of Aveiro, Portugal
Francesco Amigoni	Politecnico di Milano, Italy
John Anderson	University of Manitoba, Canada
Sven Behnke	University of Bonn, Germany
Andreas Birk	Jacobs University, Germany
Andrea Bonarini	Politecnico di Milano, Italy
Ansgar Bredenfeld	Fraunhofer IAIS, Germany
Ramon Brena	Tecnologico de Monterrey, Mexico
Hans-Dieter Burkhard	Humboldt University Berlin, Germany
Vincenzo Caglioti	Politecnico di Milano, Italy
Stefano Carpin	University of California, Merced, USA
Riccardo Cassinis	University di Brescia, Italy
Weidong Chen	Shanghai Jiao Tong University, China
Xiaoping Chen	University of Science and Technology of China, China
Silvia Coradeschi	Orebro University, Sweden
Paulo Costa	University of Porto, Portugal
M. Bernardine Dias	Carnegie Mellon University, USA
Amy Eguchi	Bloomfield College, USA
Alessandro Farinelli	University of Southampton, UK
Emanuele Frontoni	Univ Politecnica delle Marche, Italy
Giuseppina Gini	Politecnico di Milano, Italy
Alexander Hofmann	Technikum Wien, Austria
Dennis Hong	Virginia Tech, USA
Harukazu Igarashi	Shibaura Institute of Technology, Japan
Giovanni Indiveri	University of Lecce, Italy
Luca Iocchi	Sapienza University of Rome, Italy
David Jahshan	University of Melbourne, Australia

Vitor Santos	University of Aveiro, Portugal
Stefan Schiffer	RWTH Aachen University, Germany
Ali Shahri	Iran University of Science and Technology, Iran
Domenico G. Sorrenti	Università di Milano - Bicocca, Italy
Mohan Sridharan	University of Birmingham, UK
Tomoichi Takahashi	Meijo University, Japan
Yasutake Takahashi	Osaka University, Japan
Ahmed Tawfik	Univ of Windsor, Canada
Tijn van der Zant	University of Groningen, The Netherlands
Sergio Velastin	Kingston University, UK
Igor Verner	Technion - Israel Institute of Technology, Israel
Ubbo Visser	University of Miami, USA
Nikos Vlassis	Technical University of Crete, Greece
Oskar von Stryk	Technische Universität Darmstadt, Germany
Thomas Wagner	Bremen University, Germany
Alfredo Weitzenfeld	Inst Tecnologico Autonomo de Mexico, Mexico
Mary-Anne Williams	University of Technology, Sydney, Australia
Andrew Williams	Spelman College, USA
Martijn Wisse	Delft University of Technology, The Netherlands
Franz Wotawa	Technische Universität Graz, Austria
Guangming Xie	Peking University, China
Andreas Zell	University of Tuebingen, Germany
Mingguo Zhao	Tsinghua University, China
Changjiu Zhou	Singapore Polytechnic, Singapore
Stefan Zickler	Carnegie Mellon University, USA
Vittorio Ziparo	Sapienza University of Rome, Italy

Additional Reviewers

Markus Bader	Balajee Kannan
Oliver Birbach	Fabio Dalla Libera
Domenico Daniele Bloisi	Gil Lopes
João Certo	Hartmut Messerschmidt
Hon Keat Chan	Hugo Picado
Andrea Cherubini	Dietmar Schreiner
Ertan Dogrultan	Armando Sousa
Christian Dornhege	Matteo Taiana
Francesco Maria delle Fave	Stephan Timmer
Thomas Gabel	Jingchuan Wang
Kai Haeussermann	Yong Wang
Roland Hafner	Feng Wu
Zina M. Ibrahim	Marco Zaratti
Giovanni Indiveri	Yue Zhang
E. Gil Jones	

Table of Contents

Coordinated Action
in a Heterogeneous Rescue Team

Fares Alnajar, Hanne Nijhuis, and Arnoud Visser

Intelligent Systems Laboratory Amsterdam, Universiteit van Amsterdam,
Science Park 107, NL 1098 XG Amsterdam, The Netherlands
f.alnajar@student.uva.nl, h.nijhuis@student.uva.nl, a.visser@uva.nl

Abstract. In this paper we describe a new approach to make use of
a heterogeneous robot team for the RoboCup Rescue League Virtual
Robot competition. We will demonstrate coordinated action between a
flying and a ground robot. The flying robot is used for fast exploration
and allows the operator to find the places where victims are present in
the environment. Due to the fast aggregation of the location error in the
flying robot no precise location of the victim is known. It is the task of
the ground robot to autonomously go the point of interest and to get
an accurate location of the victim, which can be used by human rescue
workers to save the victim. The benefit of this approach is demonstrated
in a small number of experiments. By integrating the abilities of the two
robots the team's performance is improved.

1 Introduction

Since long it has been indicated that a heterogeneous robot team should have
operational benefits [1]. For Urban Search and Rescue operations, the benefits
seem even more promising [2]. Many teams [3,4] have indicated the possibility of heterogeneous team consisting of an aerial and ground robot, but actual
demonstrations are sparse [5]. Here, the benefit of coordinated action between a
teleoperated robot and a semi-autonomous ground-robot is demonstrated.

1.1 Relevance

In situations where a disaster like an earthquake has occurred, searching for
survivors in the area could be dangerous due to (partly) collapsed buildings that
are unstable. It could also be difficult for humans to search in such a collapsed
building if the available room is too small to crawl through. In such situations
robots could be deployed to search the area and hopefully supply some useful
information on the location and status of possible survivors. These robots could
be operated by humans (by remote control), but if they are able to explore
(semi) autonomously one could deploy a whole team of robots simultaneously
to cover a bigger area. To investigate the possibilities of multiple robots in these
situations, the Virtual Robot competition of the RoboCup Rescue League was
introduced [6].

J. Baltes et al. (Eds.): RoboCup 2009, LNAI 5949, pp. 1–10, 2010.

Because deploying actual robots in actual disaster-situations is a very complex task, the *virtual robots* competition is done in a simulated world. Simulation should not replace experiments with real robots, but a simulated world makes complex experiments reproducible and controllable. This provides the opportunity for rapid development by focusing on only a relevant aspect. Researchers can use this simulated world as rapid development environment, and can explore the design space for behaviors and co-operation between multiple robots before they are validated on real platforms.

1.2 USARSim

Since 2006 there have been annual world competitions in a simulation environment called USARSim. It's based on the Unreal Tournament 2 engine [7] and provides the ability to have robots operate in a 3D world with the laws of physics (like gravity) already implemented.

For our research the CompWorldDay1 map[1] is used as 3D map. This map supplies a large outdoor- as well as indoor-environment (an overview of the outdoor area we operate in can be seen in Figure 1(c)). Many obstacles like cars, buildings and construction are present, but the sky is fairly empty. This will give the aerial robot the opportunity to fly around without excessive need for 'obstacle avoidance'.

Robots are recreated in the virtual world, based on real machines that are used in the RoboCup Rescue League. For this research the AirRobot® and the Pioneer 2-AT (P2AT) were used, as are depicted in figure 1(a) and 1(b).

1.3 Performance Metrics

To be able to compare the performance of the participating teams some metrics were defined. The initial 2006 metrics were specified with the following formula:

$$S = \frac{V_{ID} \times 10 + V_{ST} \times 10 + V_{LO} \times 10 + t \times M + E \times 50 - C \times 5 + B}{(1 + H)^2}$$

Where V_{ID} is the number of victims identified, V_{ST} is the number of victims for which a status was reported, V_{LO} is the number of properly localized victims, B is an optional amount of bonus points rewarded by a referee for additional information on victims, t is a scaling factor for the accuracy of the map, M is the points assigned by a referee for the quality of the map, E is the points assigned by a referee for the exploration efforts, C the collisions between a robot and a victim and H the number of operators.

Over the years these metrics have changed [8] into the latest 2009 metrics:

$$S = \frac{E * 50 + M * 50 + V * 50}{(H)^2}$$

[1] Available for download on: http://downloads.sourceforge.net/usarsim/

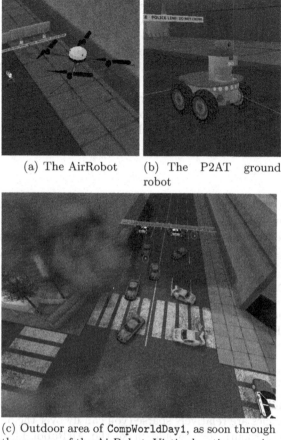

(a) The AirRobot   (b) The P2AT ground robot

(c) Outdoor area of CompWorldDay1, as soon through the camera of the AirRobot. Victim locations are indicated by red circles.

Fig. 1. Images from the USARSim environment

where V is an aggregate number indicating the effectiveness of victim detection and $H \geq 1$ is guaranteed. Part of V is the accuracy of the localization of the victim; if the victim is reported more than 2.5 meters from its actual location it is counted as false alarm with a corresponding penalty.

Although the AirRobot can be used to detect victims, initial tests indicated that the accuracy of localization of the AirRobot drops fast after a few turns. Although not impossible, building a map on visual clues alone is quite challenging [9]. Due to the limitations of the AirRobot, the performance of heterogeneous team will be based on the map of the ground-robot alone.

1.4 Related Work

Experiments with a heterogeneous team in rescue applications have been tried before. The 2nd place of the 2008 Virtual Robot competition was actually rewarded to a heterogeneous team consisting of one aerial and one ground robot [5]. Note that both robots were manually operated. Our approach incorporates semi-autonomous behavior as explained in section 3.1. Semi-autonomy by following an online planned path is demonstrated by many teams in the competition (e.g. [4,3,10,11]). To coordinate the actions of multiple robots by dropping RFID tags is inspired by the approach introduced by the Freiburg team [12]. The Freiburg robots avoided places where RFID tags were dropped, to optimize the exploration efforts E. Our ground robot actively searches for RFID tags, to optimize the victim detection V. In both approaches the location of the RFID tags is used to ensure that all robots use the same global locations.

1.5   Outline

In the following section we will explain the approach we took on showing improvements by using a heterogeneous team. Next we will cover the methods we used to implement this approach in section 3. We will discuss the experiments and their results in section 4 to round of with a discussion and a conclusion.

2   Approaches

We discuss several approaches to make the heterogeneous team operate (semi) autonomously. For the moment our approach is tested on outdoor situations only.

2.1   Exploration

The main advantages of the AirRobot are its speed and its ability to move at such height that obstacles are relatively sparse (you can fly over the obstacles instead of driving around them). This would make the AirRobot ideal for exploring a large area quickly. Furthermore the ability to explore from a high position provides a higher probability of having an unobstructed view on victims.

Unfortunately the pay-load of the robot is very limited, so in real-life situations we can not supply the robot with heavy gear. This limits not only the amount and type of sensors the robot could carry, but also the computational power available on board. For this research the payload is limited to a camera, a victim-sensor, a RFID dispenser and an inertia navigation system.

2.2   Victim Localization

Since the main goal of the Rescue League is to find victims in a disaster-area we want to use the AirRobot to search for them, because this can be done much faster from the sky than by a slow ground robot. The AirRobot uses its inertia navigation system (acceleration sensor) and therefore rapidly accumulates an error on the estimation of its location. The threshold on accurate victim localization is so sharp that GPS is also no option. This means that a victim report of the AirRobot could not be trusted, and should be verified by a ground robot.

2.3   Sharing Knowledge

Each member of the team (the ground robot and the aerial one) makes an estimation of its position. Both robots have an inertial navigation system, which gives a relative position estimate. This relative position estimate is used by the ground robot as initial estimate for a SLAM routine [13], which returns an absolute location estimate by comparison of the current range measurements against previous measurements aggregated on a map.

To unify the location estimation of the two robots, the estimations of a specific point by the two robots are computed and used to find out the shift the aerial robot makes with respect to the ground one.

There are two ways to achieve this 'unification'. In the first one, the ground robot computes this shift and sends the 'correction' to the aerial robot for the next estimations. But this method requires a consistent connection with the aerial robot which is not always available in the disaster situations, and requires the two robots to be close to each other. The second way is by storing the shift 'correction' at the ground robot as list for every victim/RFID tag it has reached. The ground robot uses this list to correct the position it is currently heading. When the aerial robot comes back again into communication range the ground robot transmit the whole list to allow the aerial to correct its traveled path as a sort of post-processing, as described in section 3.2.

2.4   Cooperation

The two robots of our team have clearly different roles. The aerial robot is used for exploring the environment and searching for victims. This robot is tele-operated. The aerial robot sends the approximated locations of the victims to the ground one, which is also equipped with a victim sensor. So as the ground robot gets close to the victim, it can detect its exact position, and make an accurate victim report. The positions sent by the aerial robot are imprecise, and only used guide the ground robot to the neighborhood of the victim. The ground robot makes further investigation for the victim in that neighborhood (the victim sensor has a range of 6 meters). There is no guarantee that the absolute position error of the AirRobot is less than 6 meters, but because the ground robot maintains a gradual 'correction' of the aerial robot localization, this strategy is only sensitive for the accumulation of the error between two victim locations.

3   Methods Used

In this section we explain the details of the semi-autonomous following-behavior of the ground robot to navigate the environment. Also we discuss how the two robots calculate the correction vector for the localization estimate of the aerial robot.

3.1 Following

The `follow-behavior` (as used by the ground robot) is a collection of 'motions' and the rules to switch from one motion to another. The switch between two motions depends on observations of certain events in the local environment around the robot (e.g. presence of obstacles, victims, and/or other teammates). A switch should indicate when a motion is not longer adequate for this situation. This behavior is built on the motions used for the fully autonomous exploration [14]. The motion contains the reactive schemes that directly couple sensor measurements to steering commands which navigate the environment. In each moment only one motion is active, and the robot behavior switches from one motion to another, depending on the robots situation. Here we give a description of these motions.

Following: This is the default motion of the `follow-behavior`, in which the robot waits for the position of the next target (the victim positions). When it receives a target-position, it plans online a path from its current position to the target, based on the occupancy grid map build so far. The path-planning in this experiment is based on a breath-first algorithm, but also an A*-algorithm is available. After the robot drove a certain distance (4 meters) it re-plans the path again on the new occupancy grid map, to incorporate information about the environment previously not visible.

Avoid Teammate: This motion is called when there is potential risk of collision between two robots (when the AirRobot is still on the ground and the distance between them is less than 1 meter). There are four states the robot can have while trying to avoid a teammate:
  - The teammate is facing the robot from the front side: in this case the robot should turn right or left to get out of the way.
  - The teammate facing the robot from the back side: the same as the previous state.
  - The teammate is in front of the robot but not facing it: the robot waits till the teammate move a way
  - The teammate is behind the robot but not facing it: the robot keeps its normal behavior (return the control to the 'following' motion).

Avoid Victim: This motion is used when the robot gets closer to the victim less than 1 meter. After detecting the closest part of the victim, the motion keeps the robot away of the victim. When the robot is far (more than 2 meters), the control is returned to the following motion.

As we can see the flowing of control from one motion and another depends on the situation of the robot, this is depicted in figure 2.

3.2 Updating Location Estimation Shift

Essential in our approach is calculation the difference in location estimation between the team members. Our method is by based on dropping RFID tags,

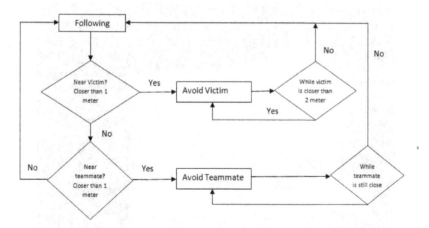

Fig. 2. Diagram of the `follow-behavior`

equivalent with the method used by [12], but now to find target points back. An RFID tag is small and cheap radio wave responder, which is often used for tracking and identification of logistic goods. These tags can be used as 'smart dust'; at some points the AirRobot stores its current location estimate in a RFID tag and drops this tag. When the ground robot comes close enough to this RFID tag, it can read the content the tag, and its current absolute location estimate and the read position as read from the tag (which was the location previously estimated by the aerial robot). Based on this comparison the new shift vector is calculated.

4   Experiments and Results

We tested our heterogeneous team on outdoor environment. We used the "P2AT"-model for the ground robot. The ground robot visits the places reported by the aerial robot and detect the victim locations accurately. At end of the test run, there were even some victims that the ground robot didn't have time to visit and detect their exact position.

To map illustrated in Fig. 3 is the result of a short flight of 8 minutes. Full competition runs can last 20 to 40 minutes. In this experiment three tags are dropped by the AirRobot, two near a victim, and one as intermediate point. Victims are reported as small red crosses. As can be seen, for each victim two crosses are given: the location as estimate by the AirRobot and the position as estimated by the P2AT. The difference is between the estimates is nearly 3 meters, above the threshold of the competition. Only by the ground-robot the victims are reported at the right location. As supplementary material also a video is available[2], from the view of the ground-robot. In the video can be seen that the ground-robot is constantly reading RFID tags. Most of them are the

---

[2] http://www.science.uva.nl/~arnoud/publications/Alnajar_CoordinatedAction.wmv

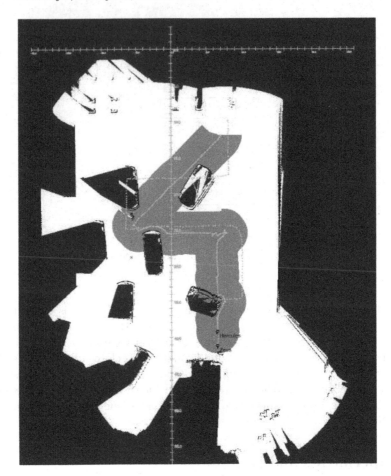

Fig. 3. Map generated by the ground robot in the outdoor area of `CompWorldDay1`

RFID tags with only an ID always present in the 2006 competition world[3]. Only near the male and female victim RFID tags are dropped by the AirRobot.

5   Discussion

Our team works well in the outdoor environment, but care must be taken when testing the team in indoor environments. The exploration of the aerial robot becomes very restricted due to the small free space and the many obstacles around the robot. As long as the AirRobot has no obstacle sensors, it has to remain in the area already explored by the ground robots. When the ground robots are equipped with sophisticated range scanner, this is not a severe restriction,

---

[3] Used at that time to estimate the accuracy of the maps, currently replaced by georeferenced maps.

because this range scanner can have a range of 80 meters. The remaining danger are obstacles which are not straight (as most buildings), but have a larger volume at the flight height of the AirRobot than at the scan-height of the ground-robots (as most trees). These obstacles can only be avoided with obstacle sensors on board of the robot. Due to the limited payload of the AirRobot such obstacle sensors have to be carefully chosen. A possible solution could be obstacle detection based on visual clues.

Another topic of future work is to achieving the relative position estimation update via visual tracking. Currently the relative position between the two robots is quickly lost, something that is partly corrected when a RFID tag is found. If the ground-robot could visually track the AirRobot, a high frequency update of this correction could be made (when an accurate height position is available). Using a bright color for the AirRobot (e.g. orange) can make the tracking easier.

Another possibility is extending the team size further. Many questions could be studied: How to divide the work of the ground robots (going to the victims)? What's the difficulty in adding more than one flying robot? What is the optimal ratio between the ground robots and the aerial ones? How can teams of aerial and ground robots automatically be formed?

6 Conclusion

In this paper a new coordinate action between an AirRobot and a ground robot is described. The two robots coordinate their behavior partly by communicating target points and partly through the environment by dropping and finding RFID tags. Each robot has its own role. The AirRobot is used for fast and course exploration, while the ground robot automatically inspects the regions of interest and needs no further attention of the operator. Together the heterogeneous team demonstrates an attractive way to perform a rescue mission.

References

1. Jennings, J., Whelan, G., Evans, W.: Cooperative search and rescue with a team of mobile robots. In: Proc. of the IEEE Int. Conf. on Advanced Robotics (ICAR), pp. 193–200 (1997)
2. Murphy, R., Casper, J., Micire, M., Hyams, J.: Mixed-initiative Control of Multiple Heterogeneous Robots for USAR. Technical Report CRASAR-TR2000-11, Center for Robot Assisted Search & Rescue, University of South Florida, Tampa, FL (2000)
3. Pfingsthorn, M., Rathnam, R., Stoyanov, T., Nevatia, Y., Ambrus, R., Birk, A.: Jacobs Virtual Robot 2008 Team - Jacobs University Bremen, Germany. In: Proceedings CD of the 12th RoboCup Symposium (2008)
4. Calisi, D., Randelli, G., Valero, A., Iocchi, L., Nardi, D.: SPQR Rescue Virtual Robots Team Description Paper. In: Proceedings CD of the 12th RoboCup Symposium (2008)
5. Balaguer, B., Carpin, S.: UC Mercenary Team Description Paper: RoboCup 2008 Virtual Robot Rescue Simulation League. In: Proceedings CD of the 12th RoboCup Symposium (2008)

6. Nourbakhsh, I.R., Sycara, K., Koes, M., Young, M., Lewis, M., Burion, S.: Human-Robot Teaming for Search and Rescue. Pervasive Computing, 72–78 (2005)
7. Balakirsky, S., Scrapper, C., Carpin, S., Lewis, M.: USARSim: Providing a Framework for Multi-robot Performance Evaluation. In: Proc. of the Performance Metrics for Intelligent Systems (PerMIS) Workshop, pp. 98–102 (2006)
8. Balakirsky, S., Scrapper, C., Carpin, S.: The Evolution of Performance Metrics in the RoboCup Rescue Virtual Robot Competition. In: Proc. of the Performance Metrics for Intelligent Systems (PerMIS) Workshop, pp. 91–96 (2007)
9. Booij, O., Terwijn, B., Zivkovic, Z., Kröse, B.: Navigation using an appearance based topological map. In: Proceedings of the International Conference on Robotics and Automation (ICRA 2007), Roma, Italy, pp. 3927–3932 (2007)
10. Velagapudi, P., Kwak, J., Scerri, P., Lewis, M., Sycara, K.: Robocup Rescue - Virtual Robots Team STEEL (USA) - MrCS - The Multirobot Control System. In: Proceedings CD of the 12th RoboCup Symposium (2008)
11. Visser, A., Schmits, T., Roebert, S., de Hoog, J.: Amsterdam Oxford Joint Rescue Forces - Team Description Paper - Virtual Robot competition - Rescue Simulation League - RoboCup 2008. In: Proceedings CD of the 12th RoboCup Symposium (2008)
12. Ziparo, V.A., Kleiner, A., Marchetti, L., Farinelli, A., Nardi, D.: Cooperative Exploration for USAR Robots with Indirect Communication. In: Proc. of 6th IFAC Symposium on Intelligent Autonomous Vehicles, IAV 2007 (2007)
13. Pfingsthorn, M., Slamet, B.A., Visser, A.: A Scalable Hybrid Multi-Robot SLAM Method for Highly Detailed Maps. In: Visser, U., Ribeiro, F., Ohashi, T., Dellaert, F. (eds.) RoboCup 2007: Robot Soccer World Cup XI. LNCS (LNAI), vol. 5001, pp. 457–464. Springer, Heidelberg (2008)
14. Visser, A., Slamet, B.A.: Balancing the Information Gain Against the Movement Cost for Multi-robot Frontier Exploration. In: Bruyninckx, H., Přeučil, L., Kulich, M. (eds.) European Robotics Symposium 2008. Springer Tracts in Advanced Robotics, vol. 44, pp. 43–52. Springer, Heidelberg (2008)

Concept Evaluation of a Reflex Inspired Ball Handling Device for Autonomous Soccer Robots

Harald Altinger[2], Stefan J. Galler[2], Stephan Mühlbacher-Karrer[1],
Gerald Steinbauer[2], Franz Wotawa[2], and Hubert Zangl[1,*]

[1] Institute of Electrical Measurement and Measurement Signal Processing,
Graz University of Technology, 8010 Graz, Austria
[2] Institute for Software Technology, Graz University of Technology, 8010 Graz,
Austria
team@robocup.tugraz.at
http://www.robocup.tugraz.at

Abstract. This paper presents a concept evaluation for a passive *ball handling device* for autonomous robots that enables the robot to "feel" the ball. A combination of a capacitive and a pressure sensor delivers accurate information of the ball position and movement within the *ball handling device*- even without touching it. Inspired by the human *reflex* the sensor values are evaluated to implement a low-level based control loop. This should enable the robot to make minor movement corrections to the overall path calculated by the high-level control system.

1 Introduction

Playing soccer is not a trivial task. A soccer player has to be fast with good stamina, should be a team-player and should be able to play the ball precisely in the desired direction. This is true for human beings as well as mobile autonomous soccer robots.

Robots do not have any problems with their stamina, but handling the ball is a very difficult task for them, since they do not have a vision and coordination system as humans. For example, an omni-directional vision system of autonomous soccer robots usually has a limited resolution with respect to the localization of the ball in the vicinity of the robot, e.g., an accuracy of about five centimeter. Therefore, a separated *ball handling device* is important.

Currently, teams in the RoboCup Middle-Size League use two different kinds of ball handling systems: passive and active ones. Active ball handling devices use some kind of actuators to prevent the robot from losing the ball [1]. Passive devices do not use any actuators. Both may use sensors to detect the ball.

We propose a *ball handling device* that is inspired by humans. We re-build the human *reflex* system for autonomous robots with high-speed sensors and a micro controller. Instead of manipulating the ball movement we are able to detect and react on the moving ball within milliseconds. Based on the path commanded by the high-level software the micro controller is allowed to adapt the motion

---

* Authors are listed in alphabetical order.

J. Baltes et al. (Eds.): RoboCup 2009, LNAI 5949, pp. 11–22, 2010.

command within limits to prevent the robot from losing control over the ball. The objective of the paper is to evaluate different sensors to be used in the *ball handling device*.

The paper is structured as follows: Section 2 describes the basic idea of our *reflex* inspired *ball handling device*. Following the concept one can find the test setup and the results in Sect. 3. Related work is summarized in Sect. 4.

2   Concept

A typical control cycle of a Middle-Size League robot starts with data acquisition using a directed or omni-directional camera system and perhaps some other sensor input transformation. High level software components perform multiple tasks such as self-localization, AI planning, path planning, obstacle avoidance and generate commands for the motion control of the robot hardware. As many high level tasks are involved in this process and huge amounts of data have to be processed, this control loop has a limited reaction time and typically achieves rates of 20 to 50 frames per second.

Inspired by humans we propose a concept of *reflexes* for mobile autonomous robots. Based on the information obtained by tactile sensors, a underlying loop - a *reflex*- is introduced. This loop makes minor corrections to the commanded path of the high level system in order to keep the ball within the handling device of the robot. As the complexity of the *reflex*-loop is much lower than for level control, the reaction time can be significantly reduced.

As an example, if a human soccer player sees a chance to get a ball he or she starts running towards it. This decision includes commands to our muscles to move the legs. In addition, a human has *reflexes*. When the soccer player steps onto a small stone, *reflexes* prevent him from falling. *reflexes* are fast movements not triggered by our brain but some kind of pre-trained actions to be able to react fast on unpredicted situations.

We propose such a high-speed control loop for a ball handling device similar to *reflexes* for autonomous mobile robots. Figure 1 shows both the high-level cognitive decision loop and the low-level *reflex* loop. Realizing this device requires

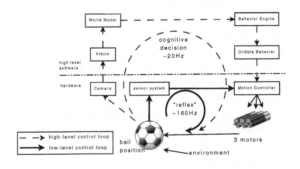

Fig. 1. High-level and low-level control loop

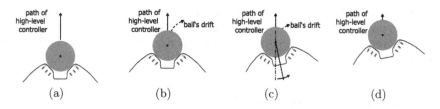

Fig. 2. Low-level control loop: Within Fig. 2(a) the robot is on the path of the high-level controller. Figure 2(b) shows a ball trying to drift off which is detected with our sensors mounted on the *ball handling device*. During Fig. 2(c) the low-level control loop tries to absorb the ball's drift with minor movement corrections. Finally at Fig. 2(d) the robot is back on the path of the high-level controller.

accurate and high-speed sensors that emulate the human capabilities of sensing its environment, and low-level computer power for reacting on unexpected changes in the environment. Implementing this *ball handling device* allows the robot to calculate minor movement corrections in order to avoid losing the ball.

The following subsections point out two essential components for the proposed low-level *reflex*-loop. The rest of the paper focuses on the evaluation of the most appropriate sensor technologies for the implementation of this kind of *ball handling device*.

2.1   Sensor System

Comparing the proposed device with human capabilities requires the sensor system to accurately detect the *ball position* within the *ball handling device* and nearby the robot. While catching the ball with the device it is also interesting to determine the *force a ball applies* to the robot. This information in combination with the exact contact point allows the robot to predict the bouncing direction and speed of the ball in an adequate way.

A *high sampling rate* is the basis for the proposed *reflex* system. In the RoboCup Middle-Size League robots are moving with up to five meters per second, the ball is kicked with up to eight meters per second. When sampling the RoboCup environment with 100Hz the ball may have moved about 13 centimeters between two sensor readings.

2.2   Measurement Evaluation

The heart of the *reflex* based *ball handling device* is the micro controller reading all sensor data and adapting the motion command of the robot. The high-level software provides information if the robot is currently in dribbling mode and the motion command - consisting of two directional velocity values and a rotational velocity value. If the robot is not in dribbling mode, the motion command is directly passed to the motion controller. The motion controller is responsible for transforming those three velocity values to three actual motor speeds.

Within dribbling mode the *reflex* mechanism adapts the motion command provided by the high-level software to prevent the robot from losing the ball but without completely overwriting the overall moving direction. Basically, we propose to only adapt the robots relative position to the ball to ensure a contact point between the ball an the center of the *ball handling device* as can be seen in Figure 2(d).

3   Experiment Setup and Results

To find the best combination of sensors we experimented with four different sensor types: two contact sensors and two non-contact sensors. The following subsections describe the test setup and the result for each of those sensors in detail.

To compare our results we developed a test setup that allows us to measure all values at the same time. Figure 3(a) shows the *ball handling device* and the location of all evaluated sensors. We distinguish between a static test case and a dynamic test case. Within the static test case we repositioned the ball on a predefined path through the *ball handling device*. The dynamic test case emulates a real playing situation. The ball is rolled down a ramp to ensure a predefined velocity when it hits the *ball handling device* (see Figure 4).

A micro controller periodically reads the measurements of the used sensors. The result is sent via controller area network (CAN) protocol to a computer that logs and evaluates the data. Figure 5 shows a general block diagram of the measurement setup.

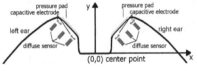

(a) Top view of our *ball handling device*, which shows the mounting position of three sensors.

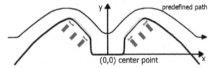

(b) Location of the "center point" and an illustration of the predefined path we use for the static experiments.

Fig. 3. Test setup

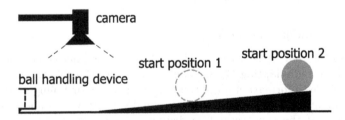

Fig. 4. Setup for dynamic experiments: The ball is placed at 2 different start positions. The first position results in a ball velocity of 0,22 m/s when hitting the *ball handling device*. Start position 2 results in a ball velocity of 0,31 m/s. The camera is used to determine the velocity.

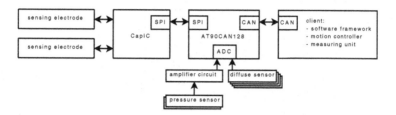

Fig. 5. Block diagram of our measurement setup: The micro controller triggers all sensor readings and sends the pre-processed values (filtering, amplifying, averaging) via CAN to all interested clients. For the test setup the client is a logging and evaluating unit on a computer.

3.1   Micro Switch

Humans can precisely detect which parts of their hands have contact with an object. To rebuild this for the robot, multiple micro switches could be used. Therefore, multiple micro switches are mounted on the *ball handling device*, which allow the robot to detect where the ball hits the device. We use two different arrangement on each side, one with a 6x1 grid, one with a 3x2 grid adopted to the contact line between a ball and the *ball handling device*.

All evaluated commercial micro switches are not able to detect when the ball hits the *ball handling device* due to the low force a ball applies onto it. Our experiments show a gliding friction of about 1 N and a rolling friction of about $\frac{1}{10}$ N. Therefore, this sensor is not applicable for the *reflex* based *ball handling device*.

3.2   Pressure Sensor

Similar to micro switches the pressure sensor is a representative of a contact sensor. It consists of a pressure pad filled with air and a silicon piezo resistive absolute pressure sensor SCCP30ASMTP [2] from Honeywell. Applying force onto the pressure pad compresses the air inside which can be detected by the connected sensor. In contrast to micro switches the signal is analog and proportional to the applied force.

Sommer [3] uses pressure sensors to detect collisions of the robot with objects and other robots. In our case we mount two pressure pads on the *ball handling device*. One on the left and one on the right side of the *ball handling device*, which enables us to determine on which side the ball touches the robots (see Figure 3(a)). Due to little dribbling energy of the ball the signal has to be pre-processed by an amplifier circuit. The amplifier consists of three stages. After the first amplifying phase, the signal is filtered and once more amplified. Corresponding to the amplitude of the analog signal the velocity of the ball hitting the *ball handling device* can be determined up to 0.25 m/s, above the amplifier circuit is designed to deliver maximum scale. Due to the characteristic of the amplifier circuit we get a signal even if the ball hits the pressure pads very

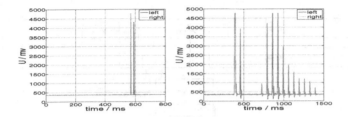

Fig. 6. Left picture, sensor signal from the pressure pad dynamic case: The analog signal range is in-between 0 and 5 V. The ball hits the right short before the left pad, the noise is negligible. Due to this measurement values the ball's position can be distinguished clearly between left and right. Right picture, sensor signal from the pressure pad bouncing: To simulate dribbling we bounced the ball with a cam lever. At the beginning both pressure sensors triggers at the same time which means that the ball's position is in the center of the *ball handling device*. Between 900 ms and 1150 ms it is clearly detectable that the balls position drifts from the center point to the right ear while the applied cam lever force decreases.

smoothly. The micro controller samples this signal with 1 kHz. This information can be used to detect if the ball is bouncing.

3.3 Diffuse Sensor

Diffuse sensors belong to the non-contact sensor category. We use the OHDK 10P5101 sensor [4] from Baumer Electronics Ltd, which is a diffuse sensors with background suppression and a binary output signal. Either an object is within the pre-defined trigger distance or not. The OHDK 10P5101 can be configured to trigger at distances between three and 130 mm with an accuracy of 0.2 mm. The sensor readings are provided at 40 Hz. It is not necessary to pre-process the signal of this sensor.

Combining multiple diffuse sensors with different distance configurations allows the robot to locate the ball position. Figure 3(a) shows the setup of all six diffuse sensors used in our test setup.

To emulate a similar behavior as the capacitance sensor (see Sect. 3.4) provides we adjust the trigger distances of all six sensors to three different threshold values (0.5 cm, 1.5 cm and 5.5 cm) on each side of the *ball handling device*. Due to the fact that those sensors use laser class two diodes as their light source we needed to adjust the height of the sensor mounting to different positions for each side to prevent interferences. Figure 7 visualizes the resulting sensitivity of the test setup.

3.4 Capacitance

Capacitive sensors are capable to determine measurands that, in some way, affect the coupling capacitance between two or more sensing electrodes [5]. The simplicity of the sensor elements is unparalleled: They essentially consist of two or more conductive areas. Despite this simplicity, capacitive sensing technology

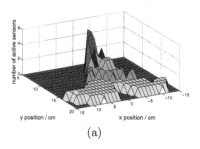
(a)

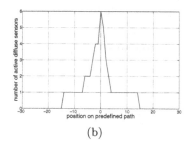
(b)

Fig. 7. Left picture, sensitivity graph of diffuse sensors: With this setup a clear decision can be made if the ball is within the center of the *ball handling device* or not. The outer sensors realizes a far field detection, the mid and inner sensors a near field. This results within an increasing sensitivity into the direction of the center. Due to the shape of the *ball handling device* the resulting sensitivity area is asymmetric. Right picture, position on a predefined path (see Figure 3(b)): It shows an increasing number of active diffuse sensors if the ball moves into the center of the *ball handling device*. Due to the asymmetric *ball handling device* the curve decreases much more on the right side. This enables us to track the balls movement. The decision between left and right can be made with the knowledge which sensor triggers first.

is very versatile due to a plurality of modes of operation and applicable materials (either conductive or dielectric). With the availability of small and fairly low cost monolithic sensor interfaces (e.g., Cypress CY8C21x34 family, Analog Devices capacitance to digital converters), the acceptance of capacitive technology has constantly increased during the last years.

The circuit used for the *reflex* based *ball handling device* is *CapIC*, a versatile interface IC for capacitive sensing, which was developed at the Institute of Electrical Measurement Signal Processing at Graz University of Technology and Infineon Austria. *CapIC* is particularly suitable for dynamic observations in various environments due to a high measurement rate and high configurability. The electrodes can be directly exposed to the environment and do not require shielding from external electromagnetic interference. The sensor front-end uses a fully differential approach for an enhancement of the sensitivity as well as the robustness with respect to disturbers and for reduced electromagnetic emissions.

Figure 8 depicts a system model of the sensor interface *CapIC*. It is based on a carrier frequency principle and uses an array of transmitter electrodes with selectable phase and a differential pair of receiver electrodes. Excitation signals (the carrier) are applied to one (or more) electrodes, the displacement currents injected to a pair of differential receivers are converted to a differential voltage signal by means of shunt impedances, which also form a filter tuned to the carrier frequency. Alternatively, the circuit can also operate in single ended mode, i.e., with only one non-differential receiver. The voltage signal is buffered and amplified in an HF amplifier stage. Besides amplification, this amplifier is also important to decouple the demodulator from the shunt impedance, as charge injection from the demodulator would lead to undesired resonance effects in the shunt impedances. The HF amplifier is implemented as a linearized

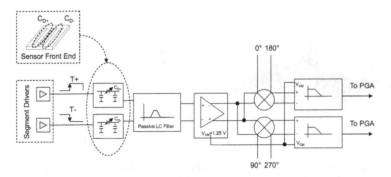

Fig. 8. System model of *CapIC* [5]: An excitation signal (the carrier) is applied to one (or more) transmitter electrodes, the receivers are connected to passive filters forming a shunt impedance for a current to voltage conversion [6]. The voltage signal is buffered and amplified and subsequently multiplied with the carrier for the I channel and with a 90° phase shifted carrier and low-pass filtered. The sensor array uses Time Division Multiple Access (TDMA): After one acquisition, other electrode(s) are excited such that all desired capacitances and conductances between the transmitter electrodes and the receiver electrode are obtained after a full sequence of excitation patterns.

differential amplifier stage. The buffered and amplified signal is mixed with the carrier for the I channel and with a 90° phase shifted carrier for the Q channel. After the signals are low-pass filtered, offset values due to offset capacitance and offset conductance can be compensated (digitally controllable). Consequently, the gain setting of the Programmable Gain Amplifier (PGA) can be adjusted to make use of the full conversion range of a 12 bit successive approximation AD converter.

In our test setup we used two pairs of electrodes. One is mounted on the right ear, the other on the left ear of the *ball handling device* (see figure 3(a)). Both are connected to the same *CapIC* circuit which itself is connected to an AT90CAN128 processor board which can be found in [7]. The connection between those two components is established via serial peripheral interface (SPI). Based on our *CapIC* configuration and after averaging we obtain readings at 160 Hz.

The capacitive measurement can be used to determine the ball's position between the left and right side of the *ball handling device*, however compared with the diffuse sensors it is not possible to ascertain a precise position of the ball based on the current measurement value from one differential pair of electrodes alone. Therefore, a history based signal processing is necessary. Due to the fact that every object (conductive or dielectric) can influence the measured capacitive value the *CapIC* system can detect a wide variety of objects. Within the RoboCup environment the possible objects are limited to balls, humans and robots which reduces complexity for classification. The ball shows the smallest effect, which causes a short sensing range up to 4 cm. The dead center of the sensing elements can cause wrong measurements. To avoid this, the electrode pairs are mounted 1 cm behind the surface of the *ball handling device*.

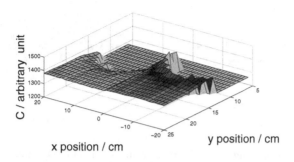

Fig. 9. Sensing area of two electrodes connected to *CapIC*: The flat area in the back represents the ball handling device. According to the differential measurement the signal within this sensitivity plot shows a valley on the right and a peak on the left side (robots view). The smooth change in the middle is caused by a longtime offset drift from *CapIC*. Those resulting potential areas are used to localize the ball within the ball handling device.

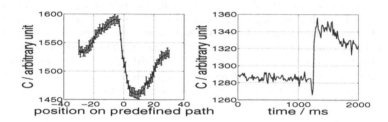

Fig. 10. Left picture, static case: The Ball is repositioned along a predefined path trough the *ball handling device*. The plotted values represent an averaging the error bars the maximum and minimum before averaging. Combined with the sensitivity plot the first peak indicates that the ball enters the *ball handling device* on the left side. Second the lines decline which means that the ball goes through the middle and leaves the *ball handling device* to the right side when the bottom is reached. The starting signal level is the same as the end level which means that the ball is completely out of the sensing area. Right picture, dynamic case: Sensor signal while the ball is rolling down a ramp - to ensure a predefined velocity - straight forward into the *ball handling device*. It is obvious that the ball hits the right side, bounces on the left and comes to a still stand nearby the right ear of the *ball handling device*, which can be detect with a higher signal level at the end of the plot compared to the beginning.

3.5   Comparision of Evaluated Sensors

Table 1 summarizes all important attributes of the evaluated sensors and the test results. All required values are adjusted to the visions performance. In our test setup the resolution within the x-y plane can not be determined for contacting sensors because it is not possible to determine the position of the action point on the pressure pad if the ball hits the *ball handling device*. Within the rough RoboCup environment it is necessary to protect all sensible components. The

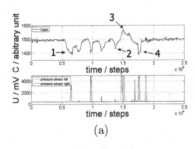 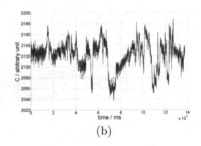

(a)   (b)

Fig. 11. Robot is moving on the pitch, trying to dribble the ball: (a): The robots operator can see the robot and the ball. The ball first touches the right side at mark 1, which can be seen within a decreasing *CapIC* value and a short peak from the pressure sensor. The ball starts to bounce on the right hand side of the *ball handling device*, until mark 2. At mark 3 the ball bounces to the left hand side, which can be seen when the *CapIC* value increases. Finally the ball is rebound to the right hand side (at mark 4) and leaves the *ball handling device*. There is a good correspondence between the *CapIC* measurement an the pressure measurement which can be seen every time the ball touches the *ball handling device*. (b): Control loop evaluation with a robot operator who neither sees the robot nor the ball. The operator's view is limited to the measurement values displayed on the evaluating computer when making minor movement corrections. An upward trend of the measurement curve indicates the ball is moving to the left side of the *ball handling device* which requires a movement correction to the left side. A reaction to the right side is required in case of a downward trend. A video analysis shows a good correlation between the measurement curve and the necessary movement corrections to dribble the ball successfully.

Table 1. Comparison of evaluated sensors: The micro switch can not be chosen, see Section 3.1, therefor the pressure sensor is used as contact sensing element. The *CapIC* sensor satisfy all defined requirements for our *ball handling device*. In comparison the diffuse sensor has several disadvantages, mainly a low sampling rate, which disqualifies this sensors to be used as the non-contact sensing element.

	sampling rate		resolution x-y plane		exposed	mountable	costs
sensor	required	realized	required	realized	required	realized	calculated
	Hz	Hz	cm	cm			€
micro switch	> 40	1 k	< 5	X[1]	✔	✔	12· € 0.2
pressure	> 40	70	< 5	X	✔	✔	2· € 25
diffuse	> 40	40	< 5	0.02	✔	X	6· € 163
CapIC	> 40	160	< 5	1	✔	✔	1· € 100

CapIC electrodes and the pressure pads can be mounted on the surface of the robot their electronics can be distributed to protected areas. The diffuse sensors' electronic can not be separated from its sensing elements.

---

[1] Not determined.

4   Related Work

Typical applications of capacitive sensors are position sensing, material property monitoring such as moisture content or oil quality, proximity switching, occupancy detection, fill level detection etc. Furthermore, capacitive sensing is very common in biomedical and chemical applications. A comprehensive presentation on capacitive sensing can be found in [8].

The *CapIC* circuit has been used in a variety of applications and research projects. In [9] a rapid prototyping environment based on *CapIC* is described. This prototyping environment allows easy configuration of the IC and quick implementation of stand-alone demonstrators. In [6] the circuit has been used for a seat occupancy sensor, with a focuses on the reduction of electromagnetic radiation at higher frequencies. Moisture measurements using the circuit are discussed in [10]. The application of the circuitry for personal safety for chain saw users is presented in [11]. A more detailed description of the architecture of *CapIC* and its application to chemical sensing can be found in [5].

Pressure sensors are used in a wide variety of applications such as measurement of fluid/gas flow, liquid level, altitude, chemical process control, and for meteorological observations. Most pressure sensors evaluate a mechanical deformation due to a force associated with the pressure. Examples are strain gauges, piezoelectric and capacitive pressure sensors. However, there are other effects that are also used for pressure sensors, e.g., such as thermal conductivity, ionization and viscosity [12].

Optical sensors are widely used for proximity detection, e.g., in industry for object and material detection, in packaging machines, in robotics, and in laboratory automation. Multiple light sources or detectors are often used to suppress the influence of background light and inclination.

5   Conclusion and Outlook

We proposed a *reflex* based *ball handling device* that is mainly inspired by humans.

The fact that the high level ball detection with the omni-directional camera only reaches 20 Hz to 40 Hz, which is too slow for successful dribbling, lead us to a new approach with non-contact and contact sensors. We focused on decreasing reaction time which can be implemented with a low-level control loop. Within this paper we showed an evaluation of four different types of sensors.

However, to get a reliable localization of the ball we decided to use a combination of one contact and one non-contact sensor type. Two pressure sensors to detect which side the ball is hitting within the *ball handling device* and the *CapIC* to realize a 'far field' sensing area for localizing the ball. The introduced sensors facilitate a sampling rate at minimum 160 Hz which is 8 times faster than our current vision based implementation. The *CapIC* sensor is able to localize the ball within 1 cm accuracy, which is enough for improved dribbling. All required sensing elements can be mounted at mechanical exposed positions - so the application in RoboCup is possible. Various experiments as described within

Figure 11(b) show a realizable concept of the *reflex* loop to be implemented on a micro controller which will be part of future work.

We will concentrate on those experiments and different implementations of the *reflex* loop and its usability during a RoboCup soccer tournament. Another research focus is on the usability of the capacitance sensor for near-by object detection and classification. Similar to a human who feels when a person comes close to him or her - even without physical contact.

References

1. Lunenburg, J., v.d. Ven, G.: Tech united team description. In: International RoboCup Symposium 2008, Netherlands, Eindhoven University of Technology (2008)
2. Honeywell: SCC SMT - Absolute pressure sensor (2009)
3. Sommer, U.: Roboter selbst bauen. Franzis (2008)
4. Baumer Electric: OHDK 10P5101 - Diffuse sensors with background suppression (2009)
5. Zangl, H., Bretterklieber, T., Holler, G.: On the design of a fully differential capacitive sensor front-end interfacecircuit. In: Microelectronics Conference ME 2008, Vienna, Austria, October 15-16, pp. 132–136 (2008)
6. Zangl, H., Bretterklieber, T., Werth, T., Hammerschmidt, D.: Seat occupancy detection using capacitive sensing technology. In: SAE World Congress, Detroit, MI, USA, April 14-18 (2008)
7. Krammer, G.: Elektronisches Gesamtkonzept fuer einen mobilen Roboter. Master's thesis, Graz Universtiy of Technology (2007)
8. Baxter, L.: Capacitive Sensors, Design and Applications. IEEE Press, Los Alamitos (1997)
9. Hrach, D., Zangl, H., Fuchs, A., Bretterklieber, T.: A versatile prototyping system for capacitive sensing. Sensors and Transducers Journal 89(3), 117–127 (2008)
10. Fuchs, A., Zangl, H., Holler, G., Brasseur, G.: Design and analysis of a capacitive moisture sensor for municipal solidwaste. Meas. Sci. Technol. 19(2), 9 (2008)
11. George, B., Zangl, H., Bretterklieber, T.: A warning system for chainsaw personal safety based on capacitive sensing. In: IEEE International Conference on Sensors, Leece, Italy, October 26-30, pp. 419–422 (2008)
12. Webster, J.G. (ed.): The Measurement, Instrumentation, and Sensors Handbook. CRC Press LCC, Boca Raton (1999); Springer Verlag, Heidelberg, Germany and IEEE Press

Development of a Realistic Simulator for Robotic Intelligent Wheelchairs in a Hospital Environment

Rodrigo A.M. Braga[1,2], Pedro Malheiro[1,2], and Luis Paulo Reis[1,2]

[1] FEUP – Faculty of Engineering of the University of Porto, Rua Dr. Roberto Frias,
s/n 4200-465, Porto, Portugal
[2] LIACC - Artificial Intelligence and Computer Science Lab., University of Porto, Portugal
rodrigo.braga@fe.up.pt, pedro.m.malheiro@gmail.com,
lpreis@fe.up.pt

Abstract. Nowadays one can witness the increase of the world population carrying some kind of physical disability, affecting locomotion. With the objective of responding to numerous mobility problems, various intelligent wheelchair related projects have been created in the last years. The development of an Intelligent Wheelchair requires a lot of testing due to the complexity of the algorithms used and the obligation of achieving a failproof final product. This paper describes the some need for an Intelligent-Wheelchair specific simulator as well as the requirements of such a simulator. The simulator implementation, based on "Ciber-Mouse" simulator, is also described with emphasis on analyzing the limitations concerning intelligent wheelchair simulation using this adapted simulator. The changes applied on the existing software and the difficulties of robotic simulation development are described in detail. Experimental results are also presented showing that not only the simulator reveals flexible simulation capabilities but, also, enabled to validate the algorithms implemented in the physical intelligent wheelchair controlling agent.

Keywords: Simulation, Robotics, Intelligent Wheelchair.

1 Introduction

Although not obvious, conventional electric wheelchairs have limitations that difficult and may even prevent its normal usage for some people, depending on their disability [1] [2]. Afflictions that affect arm and hand coordination or even vision are sure to benefit from intelligent wheelchairs, which will give some independence back to these patients.

This scenery enables the need for Intelligent Wheelchairs (IWs). An IW will, through computational capabilities, sensors, communications and motor control, automatically (and in an intelligent fashion) provide services for its user [3] [4]. A few examples are: Multiple user interfaces for control and ordering (voice, facial recognition, joysticks, keyboard, etc.), aided driving (in case of miscalculation of the patient's manual drive, the IW will disable human control: e.g.: if a collision is imminent) or even opening a door, as well as many other domotics applications, can be dealt within the IW's communication capabilities.

J. Baltes et al. (Eds.): RoboCup 2009, LNAI 5949, pp. 23–34, 2010.
© Springer-Verlag Berlin Heidelberg 2010

After applying the hardware add-ons – such as sensors – an IW's capabilities and reliability lie on its control software. The control algorithms must be thoroughly tested. The control software should undertake Multi IW scenario tests with dynamic obstacles, in order to ensure reliability on the developed algorithms. Creating a real life hospital-like situation is not conceivable therefore this is the point where a simulator will be required. It can model not only any kind of map but it can even provide the IW with all sensor information as a real situation would.

In a generic definition [5], simulation is the imitation of a real system, in function of the time. It is used to describe and analyze the behavior of a system and can draw conclusions from "what-if" scenarios. Among the advantages of simulation, we can refer the possibility of testing every aspect of a proposal for modification or addition without endangering real resources; time compression, in order to obtain all the consequences of a modification in a shorter lapse of time; Finding errors and problems on a system becomes simpler. Finding the cause of a certain occurrence is facilitated for it is possible to isolate the simulation to the occurrence and analyze it in detail; Lower costs: typically the cost of a simulation compared to a real test is around 1%; Requirement identification: it is possible to identify the needs in terms of hardware. For example: one can, through simulation, find what will the necessary resolution be for a certain sensor.

Having justified the need for a simulator, the challenge is to find what needs to be simulated and how. An IW-specific robotic simulator must be able to realistically mimic the physics of every aspect of the wheelchair's environment and its own characteristics namely its form, dimensions, motor dynamics, sensors and communications. Moreover, the simulation must take care of all interaction between the wheelchair and the world: detect collisions, return sensor values and calculate IWs' positions.

Additionally, being visualization one of the main objectives of an IW simulation, it is important to have a simulation viewer that can show the main information: IWs' sizes, locations and sensors' values as well as the environment: map with its walls and doors.

2   Related Work

Two projects stood out during literature review of intelligent wheelchair simulation: the Bremen Autonomous Wheelchair [6] (BAW) and the *Vehicle Autonome pour Handicapés Moteurs* [7] (VAHM).

Combined, both BAW and VAHM were motivated by reasons shared with Intellwheels Simulator project: the need to test hardware platform and its performance, without submitting patients to dangerous situations. Despite this, there are conceptual differences in architecture and on the objectives. While these projects intend only to aid the development of a single isolated wheelchair, Intelwheels is multi-agent based, in the sense that a dynamic, more complex environment with multiple intelligent and collaborative objects is intended. Furthermore, BAW simulation segregates completely real and virtual worlds, leaving no room for augmented reality model (which is a critical conceptual design of Intellwheels).

3   Intellwheels Simulator

The original "Ciber-Mouse" Simulator[7][8] is a software developed for a robotic competition, held at University of Aveiro and it was the base of this project. The objective of the participants is to create a robotic agent that would control their "mouse" (a virtual robot) through a maze to find a beacon (the objective).

This software already had various characteristics very usefull for IW simulation such as virtual differential robots (two wheels) and numerous sensors (e.g. compass and proximity sensors). It also contained a 2D simulation viewer, which was specific for the competition[9].

Being based on "Ciber-Mouse" the new simulator has a similar conceptual architecture, consisting of a central simulation server, to which every agent (robotic or viewing agents) must connect to (Figure 1).

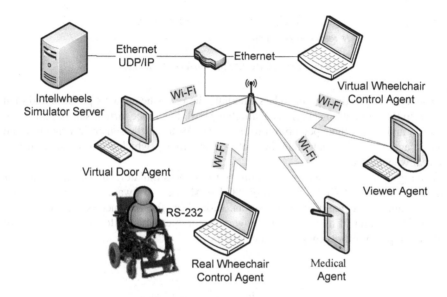

Fig. 1. Intellwheels Simulator Architecture

The most relevant work was made on the Robot Modeling and on the Collision Detection modules which will be detailed further ahead. Evident in the image is the connection of agents for controlling virtual wheelchairs and agents that control real wheelchairs. This possibility of interaction between the real and virtual worlds creates a mixed reality (MR) environment[10]. The definition of MR lies in a mid-stage point between a purely real world and a virtual one.

It is this MR support that stretches this simulator's capabilities beyond single algorithm testing, as it is now possible to see a real IW react to a more dynamic scenario (with moving obstacles, complex maps and other intelligent agents). It all depends on which and how many agents connect, for the simulator itself will not limit their size or movement.

The simulator expects a real wheelchair agent to provide real world data concerning its position, as illustrated in Figure 2.

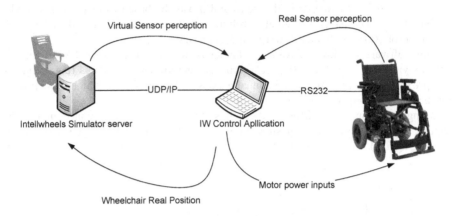

Fig. 2. Mixed Reality information exchange

The agent must, through encoders calculate the real wheelchair's position and then send it to the simulation server. The simulator will then place virtual sensors onto that position and return their perception to the agent. With all the sensor information, the robotic agent will then calculate the motor power inputs which are to be sent only to the real wheelchair.

Having defined the conceptual architecture of the simulator, the starting point for the adaption of the DCOSS version of "Ciber-Mouse" software was the conversion from circular to rectangular body robots. Moreover, the parameters that define the size and the center of movement are now configurable, which allows the modeling of different types of robots (not only wheelchairs).

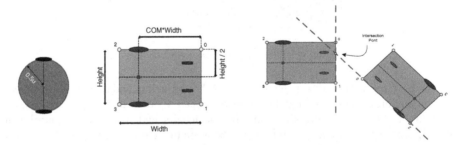

Fig. 3. Circular to rectangle shape modifications   **Fig. 4.** Robot-Robot collision check

The main usage of the robot's body is in the collision detection verification. A robot's shape has to be defined in mathematical equations that will enable the detection of intersection with other objects. In this simulator there will only be modeled 2 types of objects: walls and robots. Therefore collision checking will only have to be performed with these two types.

Originally, for Robot-Robot collision checking, the "Ciber-Mouse" simulator checked whether the distance that separates the robots' centers (thought X and Y coordinates) was smaller than two times a robot's radius (all robots were circular with the same radius). This simple algorithm is not applicable for different radius robots neither for rectangle shaped robots. The wheelchairs' size and position on the map were now modeled using four parameters: Center of movement point, the wheelchair's orientation angle, the width and height. Through this information all the robots' corners coordinates can be calculated. Using this information the new collision detection algorithm is as follows:

1. Using pairs of corners as line segment defining points, it is calculated an equation for one line segment for each robot.
2. The intersection point of the two lines is calculated. If lines are parallel no point is calculated for there is no intersection.
3. Both X and Y coordinates are checked to find whether they are located within each robot line segment. If so, then there is a collision between the two robots.
4. This process is repeated until the 4 lines of each robot are checked with the lines of every other robot.

In Figure 4 the lines cross at a point that only belongs to one of the wheelchairs, therefore no collision is detected.

An additional modification was made on the dynamic characteristics of the motors' acceleration curves. Since the original software was designed to ensure all robots were equal, every robot connected had to had the same dynamic characteristic. In this IW simulation environment it is expected that different robots connect. Equation (1), show below, was applied to allow the curve definition.

$$output_n = (1 - AccelerationCurve)*input_n + AccelerationCurve *output_{n-1} \qquad (1)$$

where $output_n$ is the new motor power output, *AccelerationCurve* is a value between 0.00 and 1.00, defining the slope of the acceleration curve, and $output_{n-1}$ is the power value from the previous period.

Similarly to size and center of movement characteristics, a new robot registration parameter was implemented to allow each robot to define their curve. This value will affect the robot speed calculation consequently affecting the robot's next position calculation by the simulator.

The proximity sensor positioning was the next functionality to be adapted. Originally, the infra red sensor could only be positioned in the perimeter of a circle, with a fixed cone of sight and a fixed direction, radial to the robot. To be true to the rectangular form, the sensor definition was now made by X and Y coordinates, relatively to the robot center, and both the cone of sight and the direction were redefined. All these parameters are now configurable by the agent, at the time of registration with the simulator. Moreover, the sheer modification of a configurable cone enables the agent to register different proximity sensors. A wider cone (around 50-70 degrees) would resemble a sonar proximity sensor whereas as thinner cone (1-10 degrees) would be more similar to an infrared proximity sensor.

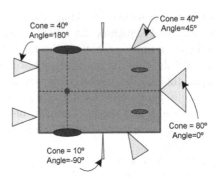

Fig. 5. Sensor definitions

An important objective of this IW simulation was to provide augmented reality scenery. In such an environment, a real IW (connected to the simulator) would be able to interact with virtual objects. Virtual wheelchairs would be able to detect its presence and virtual sensors could be attached to the real chair and provide it with additional information of the world around it. A very simple example of an application for this functionality is the correction of real sensors' errors. A Sonar proximity sensor is likely to have difficulty in distinguishing a table despite being in its normal range of sight. Depending on the shape of the legs, the height of the table and even on its color, acknowledging the presence of an object may fail. If the simulator has the table object drawn on the world, virtual sensors will report information about its location for they are not bound to the physical restrains as real sensors. The identification of a real chair is now made through a new XML tag, defined at the robot registration. A "Type" tag (later stored onto each robot's information within the simulator) will indicate whether the chair is real or simulated. Finaly, in terms of algorithms inside, the main change for mixed reality support is in the section of the code where the simulator calculates and commits the next robot position. Unlike a simulated IW, the real chair's position is not calculated by the simulator (through the left and right power orders given by the agent). Instead, the real IW will provide its current position and orientation, ideally once every simulation cycle, as previously illustrated on Figure 2.

4   Simulation Visual Representation

Visualization is the foundation for human understanding, as we process graphic information in a preconscious, involuntary fashion, similar to breathing[11]. In spite of its importance it is critical to ensure quality in a few elements, when developing simulation graphics.

Keeping this in mind, a visualization application was developed – Intellwheels Viewer. On a conceptual sense, it consists of five modules:

- Main Form – Responsible for the aggregation of all the information, allowing it to be transferred between the other modules. It also handles the initial interaction with the user, including configuration parameters of the other modules;

- Communications – This module implements the IP and UDP protocols, for physical connection with the simulator, and XML message parsing, for simulation data extraction;
- Robot – This module will store information concerning each robot: size, center of movement, position, orientation and collision status;
- 2D Visualization – This module will, through the map and robots' information, draw a two dimensional representation of the simulation;
- 3D Visualization – To represent a three dimensional environment, this module uses OpenGL technology [12] realistically displays the simulation. It has various camera options, from free camera movement to a 1st person fixed point, similarly to how a real driver of a wheelchair would see the world around.

Every simulator step, new world state data, including robot information, is sent to the viewers and is at that time that the graphical representation is updated. Figure 6 shows a simulation with three wheelchair agents, a table agent and a and door agent, on free camera viewing.

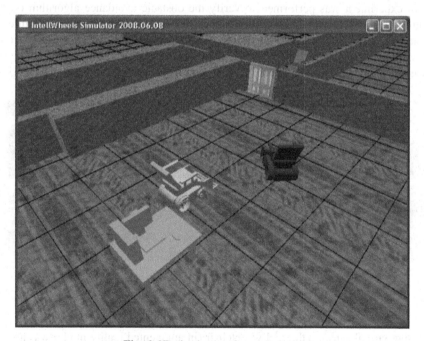

Fig. 6. 3D viewing, on free camera mode

5 Simulation Tests

To validate the performance of the simulator and confirm its importance for intelligent wheelchair development, a series of tests was performed. These tests were based on driving analysis from real, virtual and augmented reality runs, which were compared against each other. The floor of the building were (Deleted for blind review) is

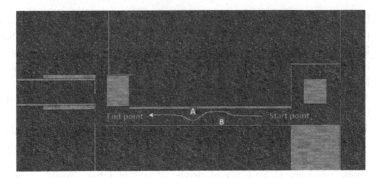

Fig. 7. Intellwheels modeled map of floor with test obstacles, in 2D mode of new viewer

set was modeled into Intelwheels XML map format, to allow a direct comparison between real and virtual tests. Figure 7 shows the simulated scenario (drawn from the CAD file) of the floor where the testes were performed.

An experiment was performed to verify the obstacle avoidance algorithm on the controlling software as well as the augmented reality environment as well (Figure 8). Obstacles (A and B block illustrated in Figure **7**) were placed both on the real and on the virtual environments, in the exact same space, with the same dimensions. The wheelchair should drive 15 meters in the X coordinate while avoiding the obstacles.

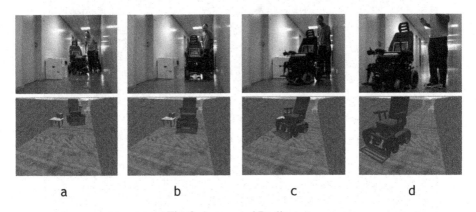

a   b   c   d

Fig. 8. Augmented Reality test

Using virtual sensors, the real wheelchair (in augmented reality mode) was able to autonomously avoid an obstacle placed in the virtual world. The simulator placed the wheelchair in its true position and the controlling agent correctly used the virtual sensors for motor power orders.

In this same experiment set-up, additional tests were taking, with both real and virtual wheelchairs and using manual and autonomous control. The results of the tests are illustrated in Table 1.

Table 1. Obstacle avoidance test results

Operation mode	Performance	Collision Count
Real/Manual	⚠	0
Virtual/Manual	✓	0
Real/Autonomous	⚠	1
Virtual/Autonomous	✓	0
Augmented/Autonomous	✓	0

Performance of these tests was based on whether the wheelchair reached the final point, if it collided and on position logs of the controlling software. The results showed that simulation-aided algorithm testing performed better than purely real tests. The main reason for this was the errors in odometer-based positioning (which accumulates error) and sonar noise which sets the chairs' controlling software into erratic decisions. It is also noticeable that the behavior of the wheelchair is almost equal in virtual and augmented reality modes.

Despite problems with real sensors, this simulation's value was proved with the success of the virtual environment tests. The control algorithm is correctly implemented, which is a conclusion that, without simulated testing, could not have been reached.

In order to test the flexibility of the simulator and of the new viewer and to verify the correct implementation of a developed door agent a new experiment was prepared: ordering a chair to move straight forward, through the virtual door. The door agent itself was configured with height=0.1m, width=1.0m and COM=0.99. Two proximity sensors were defined at each side of the door, to detect approaching objects, as illustrated in Figure 9.

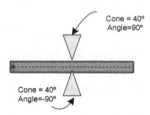

Fig. 9. Representation of the modeled door

Using a robotic agent, a virtual wheelchair was connected to the simulator and the test of door automatic opening was executed. Figure 10 is a series of print screens of the Intellwheels 3D viewer, during this automatic door test. The IW agent ordered the chair to move forward, regardless of what its own proximity sensors detect. On the other hand, the door agent was programmed to open if an object was detected and close only when the sensors stop detecting.

| A | b | c | d |

Fig. 10. Automatic opening of the door

Having the capability of agent to agent communication, through very simple and effective XML messaging, IWs could autonomously "tell" each other to follow. Figure 11 shows an example where one chair is leading the way. The other IWs receive the first chair's communication of last point and follow the orders.

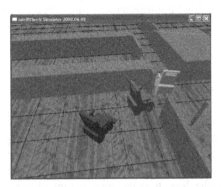

Fig. 11. Intelligent wheelchair communication

The following XML message is exchanged between the chairs:

```
<Actions …..
<Say><![CDATA[msg]]></Say> </Actions>
```

At a later stage of system developing, we have assembled a simple Mixed Reality Rehabilitation Lab. Using this setup, the virtual and augmented reality tests and simulations could be performed with increased realism, as shown in Figure 12.

Fig. 12. Mixed Reality Rehabilitation Laboratory

With such a setup it is now possible to supervise and monitor real and augmented reality tests and also provide a realistic simulation, without subjecting people or the wheelchairs' themselves to stress.

6 Conclusions and Future Work

The final product of this work was a simulation engine and a 2D/3D viewer that is able to test and validate control algorithms for intelligent wheelchairs. What was achieved during experimental testing demonstrated not only that the developed application is capable of simulating maps and wheelchairs but also showed its flexible characteristics. In fact, this simulator is capable of modeling a very wide variety of robots, due to implemented functionalities for configuring the size, center of movement, top speed and acceleration curve of each robotic agent that connects to it.

In this sense, the developed application's capabilities stretch beyond its original intent as a test board for intelligent wheelchairs. Car and pedestrian simulation can be performed, including their physical interaction (collisions). Direct communication between agents is available and, with further developments in agent conception, it could be used for emotional interaction (movement conditioned by attraction or repulsion). Having a distributed architecture, Intellwheels Simulator expects the agents to be external applications that connect through UDP. Because of this attribute, it is able to involve in a single simulation a vast number of intelligent agents, adding the possibility of testing algorithm results in a dynamic, complex environment.

This topic links with the idea of future work, which could be done on development of new robotic agents. The simulator software itself is ready and capable of receiving any kind of control for the virtual motors, therefore the challenge is now to create the various controlling agents themselves.

Concerning the simulator, an interesting addition would be to include other sensors. A good example are encoders, widely used in robot speed and position calculations.

As for the visualization application, the main modification would be to add a "drag and drop" capability. Introducing, in real time, objects and obstacles onto the simulation environment, while seeing where they will be placed is intuitive and allows a more interactive simulation.

Final words to an overview of the contribution of this work on how artificial intelligence and robotic systems can truly aid life of people. Although still a prototype, the Intellwheels project can effectively reduce the limitations and one's dependence on others when faced with physical disabilities.

Acknowledgements

This work was partially supported by FCT Project PTDC/EIA/70695/2006 "ACORD: Adaptative Coordination of Robotic Teams" and LIACC – Artificial Intelligence and Computer Science Lab. of the University of Porto, Portugal. The first author would like to thank for CAPES-Brazil for his PhD scholarship funding.

References

1. Braga, R.A.M., Petry, M., Moreira, A.P., Reis, L.P.: INTELLWHEELS – A Development Platform for Intelligent Wheelchairs for Disabled People. In: 5th International Conference on Informatics in Control, Automation and Robotics, Funchal, Madeira, Portugal, vol. I, pp. 115–121 (2008)
2. Braga, R.A.M., Petry, M.R., Moreira, A.P., Reis, L.P.: Concept and Design of the Intellwheels Development Platform for Intelligent Wheelchairs. In: Cetto, J.A., et al. (eds.) Informatics in Control, Automation and Robotics. LNEE, vol. 37, pp. 191–203. Springer, Heidelberg (2009)
3. Faria, P.M., Braga, R.A.M., Valgôde, E., Reis, L.P.: Interface Framework to Drive an Intelligent Wheelchair using Facial Expressions. In: IEEE International Symposium on Industrial Electronics (ISIE 2007), pp. 1791–1796 (2007)
4. Martins, B., Valgôde, E., Faria, P., Reis, L.P.: Multimedia Interface with an Intelligent Wheelchair. In: Proceedings of CompImage 2006 – Computational Modelling of Objects Represented in Images: Fundamentals Methods and Applications, Coimbra, Portugal, pp. 267–274 (2006)
5. Banks, J.: Introduction to Simulation. In: Proceedings of the 2000 Winter Simulation Conference, Phoenix, Arizona, United States, vol. I, pp. 7–13 (2000)
6. Laue, T., Röfer, T.: SimRobot - Development and Applications. In: Amor, H.B., Boedecker, J., Obst, O. (eds.) The Universe of RoboCup Simulators - Implementations, Challenges and Strategies for Collaboration. Workshop Proceedings of the International Conference on Simulation, Modeling and Programming for Autonomous Robots (SIMPAR 2008). LNCS (LNAI). Springer, Heidelberg (2008)
7. Bourhis, G., Agostini, Y.: The Vahm Robotized Wheelchair: System Architecture and Human-Machine Interaction. Journal of Intelligent and Robotic Systems 22(1), 39–50 (1998)
8. Lau, N., Pereira, A., Melo, A., Neves, A., Figueiredo, J.: Ciber-Rato: Um Ambiente de Simulação de Robots Móveis e Autónomos, Revista do DETUA (2002)
9. Lau, N., Pereira, A., Melo, A., Neves, A., Figueiredo, J.: Ciber-Rato: Uma Competição Robótica num Ambiente Virtual. Revista do DETUA 3(7), 647–650 (2002)
10. Lau, N., Pereira, A., Melo, A., Neves, A., Figueiredo, J.: O Visualizador do Ambiente de Simulação Ciber-Rato. Revista do DETUA 3(7), 651–654 (2002)
11. Milgram, P., Kishino, F.: A Taxonomy of Mixed Reality Visual Displays. IEICE Transactions on Information Systems E77-D, 1321–1329 (1994)
12. Rohrer, M.R.: Seeing is Believing: The Importance Of Visualization in Manufacturing Simulation. In: Winter Simulation Conference, pp. 1211–1216 (2000)
13. Woo, M., Neider, J., Davis, T.: OpenGL Programming Guide, The Official Guide to Learning OpenGL, Version 1.2., 3rd edn. Addison-Wesley, Reading (1999)

Creating Photo Maps
with an Aerial Vehicle in USARsim

Heikow Bülow, Andreas Birk, and Shams Feyzabadi

Jacobs University Bremen*
Campus Ring 1, 28759 Bremen, Germany

Abstract. Photo maps, i.e., 2D grids that provide a large scale bird's eye view of the environment, are of interest for many application scenarios but especially for safety and security missions. We present a very efficient and robust algorithm for this task, which only uses registration between consecutive images, i.e., it does not require any localization. The algorithm is benchmarked in USARsim, where the video stream of a down-looking camera of an aerial vehicle, namely a blimp, is used to generate a large scale photo map.

1 Introduction

Unmanned Aerial Vehicles (UAV) are promising tools for a fast first situation assessment in Safety, Security, and Rescue Robotics (SSRR). They are ideal eyes in the sky that cooperate with Unmanned Ground Vehicles (UGV), which provide the advantages of e.g. higher payloads and of options for mobile manipulation. Figure 1 shows a typical application scenario where a Jacobs land robot and an UAV cooperating at European Land Robotics Trials (ELROB) 2007 in Monte Ceneri, Switzerland. This scenario, in which the Jacobs team won the 1st place, required that the UAV detects hazard spots like seats of fire, which the land robot has to reach [1]. The online generation of a detailed aerial photo map by the UAV is of obvious interest in according SSRR missions.

Precise localization of UAVs is very challenging for several reasons. First of all, they typically only have a limited payload for sensors. Second, commonly found solutions like GPS receivers in combination with compasses only provide coarse pose information, which is not sufficient to fuse a sequence of images into a usable photo map. An alternative approach is to completely omit the problem of localization through additional sensors, and to only use the information in two sequentially acquired images to determine the robots movement between the two frames.

There is hence related work in the computer vision community. Optical flow techniques [2,3,4,5,6,7] are also targeted at motion estimation, but they are best suited for only estimating robot states for control, especially for aerial vehicles [8]. The classic optical flow techniques namely suffer from too large errors when the state is integrated for proper localization. An other line of research is related

---

* Formerly International University Bremen.

J. Baltes et al. (Eds.): RoboCup 2009, LNAI 5949, pp. 35–45, 2010.
© Springer-Verlag Berlin Heidelberg 2010

Fig. 1. A Jacobs land robot and an aerial robot cooperating in one of the scenarios at the European Land Robot Trials 2007 (ELROB) in Monte Ceneri, Switzerland. The aerial robot has to find and locate hazard spots like seats of fire, which the land robot then has to reach.

to computer vision work on structure from motion [9,10,11,12]. The methods are characterized by various application constraints, especially on the camera motions, as they were developed for specialized 3D model acquisitions. Here, an unconstrained integration of sequential 2D images is investigated.

So, the task here is that regions of overlap between two consecutively acquired images have to be found and suitably matched. This process of finding a template in an image is also known as registration [13,14,15,16,17,18,19]. But the task at hand is more difficult than mere registration as the region of overlap is unknown and it usually has undergone non-trivial transformations due to the robot's movements. This is comparable to image stitching [20], which is for example used to generate panoramic views from several overlapping photographs. We used this idea of employing stitching methods already before for merging 2D occupancy grid maps generated by multiple robots [21]. The scale invariant feature transform (SIFT) introduced by Lowe [20,22] is at present the most popular basis for image stitching. SIFT delivers point-wise correspondences between distinctive, non-repetitive loc al features in the two images. The number of detected features is significantly smaller than the number of pixels in the image. Other methods for identifying features include local image descriptors like intensity patterns [23,24] and the Kanade-Lucas-Tomasi Feature Tracker (KLT) [25]. Based on the Fourier Mellin transform for image representation and processing [26], we have developed an improved version that outperforms SIFT and KLT with respect to processing time and robustness as shown in experiments with real world images including aerial data.

This algorithm is applied here to generating photo maps and benchmarked in the Unified System for Automation and Robotics Simulation (USARsim) [27]. USARsim has the significant advantage that it provides a high fidelity physical simulation, i.e., realistic test data, while ground truth information of the robot is given [28,29,30,31]. This facilitates experimental validations, especially as supplements to field work where ground truth is not known.

The rest of this paper is structured as follows. A Fourier Mellin based approach to image registration is presented. It can be used to generate photo maps in a fast and robust way. In section 3, the approach is tested through experiments with a aerial vehicle in USARsim. Due to the availability of ground truth data, a quantitative assessment of strengths and weaknesses is possible. Section 4 concludes the paper.

2 Photo Mapping with Improved Fourier Mellin Registration

First, some basic terms and concepts are introduced on which our algorithm for photo mapping is based. The classical Matched Filter (MF) of two 2D signals $r * (-x, -y)$ and $s(x, y)$ is defined by:

$$q(x, y) = \int \int_{-\infty}^{\infty} s(a, b) r * (a - x, b - y) da db \qquad (1)$$

This function has a maximum at (x0,y0) that determines the parameters of a translation. One limitation of the MF is that the output of the filter primarily depends on the energy of the image rather that on its spatial structures. Furthermore, depending on the image structures the resulting correlation peak can be relatively broad. This problem can be solved by using a Phase-Only Matched Filter (POMF). This correlation approach makes use of the fact that two shifted signals having the same spectrum magnitude are carrying the shift information within its phase (equ.2). Furthermore the POMF calculation is much faster than the MF because if a signal frame of size 2^N is used, the advantages of the Fast Fourier Transform (FFT) can be exploited.

The principle of phase matching is now extended to additionally determine affine parameters like rotation, scaling and afterward translation.

$$f(t - a) \circ - \bullet F(\omega) e^{i\omega a} \qquad (2)$$

When both signals are periodically shifted the resulting inverse Fourier transformation of the phase difference of both spectra is actually an ideal Dirac pulse. This Dirac pulse indicates the underlying shift of both signals which have to registered.

$$d(t - a) \circ - \bullet 1 e^{i\omega a} \qquad (3)$$

The resulting shifted Dirac pulse deteriorates with changing signal content of both signals. As long as the inverse transformation yields a clear detectable maximum this method can used for matching two signals. This relation of the two signals phases is used for calculating the Fourier Mellin Invariant Descriptor (FMI). The next step for calculating the desired rotation parameter exploits the fact that the 2D spectrum 5 rotates exactly the same way as the signal in the time domain itself (equ.4):

$$s(x, y) = r[(x \cos(\alpha) + y \sin(\alpha)), (-x \sin(\alpha) + y \cos(\alpha))] \qquad (4)$$

$$| S(u,v) |=| R[(u\cos(\alpha) + v\sin(\alpha)), (-u\sin(\alpha) + v\cos(\alpha))] | \qquad (5)$$

where α is the corresponding rotation angle.

For turning this rotation into a signal shift the magnitude of the signals spectrum is simply re-sampled into polar coordinates. For turning a signal scaling into a signal shift several steps are necessary. The following Fourier theorem

$$f(\frac{t}{a}) \circ - \bullet \, aF(a\omega) \qquad (6)$$

shows the relations between a signal scaling and its spectrum. This relation can be utilized in combination with another transform called Mellin transform which is generally used for calculations of moments:

$$V^M(f) = \int_0^\infty v(z)z^{i2\pi f-1}dz \qquad (7)$$

Having two functions $v1(z)$ and $v2(z) = v1(az)$ differing only by a dilation the resulting Mellin transform with substitution $az = \tau$ is:

$$\begin{aligned}
V_2^M(f) &= \int_0^\infty v1(az)z^{i2\pi f-1}dz \\
&= \int_0^\infty v1(\tau)(\frac{\tau}{a})^{i2\pi f-1}d\tau \\
&= a^{-i2\pi f}V_1^M(f) \qquad (8)
\end{aligned}$$

The factor $a^{-i2\pi f} = e^{-i2\pi f ln(a)}$ is complex which means that with the following substitutions

$$z = e^{-t}, ln(z) = -t, dz = -e^{-t}dt,$$
$$z \to 0 \longrightarrow t \to \infty, z \to \infty \longrightarrow t \to -\infty \qquad (9)$$

the Mellin transform can be calculated by the Fourier transform with logarithmically deformed time axis:

$$\begin{aligned}
V^M(f) &= \int_\infty^{-\infty} v(e^{-t})e^{-t(i2\pi f-1)}(-e^{-t})dt \\
&= \int_{-\infty}^{\infty} v(e^{-t})e^{-i2\pi ft}dt \qquad (10)
\end{aligned}$$

Now the scaling of a function/signal using a logarithmically deformed axis can be transfered into a shift of its spectrum. Finally, the spectrum's magnitude is logarithmically re-sampled on its radial axis and concurrently the spectrum is arranged in polar coordinates exploiting the rotational properties of a 2D Fourier transform as described before. Scaling and rotation of an image frame are then transformed into a 2D signal shift where the 2D signal is actually the corresponding spectrum magnitude of the image frame.

Here, a sketch of the overall algorithm. The POMF is calculated as follows:

1. calculate the spectra of two corresponding image frames
2. calculate the phase difference of both spectra
3. apply an inverse Fourier transform of this phase difference

The following steps are taken for a full determination of the rotation, scaling and translation parameters:

1. calculate the spectra of two corresponding image frames
2. calculate the magnitude of the complex spectral data
3. resample the spectra to polar coordinates
4. resample the radial axes of the spectra logarithmically
5. calculate a POMF on the resampled magnitude spectra
6. determine the corresponding rotation/scaling parameters from the Dirac pulse
7. re-size and re-rotate the corresponding image frame to its reference counterpart
8. calculate a POMF between the reference and re-rotated/scaled replica image
9. determine the corresponding x,y translation parameters from the Dirac pulse

The steps are used in the Fourier Mellin based photo mapping in a straightforward way. A first reference image I_0 is acquired or provided to define the reference frame F and the initial robot pose p_0. Then, a sequence of images I_k is acquired. Image I_1 is processed with the above calculations to determine the transformations T_0^M between I_0 and I_1 and hence the motion of the robot. The robot pose is updated to p_1 and I_1 is transformed by according operations T_0^F to an image I_1' in reference frame F. The transformed image I_1' is then added to the photo map. From then on, the image I_n', i.e., the representation of the previous image in the photo map, is used to determine the motion-transformations T_n^M in the subsequent image I_{n+1}, which is used to update the pose p_{n+1} and the new part I_{n+1}' for the photo map.

3   Experiments and Results

Figures 2 and 3 show results of using the algorithm with real world UAV data for photo mapping. No localization information for the UAV is given, not even GPS data, only the raw video data is used. This real world data is well suited for a comparison with alternative approaches like SIFT, but only on the basis of a comparison of the performance between two consecutive images, i.e., single registration steps. In doing so, it can be shown that SIFT performs poorly on scenes with few distinct features and that it is in addition computationally expensive.

   The real world data is much less suited for assessing the quality of the photo mapping approach in a quantitative way as ground truth information is not available. The significant advantage of the following experiments within USARsim is that this restriction does not hold there. The presented results were processed by

Fig. 2. An image map generated with iFMI in real-time from about 300 images acquired with an UAV. The scene involves several challenges, especially large featureless areas.

a MATLAB implementation. A first C/C++ implementation were tested with 4 different frame sizes: (192x192) = 150msec, (256x256) = 230 msec, (320x320) = 260 msec and (480x480) = 430 msec on a standard PC with a 1.7 GHz CPU. The processing times already include data acquisition and overview display using the INTEL OpenCV library.

As one can see in figure 7, the approach works also well in USARsim to generate interesting overviews of a scenario - like a disaster scene - from the video stream of an aerial vehicle. The visual quality of the photo map is fully sufficient for mission planning and other qualitative tasks. The next question is how the approach performs in quantitative terms. Figures 4, 5, and 6 show a direct comparison of the estimated versus the ground truth positions. The x-axis shows time in terms of steps for 180 consecutively acquired images. The y-axis shows absolute coordinates in meter, respectively pixels for global x- and y-coordinates. The scaling shown in figure 6 shows the estimated, respectively ground truth scaling effects of the elevation of the aerial vehicle above ground; the higher the vehicle flies above ground the smaller the images its camera delivers. As mentioned, the generated photo map has a sufficient quality for performing high level tasks with it. But the quantitative analysis with USARsim allows to study the effects of the cumulative errors in the localization. Concretely, it can be shown that there is a severe drift in the real versus the estimated position, i.e., that the approach has its limits in the current form for localization. This is of interest for further work

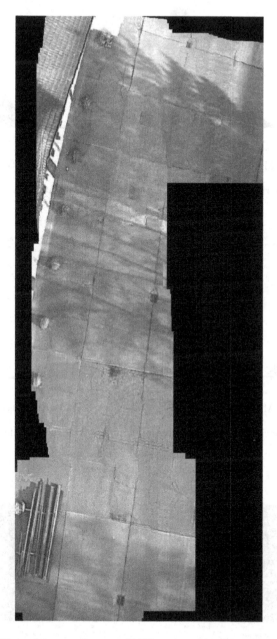

Fig. 3. About 600 areal images from an UAV are combined by iFMI in real-time into an image map

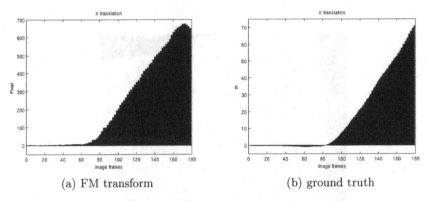

(a) FM transform   (b) ground truth

Fig. 4. Comparison of the estimated and the real translations along the x-axis

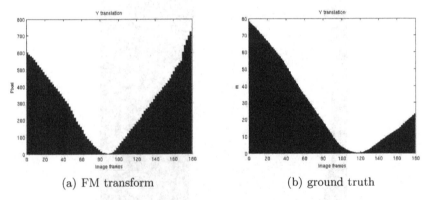

(a) FM transform   (b) ground truth

Fig. 5. Comparison of the estimated and the real translations along the y-axis

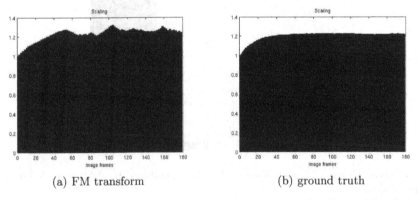

(a) FM transform   (b) ground truth

Fig. 6. Comparison of the estimated and the real scaling, i.e., the effects of the elevation of the blimp above ground

Fig. 7. Photo map M_1 generated from a sequence of 180 images in USARsim from a blimp without known poses

along this line of research, especially for incorporating this new visual odometry approach into a Simultaneous Localization and Mapping (SLAM) framework.

4   Conclusion

A Fourier Mellin Transform based approach to photo mapping was presented. It is benchmarked with experiments with videos from a blimp in USARsim. This has the advantage over field experiments that the data is physically realistic while ground truth is known for a detailed analysis. The experiments show that the approach is robust and fast. But the ground truth comparisons in USARsim also reveal limits of the approach, especially with respect to robot localization. While the photo maps are usable and provide an interesting overview of for example an incident scenario, the ground truth analysis shows that there is a clear drift in the localization estimation.

References

1. ELROB: European Land-Robot Trial, ELROB (2007),
   http://www.elrob2007.org
2. Bruhn, A., Weickert, J., Feddern, C., Kohlberger, T., Schnoerr, C.: Variational optical flow computation in real time. IEEE Transactions on Image Processing 14, 608–615 (2005)
3. Kohlberger, T., Schnoerr, C., Bruhn, A., Weickert, J.: Domain decomposition for variational optical-flow computation. IEEE Transactions on Image Processing 14, 1125–1137 (2005)
4. Liu, H., Hong, T., Herman, M., Camus, T., Chellappa, R.: Accuracy vs efficiency trade-offs in optical flow algorithms. Computer Vision and Image Understanding 72, 271–286 (1998)
5. McCane, B., Galvin, K.N.: On the evaluation of optical flow algorithms. In: Proceedings of the 5th International Conference On Control, Automation, Robotics and Vision, Singapore, pp. 1563–1567 (1998)
6. Beauchemin, S., Barron, J.: The computation of optical flow. ACM Computing Surveys (CSUR) 27, 433–466 (1995)
7. Horn, B., Schunck, B.: Determining optical flow. Artificial Intelligence 17, 185–203 (1981)
8. Kehoe, J., Causey, R., Arvai, A., Lind, R.: Partial aircraft state estimation from optical flow using non-model-based optimization. In: American Control Conference, invited paper, Minneapolis, MN, pp. 2868–2873 (2006)
9. Fitzgibbon, A., Zisserman, A.: Automatic camera recovery for closed or open image sequences. In: Burkhardt, H.-J., Neumann, B. (eds.) ECCV 1998. LNCS, vol. 1406, pp. 311–318. Springer, Heidelberg (1998)
10. Azarbayejani, A., Pentland, A.P.: Recursive estimation of motion, structure, and focal length. IEEE Transactions on Pattern Analysis and Machine Intelligence 17, 562–575 (1995)
11. Faugeras, O.: Three-dimensional computer vision – a geometric viewpoint (1993)
12. Adiv, G.: Inherent ambiguities in recovering 3-d motion and structure from a noisy flow field. IEEE Transactions on Pattern Analysis and Machine Intelligence 11, 477–489 (1989)

13. Fitch, A., Kadyrov, A., Christmas, W., Kittler, J.: Fast robust correlation. IEEE Transactions on Image Processing 14, 1063–1073 (2005)
14. Stricker, D.: Tracking with reference images: a real-time and markerless tracking solution for out-door augmented reality applications. In: Proceedings of the 2001 conference on Virtual reality, archeology, and cultural heritage, pp. 77–82. ACM Press, New York (2001)
15. Dorai, C., Wang, G., Jain, A.K., Mercer, C.: Registration and integration of multiple object views for 3d model construction. IEEE Transactions on Pattern Analysis and Machine Intelligence 20(1), 83–89 (1998)
16. Brown, L.G.: A survey of image registration techniques. ACM Comput. Surv. 24(4), 325–376 (1992)
17. Alliney, S., Morandi, C.: Digital image registration using projections. IEEE Trans. Pattern Anal. Mach. Intell. 8(2), 222–233 (1986)
18. Lucas, B., Kanade, T.: An iterative image registration technique with an application to stereo vision. In: Proceedings DARPA Image Understanding Workshop, pp. 121–130 (1981)
19. Pratt, W.: Correlation techniques of image registration. IEEE Transactions on Aerospace and Electronic Systems AES-10, 562–575 (1973)
20. Lowe, D.G.: Distinctive image features from scale-invariant keypoints. International Journal of Computer Vision 60(2), 91–110 (2004)
21. Birk, A., Carpin, S.: Merging occupancy grid maps from multiple robots. IEEE Proceedings, special issue on Multi-Robot Systems 94(7), 1384–1397 (2006)
22. Lowe, D.G.: Object recognition from local scale-invariant features. In: Proceedings of International Conference on Computer Vision, pp. 1150–1157 (1999)
23. Mikolajczyk, K., Schmid, C.: A performance evaluation of local descriptors. In: Proceedings of Computer Vision and Pattern Recognition (June 2003)
24. Gool, L.V., Moons, T., Ungureanu, D.: Affine/photometric invariants for planar intensity patterns. In: Proceedings of European Conference on Computer Vision (1996)
25. Shi, J., Tomasi, C.: Good features to track. In: IEEE Conference on Computer Vision and Pattern Recognition, CVPR 1994 (1994)
26. Chen, Q., Defrise, M., Deconinck, F.: Symmetric phase-only matched filtering of fourier-mellin transforms for image registration and recognition. IEEE Transactions on Pattern Analysis and Machine Intelligence 16, 1156–1168 (1994)
27. USARsim: Unified System for Automation and Robotics Simulator, USARsim (2006), http://usarsim.sourceforge.net/
28. Carpin, S., Lewis, M., Wang, J., Balakirsky, S., Scrapper, C.: Bridging the gap between simulation and reality in urban search and rescue. In: Lakemeyer, G., Sklar, E., Sorrenti, D.G., Takahashi, T. (eds.) RoboCup 2006: Robot Soccer World Cup X. LNCS (LNAI), vol. 4434, pp. 1–12. Springer, Heidelberg (2007)
29. Carpin, S., Lewis, M., Wang, J., Balarkirsky, S., Scrapper, C.: USARSim: a robot simulator for research and education. In: Proc. of the 2007 IEEE Intl. Conf. on Robotics and Automation, ICRA (2007)
30. Carpin, S., Stoyanov, T., Nevatia, Y., Lewis, M., Wang, J.: Quantitative assessments of usarsim accuracy. In: Proceedings of PerMIS (2006)
31. Carpin, S., Birk, A., Lewis, M., Jacoff, A.: High fidelity tools for rescue robotics: results and perspectives. In: Bredenfeld, A., Jacoff, A., Noda, I., Takahashi, Y. (eds.) RoboCup 2005. LNCS (LNAI), vol. 4020, pp. 301–311. Springer, Heidelberg (2006)

Real-Time Hand Gesture Recognition for Human Robot Interaction

Mauricio Correa[1,2], Javier Ruiz-del-Solar[1,2], Rodrigo Verschae[1], Jong Lee-Ferng[1], and Nelson Castillo[1]

[1] Department of Electrical Engineering, Universidad de Chile
[2] Center for Mining Technology, Universidad de Chile
{jruizd,rverscha,macorrea,jolee}@ing.uchile.cl

Abstract. In this article a hand gesture recognition system that allows interacting with a service robot, in dynamic environments and in real-time, is proposed. The system detects hands and static gestures using cascade of boosted classifiers, and recognize dynamic gestures by computing temporal statistics of the hand's positions and velocities, and classifying these features using a Bayes classifier. The main novelty of the proposed approach is the use of context information to adapt continuously the skin model used in the detection of hand candidates, to restrict the image's regions that need to be analyzed, and to cut down the number of scales that need to be considered in the hand-searching and gesture-recognition processes. The system performance is validated in real video sequences. In average the system recognized static gestures in 70% of the cases, dynamic gestures in 75% of them, and it runs at a variable speed of 5-10 frames per second.

Keywords: dynamic hand gesture recognition, static hand gesture recognition, context, human robot interaction, RoboCup @Home.

1 Introduction

Hand gestures are extensively employed in human non-verbal communication. They allow to express orders (e.g. "stop"), mood state (e.g. "victory" gesture), or to transmit some basic cardinal information (e.g. "two"). In addition, in some special situations they can be the only way of communicating, as in the cases of deaf people (sign language) and police's traffic coordination in the absence of traffic lights.

Thus, it seems convenient that human-robot interfaces incorporate hand gesture recognition capabilities. For instance, we would like to have the possibility of transmitting simple orders to personal robots using hand gestures. The recognition of hand gestures requires both hand's detection and gesture's recognition. Both tasks are very challenging, mainly due to the variability of the possible hand gestures (signs), and because hands are complex, deformable objects (a hand has more than 25 degrees of freedom, considering fingers, wrist and elbow joints) that are very difficult to detect in dynamic environments with cluttered backgrounds and variable illumination.

Several hand detection and hand gesture recognition systems have been proposed. Early systems usually require markers or colored gloves to make the recognition easier. Second generation methods use low-level features as color (skin detection) [4][5],

J. Baltes et al. (Eds.): RoboCup 2009, LNAI 5949, pp. 46–57, 2010.

shape [8] or depth information [2] for detecting the hands. However, those systems are not robust enough for dealing with dynamic conditions; they usually require uniform background, uniform illumination, a single person in the camera view [2], and/or a single, large and centered hand in the camera view [5]. Boosted classifiers allow the robust and fast detection of hands [3][6][7]. In addition, the same kind of classifiers can be employed for detecting static gestures [7]; dynamic gestures are normally analyzed using Hidden Markov Models [4][23]. 3D hand model-based approaches allow the accurate modeling of hand movement and shapes, but they are time-consuming and computationally expensive [6][7].

In this context, we are proposing a robust and real-time hand gesture recognition system to be used in the interaction with personal robots. We are especially interested in dynamic environments such as the ones defined in the *RoboCup @Home league* [20] (our team participates in this league [21]), with the following characteristics: variable illumination, cluttered backgrounds, (near) real-time operation, large variability of hands' pose and scale, and limited number of gestures (they are used for giving the robot some basic information). It is important to mention that in the new RoboCup @Home league' rules gesture recognition is emphasized: *An aim of the competition is to foster natural interaction with the robot using speech and gesture commands* (2009's Rules book, pp. 7, available in [20]). For instance, in the new "Follow me" test, gesture recognition is required to complete adequately the test (2009's Rules book, pp. 23: *When the robot arrives at [...] it is stopped by a HRI command (speech, gesture recognition or any other 'natural' interaction), and using HRI the robot should either move backwards, move forward, turn left or turn right [...]. Then the robot is commanded using HRI to follow the walker.*).

The proposed system is able to recognize static and dynamic gestures, and their most innovative features include:

- The use of context information to achieve, at the same time, robustness and real-time operation, even when using a low-end processing unit (standard notebook), as in the case of humanoid robots. The use of context allows to adapt continuously the skin model used in the detection of hand candidates, to restrict the image's regions that need to be analyzed, and to cut down the number of scales that need to be considered in the hand-searching and gesture recognition processes.

- The employment of boosted classifiers for the detection of faces and hands, as well as the recognition of static gestures. The main novelty is in the use of innovative training techniques - active learning and bootstrap -, which allow obtaining a much better performance than similar boosting-based systems, in terms of detection rate, number of false positives and processing time.

- The use of temporal statistics about the hand's positions and velocities and a Bayes classifier to recognize dynamic gestures. This approach is different from the traditional ones, based on Hidden Markov Models, which are not able to achieve real-time operation.

This article is focused on the description of the whole system, and the use of context to assist the gesture recognition processes. In sections 2 the rationale behind the use of context information in the proposed gesture recognition system is described. In section 3 the whole gesture recognition system and its modules are described. Results of the application of this system in real video sequences are presented and analyzed in section 4. Finally, some conclusions of this work are given in section 5.

2 Context Integration in HRI: Improving Speed and Robustness

Visual perception in complex and dynamical scenes with cluttered backgrounds is a very difficult task, which humans can solve satisfactorily. However, computer and robot vision systems perform very badly in this kind of environments. One of the reasons of this large difference in performance is the use of context or contextual information by humans. Several studies in human perception have shown that the human visual system makes extensive use of the strong relationships between objects and their environment for facilitating the object detection and perception ([13]-[17], just to name a few). Context can play a useful role in visual perception in at least three forms [18]: (i) Reducing perceptual aliasing: 3D objects are projected onto a 2D sensor, and therefore in many cases there is an ambiguity in the object identity. Information about the object surround can be used for reducing or eliminating this ambiguity; (ii) Increasing perceptual abilities in hard conditions: Context can facilitate the perception when the local intrinsic information about the object structure, as for example the image resolution, is not sufficient; (iii) Speeding up perceptions: Contextual information can speed up the object discrimination by cutting down the number of object categories, scales and poses that need to be considered.

The recognition of static and dynamic gestures in dynamic environments is a difficult task that usually requires the use of image processing algorithms to improve the quality of the images under analysis and to extract the required features (color, movement and even texture information), and statistical classifiers to detect the hands and to classify the gestures. In HRI applications there exists a tradeoff between carrying out a detailed analysis of the images, using an image's resolution that allows recognizing gestures at a given distance of a few meters, which usually can take more than one second per image, and the requirement of real-time operation to allow a proper interaction with humans. Context can be used to deal with this situation, and to achieve, at the same time, robustness and real-time operation, even when using a low-end processing unit (standard notebook), as in the case of humanoid robots.

In this work, the main sources of context to be used are human faces appearing in the image, and the existence of a physical world with defined laws of movement. Main assumptions are:

- We are dealing with an HRI application in which a human is communicating with a robot using hand gestures. Therefore a frontal human face will be observed in some or even several frames of the video sequence.

- Frontal face detectors are much more robust than hand detectors, mainly due to the fact that a hand is a deformable object with more than 25 degrees, whose pose changes largely depending on the observer's viewpoint. In the literature it can be observed that frontal face detectors achieve a much higher detection rates than hand detectors, and they are much faster.

- The robot and the human have upright positions, and their bodies (hands, heads, main-body, etc.) move according with the physical rules (gravity, etc.). This allows (i) to make some basic assumptions about the relative position and scale of the objects, as well as about their orientation, and (ii) to track the position of detected objects (e.g. a face), and to actively determine their position in the next frames.

- Normally the human user is not wearing gloves and the hand-gesture is a part of a sequence, in which the hand is moved. Therefore, candidate hand regions can be detected using skin and motion information.

In the proposed gesture recognition system a face detector is incorporated, and the information about the detected face is used to: (i) adapt continuously the skin model using the pixels contained in a sub-region of the face's area, (ii) determine the image's region that need to be analyzed for detecting skin and movement, as well as new faces, (iii) cut down the number of scales that need to be considered in the hand-searching process, (iv) normalize the input to the dynamic gesture recognition module, so that it is translation's and scale's invariant. In addition, (v) hand-searching process is restricted to regions where a minimal amount of skin and movement is detected, and (vi) after detecting a hand for the first time, it is tracked until track is lost. Then, hand detection is restarted. In the next section all these processes are explained.

3 Proposed Hand Gesture Recognition System

The system consists of five main modules *Face Detection and Tracking* (FDT), *Skin Segmentation and Motion Analysis* (SMA), *Hand Detection and Tracking* (HDT), *Static Gesture Recognition*, and *Dynamic Gesture Recognition* (see figure 1).

The FDT module is in charge of detecting and tracking faces. These functionalities are implemented using boosted statistical classifiers [11], and the *meanshift* algorithm [1], respectively. The information about the detected face (DF) is used as context in the SMA and HDT modules. Internally the CF1 (Context Filter 1) module determines the image area that needs to be analyzed in the current frame for face detection, using the information about the detected faces in the past frame.

The SMA module determines candidate hand regions to be analyzed by the HDT module. The *Skin Segmentation* module uses a skin model that is adapted using information about the face-area's pixels (skin pixels) in order to achieve some illumination invariance. The module is implemented using the *skindiff* algorithm [9]. The *Motion Analysis* module is based on the well-known background subtraction technique. CF2 (Context Filter 2) uses information about the detected face and the human-body dimensions to determine the image area (HRM: *Hand Region Mask*) where a hand can be present in the image. Only this area is analyzed by the *Skin Segmentation* and *Motion Analysis* modules.

The HDT module is in charge of detecting and tracking hands. These functionalities are implemented using boosted statistical classifiers and the *meanshift* algorithm, respectively. CF3 (Context Filter 3) determines the image area where a hand can be detected in the image, using the following information sources: (i) skin mask (SM) which corresponds to a skin probability mask, (ii) motion mask (MM) that contains the motion pixels, and (iii) information about the hands detected in the last frame (DH: *Detected Hand*).

The Static Gesture Recognition module is in charge of recognizing static gestures. The module is implemented using statistical classifiers: a boosted classifier for each gesture class, and a multi-class classifier (J48 pruned tree, Weka's [19] version of C4.5) for taking the final decision. The Dynamic Gesture Recognition module recognizes dynamic gestures. The module computes temporal statistics about the hand's positions and velocities. These features feed a Bayes classifier that recognizes the gesture.

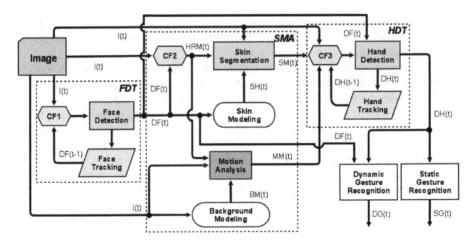

Fig. 1. Proposed hand gesture recognition system. CF*i*: *Context Filter i*. I: Image. DF: *Detected Face*. HRM: *Hand Region Mask*. SH: *Skin Histogram*. SM: *Skin Mask*: BM: *Background Model*. MM: *Motion Mask*. DH: *Detected Hand*. DG: *Dynamic Gesture*. SG: *Static Gesture*. *t*: Frame index. See main text for a detailed explanation.

3.1   Face Detection and Tracking

The FDT module is in charge of detecting and tracking faces. These functionalities are implemented using boosted statistical classifiers [11] and the *meanshift* algorithm [1], respectively. The face detector corresponds to a nested cascade of boosted classifiers, which is composed by several integrated (nested) layers, each one containing a boosted classifier. The cascade works as a single classifier that integrates the classifiers of every layer. Weak classifiers are linearly combined, obtaining a strong classifier.

The *meanshift* algorithm is used to predict the face position in the next frame. The seed of the tracking process is the detected face. We use RGB color histograms as feature vectors (model) for *meanshift*, with each channel quantized to 16 levels (4 bits) and the feature vector weighted using an Epanechnikov kernel [1]. The prediction given by *meanshift* is internally used by the CF1 (Context Filter 1) module to determine the image area that needs to be analyzed in the current frame:

$$x^v = \max\left(0, x^f - w^f\right) y^v = \max\left(0, y^f - h^f\right) w^v = \min\left(3 \cdot w^f, I_{width}\right) h^v = \min\left(3 \cdot h^f, I_{height}\right) \quad (1)$$

with x^f / y^f the *x/y* coordinates of the detected face (bounding box) upper-left corner, w^f / h^f the face's width/height, I_{width} / I_{height} the image's width/height, and x^v, y^v, w^v, h^v the coordinates, width and height of the image's area to be analyzed.

If a face is not detected in a frame, the prediction given by *meanshift* is used instead. The tracking module is reset after a fixed number of frames (about 200) in order to deal with cases such as faces incorrectly tracked or detected, or new persons entering the detection area.

3.2   Skin Segmentation and Motion Analysis

The SMA module determines candidate hand regions to be analyzed in the HDT module. *Skin Segmentation* is implemented using the *skindiff* algorithm [9]. *Skindiff* is a fast skin detection algorithm that uses neighborhood information (local spatial context). It has two main processing stages, pixel-wise classification and spatial diffusion. The pixel-wise classification uses a skin probability model G_t, and the spatial diffusion takes into account neighborhood information when classifying a pixel. The skin probability model is continuously adapted using information of the face-area's pixels (skin pixels). The adaptation is done by taking skin pixels of the face area, and updating a non-parametric skin model implemented using histograms:

$$G_t = G_{t-1}\alpha + \hat{G}_{face(t)}(1-\alpha), \tag{2}$$

where $\hat{G}_{face(t)}$ is estimated using the currently detected face, and G_o is the initial model, which can be initialized from a previously stored model (in our case the MoG model proposed in [22]).

The *Motion Analysis* module is based on the well-known background subtraction technique. CF2 (Context Filter 2) uses information about the detected face, the fact that in our system gestures should be made using the right hand, and the human-body dimensions to determine the image area (HRM: *Hand Region Mask*) where a hand can be present in the image:

$$x^w = \max\left(0, x^f - 3 \cdot w^f\right) y^w = \max\left(0, y^f - h^f\right)$$
$$w^w = \min\left(4.5 \cdot w^f, I_{width}\right) h^w = \min\left(4 \cdot h^f, I_{height}\right) \tag{3}$$

with x^w, y^w, w^w, h^w the coordinates, width and height of the image's area to be analyzed. Note that just this area is analyzed by *Skin Segmentation* and *Motion Analysis* modules.

3.3 Hand Detection and Tracking

In order to detect hands within the image area defined by the HRM a cascade of boosted classifiers is used. Although this kind of classifiers allows obtaining very robust object detectors in the case of face or car objects, we could not build a reliable generic hand detector easily. This mainly because: (i) hands are complex, highly deformable objects, (ii) hand possible poses (gestures) have a large variability, and (iii) our target is a fully dynamic environment with cluttered background. Therefore we decided to switch the problem to be solved, and to define that the first time that a hand should be detected, a specific gesture must be made, the fist gesture. The fist is detected using a boosted classifier, similar to the one used for face detection, but built specifically for that gesture. The hand detector also takes as input the skin mask and the motion mask, and only analyzes regions where at least 5% of the pixels correspond to skin and movement. The integral image representation is employed to speedup this calculation (regions of different sizes can be evaluated very fast) [12].

The hand-tracking module is built using the *meanshift* algorithm [1]. The seeds of the tracking process are the detected hands (fist gesture). We use RG color and rotation invariant LBP histograms as feature vectors (model) for *meanshift*, with each channel quantized to 16 levels (4 bits). The feature vector is weighted using an Epanechnikov kernel [1]. Rotation invariant LBP features encode local gradient information, and they are needed because if only color is used, some times *meanshift* tracks the arm instead of the hand.

As already mentioned, once the tracking module is correctly following a hand, there is no need to continue applying the hand detector, i.e. the fist gesture detector, over the skin blobs. That means that the hand detector module is not longer used until the hand gets out of the input image, or until the *meanshift* algorithm loses track of the hand, case where the hand detector starts working again. At the end of this stage, one or several regions of interest (ROI) are obtained, each one indicating the location of a hand in the image. This module is explained in detail in [10].

3.4 Recognition of Static Gestures

In order to determine which gesture is being expressed, a set of single gesture detectors are applied in parallel over the ROIs delivered as output of the HDT module (DH: Detected Hand). Each single gesture detector is implemented using a cascade of boosted classifiers. The learning framework employed for building and training these classifiers is described in [11]. Currently we have implemented detectors for the following gestures: *fist*, *palm*, *pointing*, and *five* (see figure 2).

Due to noise or gesture ambiguity, it could happen than more than one gesture detector will give positive results in a ROI (more than one gesture is detected). For discriminating among these gestures, a multi-gesture classifier is applied. The used multi-class classifier is a *J48 pruned tree (Weka's* [19] *version of C4.5)*, built using the following four attributes that each single gesture detector delivers:
- *conf*: sum of the cascade confidence's values of windows where the gesture was detected (a gesture is detected at different scales and positions),
- *numWindows*: number of windows where the gesture was detected,
- *meanConf*: mean confidence value given by *conf/numWindows*, and
- *normConf*: normalized mean confidence value given by *meanConf/maxConf*, with *maxConf* the maximum possible confidence that a window could get.
  This module is explained in detail in [10].

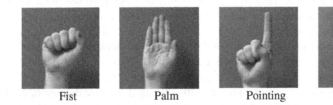

| Fist | Palm | Pointing | Five |

Fig. 2. Hand gestures detected by the system

3.5 Recognition of Dynamic Gestures

The Dynamic Gesture Recognition Module (DGRM) stores and analyzes sequences of detected hands (DH) online. The number of stored detections is fixed, so older detections that would exceed the predefined capacity are discarded as new detections arrive. Stored detections are discarded altogether when an inactivity condition is detected (still hand, hand out of camera range). Every time a new detection arrives, subsequences of the stored sequence that end with this new detection are analyzed. This analysis consists of computing a feature vector that comprises geometric and kinematical characteristics of the subsequence. Each subsequence's feature vector is fed to a Naïve Bayes classifier, which calculates a score for each possible dynamic gesture. This score represents the likelihood of the gesture in the given subsequence. In other words, every time a new detection (DH) arrives, a set of scores associated to each gesture is obtained (each score corresponding to a given subsequence). For each gesture, the highest of these scores is taken to be the best likelihood of that gesture having occurred, given the last frame. Finally, for each frame and each gesture, only this highest score is kept.

The highest scores alone could be used to determine the recognized gesture at the moment. However, in order to add robustness, the score should be consistently high during a interval of frames. So, for each gesture, the moving average of the last k highest scores is kept. In any given moment, the gesture with the best moving-average (*bma*) score is declared as the recognized gesture of that moment. Since not any frame is a real-end of a gesture, gesture segmentation becomes a problem. Thresholding the *bma* is a possible approach for gesture spotting. The thresholds can be learned from the training set. In addition, the current *bma* can be decremented in each round as a penalty for the subsequence from which it was extracted becoming older.

Each detected hand is represented as a vector (x, y, v_x, v_y, t), where (x, y) is the hand's position, (v_x, v_y) the hand's velocity, and t the frame's timestamp. In order to achieve translation and scale invariance, coordinates (x, y) are measured with respect to the face, and normalized by the size of the face. Using this vector, statistics (features) that characterize the sequences are evaluated. Some of the features are: mean hand's position in the x and y axis, mean hand's speed in the x and y axis, components of the covariance matrix of vector (x, y, v_x, v_y), area and perimeter of the convex hull of the (x, y) positions, average radius and angle with respect to a coordinate system placed on the mean (x, y) point, the percentage of points that fall on each cell of a 3x3 grid that exactly covers the positions of all detected hands, among others. Note that most of these features can be quickly evaluated using the same features evaluated in the previous frame (e.g. moving average).

4 Results

The whole gesture recognition system was evaluated in real video sequences obtained in office environments with dynamic conditions of illumination and background. In all these sequences the service robot interact with the human user at a variable distance of one to two meters (see example in figure 3). The size of the video frames is 320x240 pixels, and the robot main computer where the gesture recognition system runs is a standard notebook (Tablet HP 2710p, Windows Tablet SO, 1.2 GHz, 2 GB in RAM). Under these conditions, once the system detects the user's face, it is able to run at a variable speed of 5-10 frames per second, which is enough to allow an adequate interaction between the robot and the human user. The system's speed is variable because it depends on the convergence time of *meanshift* and the face and hands statistical classifiers. It should be noted that when the different context filters are deactivated and the complete image is analyzed, the system's speed is lower than 1 frame per second. This indicates that the use of context is essential to achieve the application requirements.

Fig. 3. Interacting with the robot in an unstructured environment

Recognition of Static Gestures. In order to evaluate this module, a database of 5 real-video sequences consisting of 8,150 frames, obtained in office environments, with variable illumination and cluttered backgrounds was built. In each sequence a single human was always interaction with our robot (altogether 4 different persons performing the 4 considered gestures). In figure 4 are shown the ROC curves of the single, static gesture detectors. Table 1 shows a confusion matrix of the multi-gesture recognition module, which consists of the four single, static gesture detectors and the multi-gesture classifier, evaluated on the same video sequences. The first thing that should be mention is that the hand detection system together with the tracking system did not produce any false negative out of the 8,150 analyzed frames, i.e. the hands were detected in all cases. From table 1 it can be observed that the gesture detection and recognition modules worked best on the *five* gesture, followed by the *pointing*, *fist* and *palm* gestures, in that order. The main problem is the confusion of the *fist* and *pointing* gestures, which is mainly due to the similarly of the gestures. In average the system correctly recognized the gestures in 70% of the cases. If the *pointing* and the *fist* gestures are considered as one gesture, the recognition rate goes up to 86%.

Recognition of Dynamic Gestures. We evaluate the proposed gesture recognition framework in the *10 Palm Graffiti Digits* database [23], where users perform gestures corresponding to the 10 digits (see example in figure 5). In the experiments the users and signers can wear short sleeved shirts, the background may be arbitrary (e.g, an office environment) and even contain other moving objects, and hand-over-face occlusions are allowed. We use the easy test set, which contains 30 short sleeve sequences, three from each of 10 users (altogether 300 sequences).

The system was able to detect and track hands in 266 of the 300 sequences (89%). In these 266 sequences, the dynamic gestures (i.e. digits) were correctly recognized in 84% of the cases. This corresponds to a 75% recognition rate (225 from 300 cases). It can be seen that this recognition rate is very similar to the one obtained in state of the art systems (e.g. [23], based on Hidden Markov Models), which are not able to operate in real-time or near real-time.

Table 2 shows a confusion matrix of the dynamic gesture recognition. It can be observed that the recognition rate of 6 digits is very high ("0"-"4", "8" and "9"). Two digits are recognized in most of the cases ("6" and "7"), and just the "5" digit has recognition problems. The "5" is confused, most of the time with the "3".

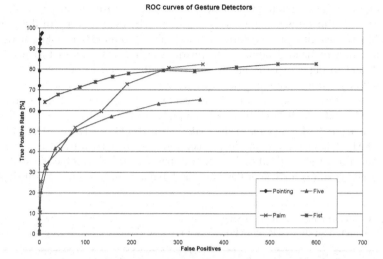

Fig. 4. ROC curves of the single, static gesture detectors

Fig. 5. Example of tracked hands in the *10 Palm Graffiti Digits* database [23]

Table 1. Confusion matrix of the final static, multi-gesture recognition module (rows: real gesture, columns: predicted gesture). RR: Recognition Rate.

	Fist	Palm	Pointing	Five	Unknown	RR (%)
Fist	1,533	2	870	9	15	63.1
Palm	39	1,196	10	659	15	62.3
Pointing	436	36	1,503	27	86	72.0
Five	103	32	6	1,446	127	84.3

5   Conclusions

In this article a hand gesture recognition system that allows interacting with a service robot, in dynamic environments and in real-time, was described. The system detect hands and static gestures using cascade of boosted classifiers, and recognize dynamic gestures by computing temporal statistics of the hand's positions and velocities, and classifying these features using a Bayes classifier. The main novelty of the proposed approach is the use of context information to adapt continuously the skin model used in the detection of hand candidates, to restrict the image's regions that need to be analyzed, and to cut down the number of scales that need to be considered in the hand-searching and gesture-recognition processes.

The system performance is validated in real video sequences. The size of the video frames is 320x240 pixels, and the robot computer where the gesture recognition system runs is a standard notebook (Tablet HP 2710p, Windows Tablet SO, 1.2 GHz, 2 GB in RAM). Under these conditions, once the system detects the user's face, it is able to run at a variable speed of 5-10 frames per second. In average the system recognized static gestures in 70% of the cases, and dynamic gestures in 75% of them.

Table 2. Confusion matrix of the dynamic gesture recognition module (rows: real gesture, columns: predicted gesture). TP: True Positives. FP: False Positives. RR: Recognition Rate.

	0	1	2	3	4	5	6	7	8	9	TP	FP	RR (%)
0	20	0	0	0	0	0	0	0	1	0	20	1	95
1	0	30	0	0	0	0	0	0	0	0	30	0	100
2	0	0	22	0	0	0	0	0	0	0	22	0	100
3	0	0	0	26	0	0	0	0	0	0	26	0	100
4	0	0	0	0	30	0	0	0	0	0	30	0	100
5	0	0	0	22	0	3	2	0	0	0	3	24	11
6	4	0	0	0	0	0	23	0	0	0	23	4	85
7	0	9	0	0	0	0	0	18	0	1	18	10	64
8	0	0	0	0	0	0	0	0	28	0	28	0	100
9	0	0	0	2	0	0	0	0	0	25	25	2	93

Acknowledgements

This research was partially funded by FONDECYT project 1090250, Chile.

References

1. Comaniciu, D., Ramesh, V., Meer, P.: Kernel-Based Object Tracking. IEEE Trans. on Pattern Anal. Machine Intell. 25(5), 564–575 (2003)
2. Liu, X., Fujimura, K.: Hand gesture recognition using depth data. In: Proc. 6th Int. Conf. on Automatic Face and Gesture Recognition, Seoul, Korea, pp. 529–534 (2004)
3. Kolsch, M., Turk, M.: Robust hand detection. In: Proc. 6th Int. Conf. on Automatic Face and Gesture Recognition, Seoul, Korea, pp. 614–619 (2004)
4. Dang Binh, N., Shuichi, E., Ejima, T.: Real-Time Hand Tracking and Gesture Recognition System. In: Proc. GVIP 2005, Cairo, Egypt, pp. 19–21 (2005)

5. Manresa, C., Varona, J., Mas, R., Perales, F.: Hand Tracking and Gesture Recognition for Human-Computer Interaction. Electronic letters on computer vision and image analysis 5(3), 96–104 (2005)
6. Fang, Y., Wang, K., Cheng, J., Lu, H.: A Real-Time Hand Gesture Recognition Method. In: Proc. 2007 IEEE Int. Conf. on Multimedia and Expo, pp. 995–998 (2007)
7. Chen, Q., Georganas, N.D., Petriu, E.M.: Real-time Vision-based Hand Gesture Recognition Using Haar-like Features. In: Proc. Instrumentation and Measurement Technology Conf. – IMTC 2007, Warsaw, Poland (2007)
8. Angelopoulou, A., García-Rodriguez, J., Psarrou, A.: Learning 2D Hand Shapes using the Topology Preserving model GNG. In: Leonardis, A., Bischof, H., Pinz, A. (eds.) ECCV 2006. LNCS, vol. 3951, pp. 313–324. Springer, Heidelberg (2006)
9. Ruiz-del-Solar, J., Verschae, R.: Skin Detection using Neighborhood Information. In: 6th Int. Conf. on Face and Gesture Recognition – FG 2004, Seoul, Korea, May 2004, pp. 463–468 (2004)
10. Francke, H., Ruiz-del-Solar, J., Verschae, R.: Real-time Hand Gesture Detection and Recognition using Boosted Classifiers and Active Learning. In: Mery, D., Rueda, L. (eds.) PSIVT 2007. LNCS, vol. 4872, pp. 533–547. Springer, Heidelberg (2007)
11. Verschae, R., Ruiz-del-Solar, J., Correa, M.: A Unified Learning Framework for object Detection and Classification using Nested Cascades of Boosted Classifiers. Machine Vision and Applications 19(2), 85–103 (2008)
12. Viola, P., Jones, M.: Rapid object detection using a boosted cascade of simple features. In: Proc. IEEE Conf. on Computer Vision and Pattern Recognition, pp. 511–518 (2001)
13. Torralba, A., Sinha, P.: On Statistical Context Priming for Object Detection. In: Int. Conf. on Computer Vision – ICCV 2001, vol. 1, pp. 763–770 (2001)
14. Cameron, D., Barnes, N.: Knowledge-based autonomous dynamic color calibration. In: Polani, D., Browning, B., Bonarini, A., Yoshida, K. (eds.) RoboCup 2003. LNCS (LNAI), vol. 3020, pp. 226–237. Springer, Heidelberg (2004)
15. Jüngel, M., Hoffmann, J., Lötzsch, M.: A real time auto adjusting vision system for robotic soccer. In: Polani, D., Browning, B., Bonarini, A., Yoshida, K. (eds.) RoboCup 2003. LNCS (LNAI), vol. 3020, pp. 214–225. Springer, Heidelberg (2004)
16. Oliva, A.: Gist of the Scene, Neurobiology of Attention, pp. 251–256. Elsevier, San Diego (2003)
17. Strat, T.: Employing contextual information in computer vision. In: Proc. of DARPA Image Understanding Workshop, pp. 217–229 (1993)
18. Palma-Amestoy, R., Guerrero, P., Ruiz-del-Solar, J., Garretón, C.: Bayesian Spatiotemporal Context Integration Sources in Robot Vision Systems. In: Iocchi, L., Matsubara, H., Weitzenfeld, A., Zhou, C. (eds.) RoboCup 2008. LNCS (LNAI), vol. 5399, pp. 212–224. Springer, Heidelberg (2009)
19. Witten, I.H., Frank, E.: Data Mining: Practical machine learning tools and techniques, 2nd edn. Morgan Kaufmann, San Francisco (2005)
20. RoboCup @Home Official website (January 2009), http://www.robocupathome.org/
21. UChile RoboCup Teams official website (January 2009), http://www.robocup.cl/
22. Jones, M.J., Rehg, J.M.: Statistical Color Models with Application to Skin Detection. Int. Journal of Computer Vision 46(1), 81–96 (2002)
23. Alon, J., Athitsos, V., Yuan, Q., Sclaroff, S.: A Unified Framework for Gesture Recognition and Spatiotemporal Gesture Segmentation. IEEE Trans. on Pattern Anal. Machine Intell. (in press, electrically available on July 28, 2008)

Combining Key Frame Based Motion Design with Controlled Movement Execution

Stefan Czarnetzki, Sören Kerner, and Daniel Klagges

Robotics Research Institute
Section Information Technology
Dortmund University of Technology
44221 Dortmund, Germany

Abstract. This article presents a novel approach for motion pattern generation for humanoid robots combining the intuitive specification via key frames and the robustness of a ZMP stability controller. Especially the execution of motions interacting with the robot's environment tends to result in very different stability behavior depending on the exact moment, position and force of interaction, thus providing problems for the classical replay of prerecorded motions. The proposed method is applied to several test cases including the design of kicking motions for humanoid soccer robots and evaluated in real world experiments which clearly show the benefit of the approach.

1 Introduction

As the field of robotics shifts to more complex tasks such as search and rescue or military operations, but also service and entertainment activities, robots themselves are becoming more autonomous and mobile. To fulfill tasks in the later two areas of application, robots must be capable of navigating in and interacting with environments made for humans, and of communicating with people in their natural ways. Those environments are particularly challenging for the movement of conventional wheeled autonomous robots. Normal stairs or small objects lying on the floor become insurmountable barriers. For these reasons the design of such robots tends to mimic human appearance in respect to body design, capability of gestures and facial expressions [1].

As a consequence humanoid robots are one of the major topics of robotics research and are believed to have high potential in future applications. Despite this, present humanoid robots have a substantial lack in mobility. The humanoid shaped form of a two-legged robot results in a relatively high center of mass (CoM) of the body while standing upright. As a result the stance of a humanoid robot is quite unstable, making it likely to tip over. Even basic tasks as walking on even ground without external disturbance are not a trivial challenge. Therefore stability is one of the central problems in this area at the moment, with research focusing mainly on the task of walking. The execution of interactions with the robot's environment represents an even more difficult task because of

J. Baltes et al. (Eds.): RoboCup 2009, LNAI 5949, pp. 58–68, 2010.

the necessary coordination of sensor input and actuator control for targeting and for keeping a stable posture.

This task is easier for a wheeled robot since because of its low CoM position it is less likely to fall over while executing movements designed to handle an object. Maintaining stability during the execution of similar tasks is not trivial for a humanoid robot. The unstable nature of its design makes it vulnerable to disturbances during motions such as lifting an object of unknown weight. The classic approach to motion design typically exploits the fact that most actions needed of a robot can be considered as sequences of motion primitives adding up to perform a certain interaction with its environment or being executed periodically in case of walking. Consequently the design consists of rigidly specifying these motion primitives or key frames and corresponding transition times. This results in a fixed motion sequence thereby making it impossible to adjust the movement online during execution.

While the specification of key frame motion provides an intuitive way of dealing with complex motions and adjusting them for specific looks or purposes, stability aspects are typically neglected but for the point that the resulting motion is stable on the reference robot used for the design. Differences between several robots of the same model or variances in interaction characteristics are normally handled by redesigning the motion for each case or trying to find a best fit that covers most cases. Therefore this static motion design approach appears to be ineligible for humanoid robots approaching the suggested tasks. Hence the proposed system extends the idea of key frame based motion design by controlled movement execution according to predefined stability criteria. This allows for a simplified specification process while differences between robots and deviations due to other reasons are compensated by the stability control.

The next section gives an overview of research on postural stability and related work. Then the proposed motion design and the control system applied to the executed motion are explained. Following this the system is evaluated using the experimental setup of a kicking motion. This application of the presented algorithm clearly shows the benefit of the control system.

2   Motion Generation and Stability

According to [2,3,4], the existing approaches to control the motions of walking robots can be divided into the following two categories:

Offline generation. A motion is designed before the execution resulting in the specification of a motion trajectory. The planned trajectory is executed once or periodically resulting in the desired motion.

Online generation. A motion is generated by a feedback control mechanism from a given motion objective in real time.

Offline motion generation has been applied since the beginning of robotics research. Teach-in techniques for industrial manipulators allow to design complex motions by manually moving the robot on a path leading to the desired motion during playback execution. The intuitive simplicity of this approach motivates the method of designing motions using key frames [5]. Different sets of joint positions are specified leading to key motion positions. The transition between these frames leads to motion fragments. Combined these motion primitives form the desired motions needed for application. Due to the high number of joints motions of humanoid robots are very difficult to design. Therefore the key frame procedure is particularly interesting for application in this field of robotics for its simple design of complex motions. As a downside the motion is executed without the possibility of online adaptation. So it is not possible to supervise and control the stability of movement execution rendering this approach unsuitable for tasks during which forces acting on the robot can change unpredictably.

Hence the concept class of online motion generation combines approaches capable of changing the planned motion during the execution which requires a method to generate a new trajectory movement. Normally a mathematical function and a model description of the robot is used to come up with a way to calculate the desired movement [2,3,4]. While finding a model or mathematical description of the desired movement is more complex than defining a key frame motion it offers the possibility to integrate feedback in the motion calculation and thereby adapt the motion to external influences. This advantage enables this kind of motion generation to use sensor feedback to supervise and control the stability of motions when unpredictable external forces act on the robot. Therefore a criteria to measure the stability of the robot is needed.

A robot's posture is called balanced and a gait is called statically stable, if the projection of the robot's center of mass to the ground lies within the convex hull of the foot support area (the support polygon). This kind of movement however covers only low speeds and momentums. Movements utilizing high joint torques and accelerations typically consist of phases in which the projection of the center of mass leaves the support polygon, but in which the dynamics and the momentum of the body are used to keep the gait stable. Those movements are called dynamically stable.

The concept of the zero moment point (ZMP) is useful for understanding dynamic stability and also for monitoring and controlling a walking robot [6,7]. The ZMP is the point on the ground where the tipping moment acting on the robot, due to gravity and inertia forces, equals zero. In the case of a quasi static motion this ZMP equals the ground projected CoM. Vukobratovic's classical ZMP notation [8] is only defined inside the support polygon. This coincides with the equivalence of this ZMP definition to the center of pressure (CoP) [9], which naturally is not defined outside the boundaries of the robot's foot. If the ZMP is at the support polygon's edge, any additional moment would cause the robot to rotate around that edge, i.e. to tip over. Nevertheless, applying the criteria of zero tipping moment results in a point outside the support polygon in this case. Such a point has been proposed as the foot rotation indicator (FRI) point [10] or the fictitious ZMP (FZMP) [8].

3   Motion Generation

In this paper a motion design concept is proposed which combines both methods discussed in section 2. Therefore a feedback controller is described in section 3.1, capable of controlling the CoM of a robot in a way satisfying the conditions of a quasi static motion. Section 3.2 describes the used key frame based feed forward control method, while section 3.3 describes the combination of both methods.

3.1   Quasi Static Feedback Controller

Motions such as manipulating objects normally require the robot to remain in position while moving only parts of its body resulting in rather slow joint movements. Hence a controller based on the assumption of a quasi static approximation is sufficient to control the motion.

At first the one dimensional problem of a center of mass R intended to reach the target position R' is considered. Without loss of generality $R' \geq R(0)$ is defined hereafter. To satisfy the condition of a quasi static motion, the acceleration must be bounded all the time.

$$|\ddot{R}(t)| \leq a_c$$

To generate the desired trajectory of the controlled motion the acceleration is set to its maximum value in the beginning and inverted once the target will be reached by maximal deceleration.

$$\ddot{R}(t) = \begin{cases} a_c & \text{if } t \in [0, t_1] \\ -a_c & \text{if } t \in [t_1, t_1 + t_2] \end{cases} .$$

To achieve this the remaining distance to the target must be covered during the time t_2,

$$-\frac{1}{2}a_c t_2^2 + \dot{R}(t_1)t_2 = R' - R(t_1) \tag{1}$$

and the velocity must be reduced to zero

$$-a_c t_2 + \dot{R}(t_1) = 0. \tag{2}$$

Elimination of t_2 out of equation 1 and 2 results in

$$\frac{1}{2}\frac{\dot{R}(t_1)^2}{a_c} = R' - R(t_1).$$

When this condition is met the acceleration is inverted. Therefore the acceleration is given by equation 3.

$$\ddot{R}(t) = \begin{cases} a_c & \text{if } \frac{1}{2}\frac{\dot{R}(t)^2}{a_c} < R' - R(t) \vee \dot{R}(t) < 0 \\ -a_c & \text{else} \end{cases} \tag{3}$$

The deviation between the measured and the desired CoM position and its current velocity is used as the system output and the acceleration of the CoM, calculated by equation 3, as the system input. The CoM position is computed by double integration as demonstrated in figure 1.

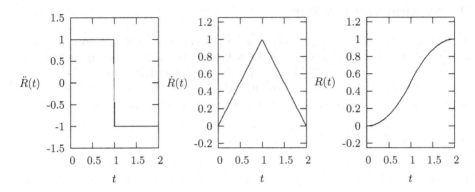

Fig. 1. Resulting position-, velocity- and acceleration curves for a quasi static controlled motion with $R(0) = 0$, $R' = 1$ and $a_c = 1$

To generalize the controller in two dimensions the control of the CoM motion is first considered to be independent and identical for both dimensions. The overall acceleration is hence bounded by the constant value $\sqrt{a_c^2 + a_c^2} = \sqrt{2}a_c$. With the target distances R_x and R_y being unequal in general the resulting movements tend to align along the axes as demonstrated in figure 2(a).

To solve this problem the coordinate system is transformed in every control step in such a way that one of its axes aligns to the current motion direction $\dot{R}(t)$ and all calculations are done is the accompanying reference system of the CoM. As in the one dimensional case the acceleration in the orthogonal direction is used to reach the target position and the orthogonal acceleration turns the movement direction towards the target. The result can be seen in figure 2(b).

3.2 Key Frame Based Feed Forward Control

Similar to the approach used in [5] a key frame based motion is modeled as a list of positions. The motion is executed by interpolating between these positions within given times. In difference to [5] the proposed algorithm uses a notation in which a key frame is not defined directly by a set of joint angles but by defining the positions of the robot's body and feet in form of coordinates in the euclidean space. The according joint angles are computed by methods of inverse kinematics. The position of the feet are either given relative to the robot coordinate system or the position of one feet is given relative the other one. This definition not only results in a more intuitive movement specification, but also is more flexible in allowing degrees of freedom in the movement to be controlled during execution to match a desired criterium.

3.3 Combining Feed Forward and Feedback Control

As discussed in section 2 classic key frame based motions are unsuitable to be controlled to meat a stability criterion during execution due to the fact that

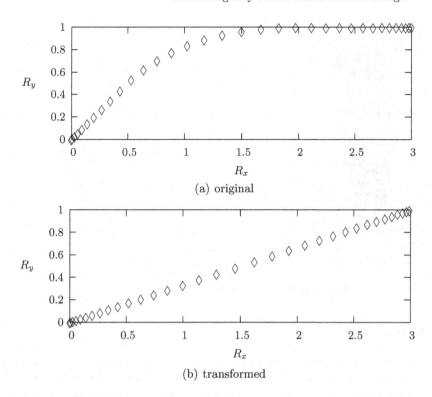

Fig. 2. Discrete two dimensional quasi static controlled motion with $R(0) = (0,0)$, $R' = (3,1)$ and $a_c = 1$

the movement of all joints is completely defined. Therefore in a novel approach both discussed methods are combined. The key frame based motion specification method discussed in 3.3 allows for a flexible motion definition without defining all degrees of freedom. While using this key frame approach to control the motion of the limbs, the orientation of the body in space, and the height of the CoM over the ground, the feedback controller presented in section 3.1 generates a motion trajectory for the horizontal components of the robot's CoM position ensuring the stability of the motion. As the desired stationary motions tend to require static stability keeping the ground projected CoM inside the robot's support polygon is sufficient to ensure this. Therefore a stable CoM trajectory is calculated in advance to match the specified motion. The fusion of the movements is then done by adding the resulting CoM position to each key frame. During motion execution the feedback controller ensures that the CoM follows the desired path by controlling the undefined degrees of freedom. The interpolation time associated to a key frame may not be equal to the time needed by the feed forward controller to reach the desired position in any case. Hence the transition from one key frame to the next can be delayed until the current CoM position is sufficiently close to the desired one and the current speed of the CoM is low enough.

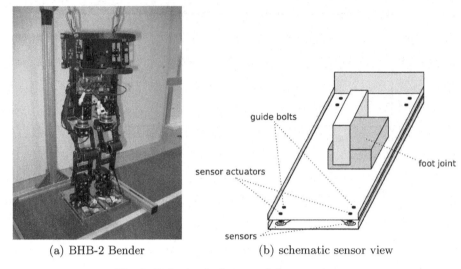

(a) BHB-2 Bender   (b) schematic sensor view

Fig. 3. Robotic platform used for experiments

4   Evaluation

To evaluate the concepts presented in this paper, experiments were conducted using a robot model of the type DHB-2 Bender (illustrated in figure 3(a)) which was designed and build by the *Dortmund University of Technology* and participated at the *German Open* and the *RoboCup* in the year 2007. In its current configuration it is 49 cm tall with a weight of 2.93 kg and a relatively high CoM of 31 cm. For more details see [11].

To measure the ZMP during the experiments the robot is equipped with sensors in the feet. Similar to the proposal of [8] four one axis force sensors of the type *FSR-149 (International Electronics Engeneering)* are integrated into the corners of each foot as illustrated in figure 3(b). As stated in [8] the measured ZMP and thereby according to section 2, in the quasi static case, the projection of the CoM to the ground can be calculated by weighted summation of the sensor values.

To calculate joint angle values from the foot and body positions, a concept of inverse kinematics using the Newton method was applied [12]. To calculate the position of the CoM a simplified model consisting of three punctual masses, one for the body and one for each leg, is used.

4.1   Application to Kicking

To proof the concept of the controller described in section 3 a quasi static motion to stand on one leg and kick a ball is described in the following. This motion is chosen because kicking a ball is an easily repeatable motion which stresses the stability aspect in two ways: First, balancing on one leg might in itself be a difficult task depending on the rigidity of the leg design and the strength of the servo motors. In addition to that the exact moment, position and force of interaction with the ball is not known in advance which might cause additional instability if not countered correctly during runtime.

In the beginning of the motion the controller is used to bring the measured CoM position over the support foot. The other foot is lifted off the ground while the controller keeps the CoM over the support foot. During the actual kicking move the lifted foot is moved forward rapidly without altering the CoM position[1]. As this part of the movement only lasts for a very short time (about 100 ms) the relatively slow quasi static controller is not fast enough to adjust the movements of the robot during this phase. After the kick the feedback controller is used again to keep the CoM over the support foot leveling out the impact of the kick. For slower movements a CoM adjustment would also be possible during the motion execution. Since slower statical movements tent to be stable by themselves a demonstration is omitted at this point. The CoM is shifted back to its original position after the kick foot is moved to the ground. The direction of the kick can be controlled by turning the kicking foot around the vertical axis before performing the actual kick move while the range of the kick can be adjusted by modifying the speed of the foot motion.

Tests have shown that due to its too flexible leg structure the robot tends to bend into the direction of the lifted foot during the phases where it stands on one foot. This effect can be minimized, although not completely avoided, by tilting the robot's body into the direction of the standing foot before lifting the leg, as thereby the angle at the hip joint is less acute. The remaining instability is compensated by the controller.

4.2 Experiments

Figure 4 shows the motion of the robot's CoM during the tested kick movements. The diagrams show the y-component of the position of the center of mass relative to the center of the right foot. The dotted lines illustrate the position as it is set by the controller and the bold line shows the position as it is measured by the foot sensors.

In figure 4(a) the movement is done under the assumption that the input CoM position is the actual CoM position therefore without utilizing the actual feedback control. As can be seen the CoM is moved over the right foot at first. After about 5 s the robot starts to lift the left foot. As the robot is no longer supported by the left foot the right leg bends and the CoM moves to the left. Accordingly the figure shows a deviation of the measured CoM position to the left. The deviation is strong enough to make the robot topple over the left side of its foot and finally fall over.

---

[1] This compensation of the CoM for the moving mass of the kicking leg is already inherent to the key frame specification if the CoM position is specified instead of the robot's coordinate system origin.

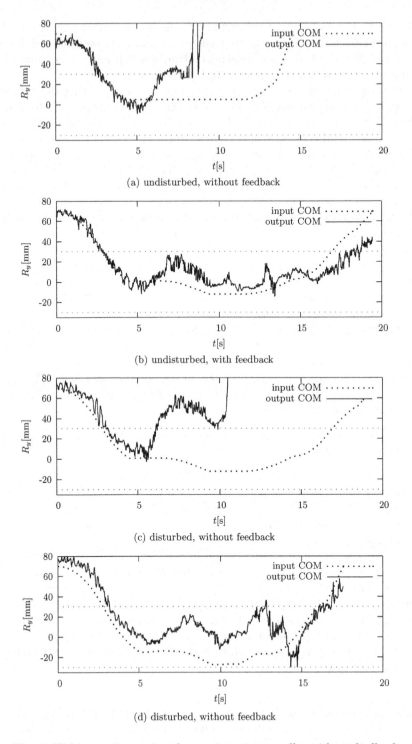

(a) undisturbed, without feedback

(b) undisturbed, with feedback

(c) disturbed, without feedback

(d) disturbed, without feedback

Fig. 4. Kicking motions using the quasi static controller without feedback

Figure 4(b) demonstrates the results of the movement utilizing the feedback control. As can be seen the controller reacts to the deviation of the CoM position by moving the CoM to the right side. So the robot is able to perform the actual kick move after about 10 s, lower the foot again and move the CoM back. Even if the disturbing movement itself is to fast to be controlled, see 4.1, the resulting disturbance can be leveled out with the help of the quasi-static controller resulting in a stable motion.

In figure 4(c) the movement resulting from the previous controlled kick is exactly repeated without the use of the feedback control. But this time an additional counterweight of 370 g is attached at the left side of the robot. The deviation of the CoM position leads again to the fall of the robot. In figure 4(d) it can be seen that the robot compensates this imbalance caused by the counterweight using the sensor feedback by shifting the whole motion to the right.

The benefit of the proposed integration of sensor feedback becomes clearly visible. Without explicit knowledge of the deviation the robot is able to adjust to unforeseen forces acting during the execution The used sensor information allows an adaptation of the defines key frame motion stabilizing the otherwise unstable motion.

5   Conclusion

This paper presents a way of combining the classical method of motion design by key frames with control algorithms for postural stability. Proof of the soundness is presented in the application to kicking motions for robot soccer. A great improvement to the robustness could be shown which even enabled the robot to perform its kick successfully with additional weights attached to it.

While uncontrolled replaying of predefined motions is still common for kicking in RoboCup leagues involving legged robots, this approach represents a far more robust and general alternative. Neither do motions need to be adapted for separate distinct robots nor do they need redesign in case of hardware wear of small decalibration of joint motors. The profit of this is obvious in the presented case and can also be of benefit in other applications involving environment interaction.

Further improvements can be achieved by introducing a model more complex than the simple quasi-static control of the robot's center of mass. Besides this, the next subject of interest is the integration of such motions directly into the robot's normal walking. This is a challenge both for motion generation and for perception accuracy that would enable faster, more fluid and natural motions and therefore faster robot soccer games.

References

1. Kanda, T., Ishiguro, H., Imai, M., Ono, T.: Development and evaluation of interactive humanoid robots. Proceedings of the IEEE 92(11), 1839–1850 (2004)
2. Sugihara, T., Nakamura, Y., Inoue, H.: Realtime humanoid motion generation through zmp manipulation based on inverted pendulum control. In: ICRA, pp. 1404–1409. IEEE, Los Alamitos (2002)

68   S. Czarnetzki, S. Kerner, and D. Klagges

3. Kajita, S., Kanehiro, F., Kaneko, K., Yokoi, K., Hirukawa, H.: The 3d linear inverted pendulum mode: a simple modeling for a biped walking pattern generation. In: IEEE/RSJ International Conference on Intelligent Robots and Systems (2001)
4. Kajita, S., Kanehiro, F., Kaneko, K., Fujiwara, K., Harada, K., Yokoi, K., Hirukawa, H.: Biped walking pattern generation by using preview control of zero-moment point. In: ICRA, pp. 1620–1626. IEEE, Los Alamitos (2003)
5. Antonelli, M., Libera, F.D., Minato, T., Ishiguro, H., Pagello, E., Menegatti, E.: Intuitive humanoid motion generation joining user-defined key-frames and automatic learning. In: Iocchi, L., Matsubara, H., Weitzenfeld, A., Zhou, C. (eds.) RoboCup 2008. LNCS (LNAI), vol. 5399, pp. 13–24. Springer, Heidelberg (2009)
6. Vukobratović, M., Borovac, B., Potkonjak, V.: Towards a unified understanding of basic notions and terms in humanoid robotics. Robotica 25(1), 87–101 (2007)
7. Czarnetzki, S., Kerner, S., Urbann, O.: Observer-based dynamic walking control for biped robots. Robotics and Autonomous Systems (in press, 2009) (accepted manuscript)
8. Vukobratovic, M., Borovac, B.: Zero-moment point – Thirty five years of its life. International Journal of Humanoid Robotics 1(1), 157–173 (2004)
9. Sardain, P., Bessonnet, G.: Forces acting on a biped robot. center of pressure - zero moment point. IEEE Transaction on Systems, Man, and Cybernetics (2004)
10. Goswami, A.: Foot rotation indicator (FRI) point: A new gait planning tooltoe-valuate postural stability of biped robots. In: IEEE International Conference on Robotics and Automation, pp. 47–52 (1999)
11. Czarnetzki, S., Hebbel, M., Nisticò, W.: DoH!Bots: Team description for RoboCup 2007. In: RoboCup 2007: Robot Soccer World Cup XI. LNCS (LNAI). Springer, Heidelberg (2008)
12. Buss, S.R.: Introduction to inverse kinematics with jacobian transpose, pseudoinverse and damped least squares methods. Department of Mathematics University of California, San Diego (2004)

Applying Dynamic Walking Control
for Biped Robots

Stefan Czarnetzki, Sören Kerner, and Oliver Urbann

Robotics Research Institute
Section Information Technology
Dortmund University of Technology
44221 Dortmund, Germany

Abstract. This article presents the application of a novel observer-based control system to achieve reactive motion generation for dynamic biped walking. The proposed approach combines a feedback controller with an online generated feet pattern to assure a stable gait. Experiments in a simulated environment as well as on real robots clearly demonstrate the robustness of the control system. The presented algorithms enable the robot not only to walk dynamically stable but also to cope with major internal disturbances like flaws of the robots internal model and external disturbances like uneven or unstable ground or collisions with objects or other robots.

1   Introduction

Humanoid robots are believed to have a high potential for future applications due to the suitability for operation in environments made for humans and due to higher acceptance by people [1], both of which are needed for service and entertainment activities. Despite this, present humanoid robots have a substantial lack of mobility. Even basic tasks such as walking on even ground without an external disturbance are not a trivial challenge. The humanoid shaped form of a two-legged robot results in a high center of mass (CoM) of its body while standing upright. As a result the stance of a humanoid robot is quite unstable, making it likely to tip over. Therefore research on stable biped walking is one of the central problems in this area at the moment. Gait planning for humanoid robots is fundamentally different from the path planning for simple robotic arms. The robots center of mass is in motion all the time while the feet periodically interact with the ground in an unilateral way, meaning that there are only repulsive but no attractive forces between the feet and ground. The movement of the center of mass cannot be controlled directly, but is governed by its momentum and the eventual contact forces arising from ground interaction. These have to be carefully planned in order not to suffer from postural instability.

This paper proposes a control system to achieve reactive motion generation for dynamic biped walking. After giving a brief overview of research on stability aspects of legged robots, the walking pattern generation and control is described. A thorough evaluation is given showing the capability of the system

J. Baltes et al. (Eds.): RoboCup 2009, LNAI 5949, pp. 69–80, 2010.

to generate stable biped walking even under difficult circumstances. The robustness is presented in experiments testing different problem settings as walking with inaccuracies and systematic errors in the model, external disturbances and on uneven or unstable ground.

2 Stability

A robot's posture is called balanced and its gait is called statically stable, if the projection of the robot's center of mass on the ground lies within the convex hull of the foot support area (the support polygon). This kind of gait however results in relatively low walking speeds. Similarly natural human gaits are normally not statically stable. Instead they typically consist of phases in which the projection of the center of mass leaves the support polygon, but in which the dynamics and the momentum of the body are used to keep the gait stable. Those gaits are called dynamically stable.

The concept of the zero moment point (ZMP) is useful for understanding dynamic stability and also for monitoring and controlling a walking robot [2]. The ZMP is the point on the ground where the tipping moment acting on the robot, due to gravity and inertia forces, equals zero. The tipping moment is defined as the component of the moment that is tangential to the supporting surface, i.e. the ground. The moment's component perpendicular to the ground may also cause the robot to rotate, but only in a way to change the robot's direction without affecting its stability, and is therefore ignored. For a stable posture, the ZMP has to be inside the support polygon. In the case when it leaves the polygon, the vertical reaction force necessary to keep the robot from tipping over cannot be exerted by the ground any longer, thus causing it to become instable and fall.

In fact, following Vukobratovic's classical notation [3], the ZMP is only defined inside the support polygon. This coincides with the equivalence of this ZMP definition to the center of pressure (CoP), which naturally is not defined outside the boundaries of the robot's foot. If the ZMP is at the support polygon's edge, any additional moment would cause the robot to rotate around that edge, i.e. to tip over. Nevertheless, applying the criteria of zero tipping moment results in a point outside the support polygon in this case. Such a point has been proposed as the foot rotation indicator (FRI) point [4] or the fictitious ZMP (FZMP) [3]. In this so-called fictitious case the distance to the support polygon is an indicator for the magnitude of the unbalanced moment that causes the instability and therefore is a useful measure for controlling the gait.

There are different approaches to generating dynamically stable walking motions for biped robots. One method is the periodical replaying of trajectories for the joint motions recorded in advance, which are then modified during the walk according to sensor measurements [5]. This strategy explicitly divides the problem into subproblems of planning and control. Another method is the realtime generation of trajectories based on the present state of the kinematic system and a given goal of the motion, where planning and control are managed in a unified system. Implementations of this approach differ in the kinematic models being used and the way the sensor feedback is handled. One group requires precise knowledge of the robot's dynamics, mass distribution and inertias of each link to generate motion patterns, mainly relying on the accuracy of the model for motion pattern generation [6,7,8]. A second group uses limited knowledge about a simplified model (total center of mass, total angular momentum, etc.) and relies on feedback control to achieve a stable motion [9,10,11]. The model used for this is often called the inverted pendulum model.

3 Motion Generation

This section describes the generation of walking patterns based on a simple inverted pendulum model using a sophisticated preview controller to generate motions resulting in a desired future ZMP movement and to be able to compensate small disturbances or unforeseen forces. The motion generation process can be seen as stages in a pipeline process, which will be described in section 3.1.

3.1 Generating the Walking Patterns

The general problem of walking can be seen as an appropriate placement of the feet and a movement of the rest of the body, both of these must satisfy the condition to keep the overall resulting motion stable. The generation of such motion patterns can be divided into separate tasks with one depending on the results of another, therefore forming a pipeline (see figure 1).

The goal of the desired walk is a certain translational and rotational speed of the robot which might change over time, either smoothly i.e. when the robot is slowing down while approaching an object or rapidly i.e. when the robot's high-level objective changes. The translational and rotational speed vector is taken as the input of the motion generation pipeline. This speed vector is the desired speed of the robot, which does not translate to its CoM speed directly for obvious stability reasons, but merely to its desired average. Thus a path is specified that the robot intends to follow. The feet of the robot have to be placed along this path to ensure the correct overall motion of the robot. Alternatively, in scenarios with uneven ground the feet placement at safe positions must be prioritized, resulting in an irregular gait dictating different changes of speed.

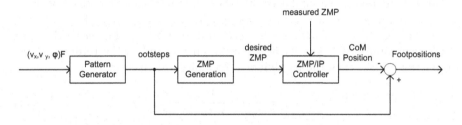

Fig. 1. Pipeline visualization of the walking pattern generation process

Once the step patterns are set, these define a region for possible ZMP trajectories to result in stable gaits, namely the support polygon at every given time. A gait can be divided into two phases, a double support phase where both feet are on the ground and a single support phase where only one foot has contact with the ground. During each single support phase the ZMP should be positioned at the center of the ground foot. Consequently in the double support phase the ZMP has to be shifted from one foot to the other. While these restrictions are sufficient to specify the stability of a gait, there is some freedom left in the specification of the exact ZMP trajectory.

The next stage of the process is the generation of a CoM trajectory in which kinematics result in the desired ZMP trajectory. As can be seen later in figures 2(c) and 4 it is not sufficient to shift the CoM at the same time as the ZMP. Instead the CoM has to start moving before the ZMP does. This is realized using a preview control described in more detail in the following section. Its output is a CoM trajectory as shown in figure 1.

All trajectories and positions calculated so far are given in a global world coordinate frame. From the step pattern the feet positions are known, and so is the position of the center of mass at a given time. If the robot's CoM relative to its coordinate frame is known (or assumed to be constant in a simple model), the difference between these directly provides the foot positions in a robot centered coordinate frame. Those can subsequently be transformed into leg joint angles using inverse kinematics.

3.2 Modeling Motion Dynamics

The main problem in the process described in section 3.1 is computing the movement of the robot's body to achieve a given ZMP trajectory. For this a simplified model of the robot's dynamics is used, representing the body by its center of mass only. In the single support phase of the walk only one foot has contact with the ground and considering only this contact point and the center of mass, the resulting motion can be described as an inverted pendulum. Its height can be changed by contracting or extending the leg, therefore allowing further control over the CoM trajectory. Restricting the inverted pendulum so that the CoM only moves along an arbitrary defined plane results in simple linear dynamics called the 3D Linear Inverted Pendulum Mode (3D-LIPM) [9]. This plane is given by its normal vector $(k_x, k_y, -1)$ and its intersection with the z-axis z_h.

For walking on an overall flat terrain the constraint plane is horizontal ($k_x = k_y = 0$) even if the ground itself is uneven. The global coordinate frame depicts the ground as the x-y-plane and the vertical direction as z. Let m be the mass of the pendulum, g the gravity acceleration and τ_x and τ_y the torques around the x- and y-axes, then the pendulum's dynamics are given by

$$\ddot{y} = \frac{g}{z_h} y - \frac{1}{mz_h} \tau_x \tag{1}$$

$$\ddot{x} = \frac{g}{z_h} x + \frac{1}{mz_h} \tau_y. \tag{2}$$

Note that even in the case of a sloped constraint plane the same dynamics can be obtained by applying certain further constraints that are not covered here [9]. According to this model the position (p_x, p_y) of the ZMP on the floor can be easily calculated using

$$p_x = -\frac{\tau_y}{mg} \qquad (3)$$

$$p_y = \frac{\tau_x}{mg}. \qquad (4)$$

Substituting equations 3 and 4 into 1 and 2 yields the following ZMP equations.

$$p_x = x - \frac{z_h}{g}\ddot{x} \qquad (5)$$

$$p_y = y - \frac{z_h}{g}\ddot{y} \qquad (6)$$

It can be seen that for a constant height z_h of the constraint plane the ZMP position depends on the position and acceleration of the center of mass on this plane and the x- and y-components can be addressed separately.

It should be noted for clarification that the ZMP notion of the 3D-LIPM [9] does not take the limitation of the ZMP to an area inside the support polygon into account. Using equations 5 and 6 for planning and controlling stable walking may result in a fictitious ZMP lying outside the support polygon. As mentioned in section 2, this is an indication of an unbalanced moment which causes instability. Since the mathematical notation of the 3D-LIPM used in the following chapters does not involve any distinction based on the support polygon, the general term ZMP will be used hereafter.

3.3   Controlling the Motion

Movement of the robot's body to achieve a given ZMP trajectory is thus reduced to planning the CoM trajectory for each direction, resulting in two systems of lesser complexity whose state at any given time is naturally represented by $(x, \dot{x}, \ddot{x})$. The ZMP position p is both the target of the control algorithm and the measurable output of the system. Equations 5 and 6 suggest that the state vector $(x, \dot{x}, p)$ is an equivalent system representation. Choosing this one incorporates the control target into the system state and significantly simplifies further derivations of the controller.

As mentioned previously, the ZMP can not be achieved correctly given its current target value alone, but the CoM needs to start moving prior to the ZMP. Hence the incorporation of some future course of the ZMP is necessary. Such data is available since part of the path to follow is already planned, as described in section 3.1. The following design of a preview controller is described in detail in [12]. It is based on the control algorithms of [13]. [14] already applied some of these to the field of biped walking but used different sensor feedback strategies.

A more natural way of using sensors is presented here. Applying elements common in control theory it is possible to directly incorporate measurements into the system using an observer model as described later in this section. The control algorithm derived here provides the basis for computing a CoM movement resulting in the desired reference ZMP. Assuming the absence of disturbances this system would be sufficient for stable walking.

The system's dynamics can be represented by

$$\frac{d}{dt} \begin{bmatrix} x \\ \dot{x} \\ p \end{bmatrix} = \begin{bmatrix} 0 & 1 & 0 \\ \frac{g}{z_h} & 0 & -\frac{g}{z_h} \\ 0 & 0 & 0 \end{bmatrix} \begin{bmatrix} x \\ \dot{x} \\ p \end{bmatrix} + \begin{bmatrix} 0 \\ 0 \\ 1 \end{bmatrix} v \tag{7}$$

where $v = \dot{p}$ is the system input to change the ZMP p according to the planned target ZMP trajectory p^{ref}. Discretizing equation 7 with time steps Δt yields

$$\mathbf{x}(k+1) = \mathbf{A}_0 \mathbf{x}(k) + \mathbf{b}v(k) \tag{8}$$
$$p(k) = \mathbf{c}\mathbf{x}(k) \tag{9}$$

where $\mathbf{x}(k)$ is the discrete state vector $[x\ \dot{x}\ p]^T$ at time $k\Delta t$. Note that A_0 describes the system's behavior according to the simplified model. This may not necessarily be identical to the real state transition of the actual robot.

The idea of previewable demand, i.e. the ZMP trajectory due to the planned step pattern, leads to a preview controller [13]. The 3D-LIPM model used to obtain the system dynamics however is only a very simplified approximation of the robot and disturbances and also the state vector itself can not be measured directly in most cases. Therefore it becomes necessary to estimate those from the system input and the measured sensor data in order for the system to work properly under realistic conditions, which leads to the introduction of an observer. The details of the derivation of the control system equations can be found in [12]. The resulting observer for the system is given by equation 10.

$$\widehat{\mathbf{x}}(k+1) = \mathbf{A}_0\widehat{\mathbf{x}}(k) - \mathbf{L}\left[p^{sensor}(k) - \mathbf{c}\widehat{\mathbf{x}}(k)\right] + \mathbf{b}u(k). \tag{10}$$

The observer-based controller designed as in equation 11 consists of integral action on the tracking error, proportional state feedback and preview action based on the future demand.

$$u(k) = -G_I \sum_{i=0}^{k} \left[\mathbf{c}\widehat{\mathbf{x}}(i) - p^{ref}(i)\right] - \mathbf{G}_x\widehat{\mathbf{x}}(k) - \sum_{j=1}^{N} G_d(j)p^{ref}(k+j) \tag{11}$$

The gains G_I, $\mathbf{G}_x$ and G_d are chosen to optimize the performance index J from

$$J = \sum_{j=1}^{\infty} \left\{ Q_e \left[p(j) - p^{ref}(j)\right]^2 + \Delta\mathbf{x}^T(k)\mathbf{Q}_x\Delta\mathbf{x}(k) + Rv^2(j) \right\} \tag{12}$$

where $\Delta\mathbf{x}$ is the incremental state vector $\Delta\mathbf{x}(k) = \mathbf{x}(k) - \mathbf{x}(k-1)$. The physical interpretation of J is to achieve regulation without an excessive rate of change in the control signal. Both the tracking error and excessive changes in state

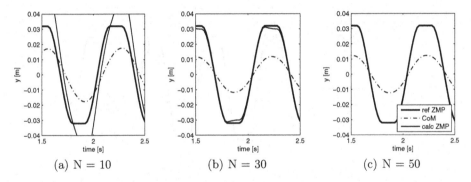

(a) N = 10   (b) N = 30   (c) N = 50

Fig. 2. Reference ZMP, calculated ZMP and CoM for different numbers N of the previewable values

and control are penalized with weights Q_e, $\mathbf{Q}_x$ and R, respectively, so that a controller optimizing J achieves a smooth regulation of the controlled system.

The effect of the availability of previewable demand is visualized in figure 2. This issue is directly related to equations 11 and 12 where the infinite horizon spanned by J is approximated by a finite preview window of size N. In case of walking, preview windows of the size of a step cycle yield near optimal approximations without adding additional motion delay to that which is already inherent in the motion generation due to step pattern planning.

The resulting control algorithm is visualized in figure 3. An intuitive illustration of this observer-based controller's performance is given in figure 4, where a constant error is added to the ZMP measurement for a period of 1.5 s. This error could be interpreted as an unexpected inclination of the ground or a

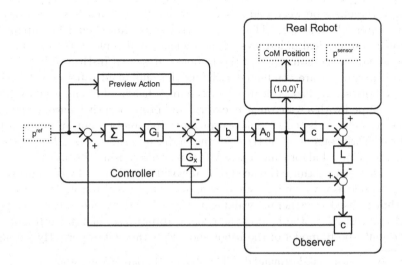

Fig. 3. Configuration of the control system

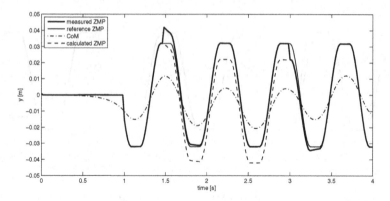

Fig. 4. Performance of the controller under the influence of a constant external disturbance resulting in an error in the measured ZMP

constant force pushing the robot to one side. The control system incorporates this
difference and compensates by smoothly adjusting the CoM trajectory, which
consequently swings more to the opposite direction.

4   Evaluation

To demonstrate the benefit of the proposed control system several experiments
are evaluated. The scenarios are designed to represent the most common reasons
of instability during gait execution, namely external disturbances and deviation
from the internal model. External disturbances occur often in the form of dis-
continuities or irregularities of the ground or external forces caused by collisions.
Even with a perfect internal model of the robot, which is nearly impossible to
obtain, divergences between real robot occur due to mechanical wearout or permanent external influences. Thus the control algorithm should be capable to
level such systematical differences of its model. A video file of the experiments
can be found on the homepage of the Robotics Research Institute[1].

The experiments are conducted utilizing the humanoid robot Nao by Aldebaran Robotics. Nao has 21 degrees of freedom, a height of 57 cm, weights 4.5 kg
and is equipped with a wide range of extero- and proprioceptive sensors including
an accelerometer in its chest. The sensor input and motion output is controlled
by a framework running at 50 Hz resulting in discrete time steps Δt of 20 ms for
the walking control algorithm. The ZMP is measured using the accelerometer.

The first test underlines the need of real world experiments by comparing the
differences between walking in simulation and with a real robot. The proposed
algorithm is used to generate walking pattern based on the calculated ZMP
trajectory. Tests with the help of the general robotics simulator SimRobot [15]
using a multi-body model of the robot show that the gait is perfectly stable in

---

[1] http://www.it.irf.uni-dortmund.de/IT/Robotics/Resource/Application-of-
Dynamic-Walking-Control.avi

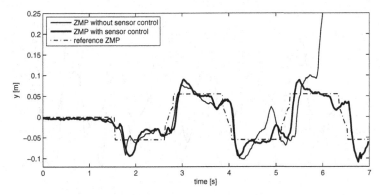

Fig. 5. Plot of the ZMP during the model error test

simulation even without sensor feedback. In comparison this gait is executed on the real robot without further adaptations. Figure 5 demonstrates the results of walking straight ahead at speed $v_x = 50 \frac{mm}{s}$ with a step duration of $t_{step} = 2.5$ s. As can be seen without sensor control the difference between the reference and the measured ZMP increases over time and the robot starts swinging at second four resulting in a fall around second six. The results of performing the same experiment with feedback control demonstrate the advantage of the proposed closed-loop system. As can be seen in figure 5 the control algorithm is able to adjust the movement according to the flaw of the model and thereby leveling the differences between the ZMPs resulting in a stabilization of the walk. As shown in figure 6 the robot keeps stable even during omnidirectional walking pattern containing substantial changes in speed and directions.

Another design target of the closed-loop system is the capability to level out unforeseen external forces. Hence as an experiment the robot is pushed during a walk at speed $v_x = 50 \frac{mm}{s}$ and a step duration of $t_{step} = 1$ s to simulate the collision with another moving object. Figure 7 illustrates the resulting controller reaction. The collision occurred around second four as can be clearly noticed

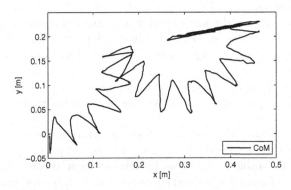

Fig. 6. CoM trajectory during omnidirectional walking

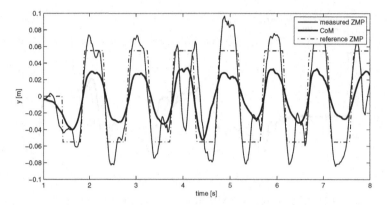

Fig. 7. Plot of the ZMP and CoM during the push test

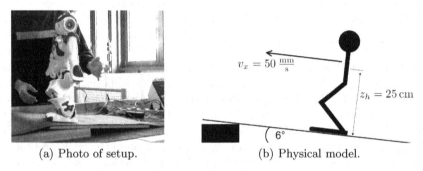

(a) Photo of setup.   (b) Physical model.

Fig. 8. Setup of the uneven floor test

by the abrupt change in the measured ZMP caused by the sudden change of acceleration. The resulting shift of the measured ZMP can be observed in the following second but is compensated by the controller with an adaptation of the desired CoM trajectory stabilizing the walk.

Walking on uneven ground or slopes leads to disturbances often resulting in a fall of the robot. Hence the last experiment, shown in figure 8(b) and 8(a), is designed to simulate these scenarios. Without adaptation the robot is walking up a slope with a gradient angle of 6°. The end of the ramp is not fixed to the block resulting in a rocking movement when walking on top of the edge. Therefore the experiment tests both the capability to level out the continuous error of the slope and to overcome the floor disturbance caused by the rocker.

Figure 9(a) illustrates the measured body orientation. Between second fourteen and fifteen the backward tilt of the robot is disturbed by the tilt of the ramp but becomes stable again in an upright position afterwards. Figure 9(b) shows the ZMP in forwards direction corresponding to this experiment. It can be seen that the robot slips due to the sloped ground shortly after second thirteen

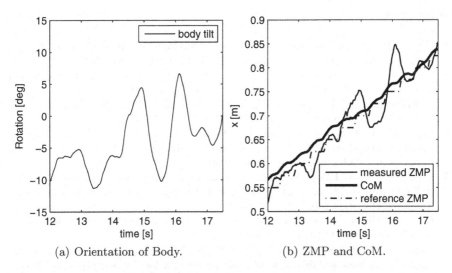

(a) Orientation of Body.   (b) ZMP and CoM.

Fig. 9. Results of the uneven floor test

and swings massively when the rocking starts, but reaches a regular walk again towards the end.

The results of the experiments clearly demonstrate the benefit of the proposed sensor feedback control. Without further adaptation to the used hardware or experimental setup the robot is able to adjust its movement to model deviations and external disturbances with the help of sensor feedback in most cases that would clearly result in a fall otherwise.

5   Conclusion

This paper presents a novel approach to biped walking based on the online generation of foot trajectories. Special focus is given to the online calculation and control of the CoM movement to achieve the desired ZMP trajectory.

The proposed control system's performance is verified using the humanoid robot Nao of Aldebaran Robotics. The approach generates dynamically stable walking patterns and performs well even under the influence of significant external disturbances. Three experimental setups are chosen as test cases. Each clearly shows the benefit of the observer-based controller for biped walking.

Further improvements can be achieved by introducing a model more complex than the 3D-LIPM. Besides this, the next subject of interest is the integration of different motion patterns other than walking into the control system. These may include planned object contact with maximized impulse, e.g. shooting a ball. A challenge both for motion generation and for perception accuracy is to enable faster and more fluid motions by integrating shooting movements into the normal step patterns.

References

1. Kanda, T., Ishiguro, H., Imai, M., Ono, T.: Development and evaluation of interactive humanoid robots. Proceedings of the IEEE 92(11), 1839–1850 (2004)
2. Vukobratović, M., Borovac, B., Potkonjak, V.: Towards a unified understanding of basic notions and terms in humanoid robotics. Robotica 25(1), 87–101 (2007)
3. Vukobratovic, M., Borovac, B.: Zero-moment point – Thirty five years of its life. International Journal of Humanoid Robotics 1(1), 157–173 (2004)
4. Goswami, A.: Foot rotation indicator (FRI) point: A new gait planning tooltoevaluate postural stability of biped robots. In: IEEE International Conference on Robotics and Automation, pp. 47–52 (1999)
5. Kim, J.Y., Park, I.W., Oh, J.H.: Experimental realization of dynamic walking of biped humanoid robot khr-2 using zmp feedback and inertial measurement. Advanced Robotics 20(6), 707 (2006)
6. Hirai, K., Hirose, M., Haikawa, Y., Takenaka, T.: The development of honda humanoid robot. In: ICRA, pp. 1321–1326 (1998)
7. Huang, Q., Kajita, S., Koyachi, N., Kaneko, K., Yokoi, K., Arai, H., Komoriya, K., Tanie, K.: A high stability, smooth walking pattern for a biped robot. In: ICRA, p. 65 (1999)
8. Yamaguchi, J., Soga, E., Inoue, S., Takanishi, A.: Development of a bipedal humanoid robot: Control method of whole body cooperative dynamic biped walking. In: ICRA, pp. 368–374 (1999)
9. Kajita, S., Kanehiro, F., Kaneko, K., Fujiwara, K., Yokoi, K., Hirukawa, H.: A realtime pattern generator for biped walking. In: ICRA, pp. 31–37. IEEE, Los Alamitos (2002)
10. Raibert, M.: Legged Robots That Balance. MIT Press, Cambridge (2000)
11. Zheng, Y.F., Shen, J.: Gait synthesis for the SD-2 biped robot to climb sloping surface. IEEE Journal of Robotics and Automation 6(1), 86–96 (1990)
12. Czarnetzki, S., Kerner, S., Urbann, O.: Observer-based dynamic walking control for biped robots. Robotics and Autonomous Systems (in press, 2009) (accepted manuscript)
13. Katayama, T., Ohki, T., Inoue, T., Kato, T.: Design of an optimal controller for a discrete-time system subject to previewable demand. International Journal of Control 41(3), 677 (1985)
14. Kajita, S., Kanehiro, F., Kaneko, K., Fujiwara, K., Yokoi, K., Hirukawa, H.: Biped walking pattern generator allowing auxiliary zmp control. In: IROS, pp. 2993–2999. IEEE, Los Alamitos (2006)
15. Laue, T., Spiess, K., Röfer, T.: SimRobot - A General Physical Robot Simulator and Its Application in RoboCup. In: Bredenfeld, A., Jacoff, A., Noda, I., Takahashi, Y. (eds.) RoboCup 2005. LNCS (LNAI), vol. 4020, pp. 173–183. Springer, Heidelberg (2006), http://www.springer.de/

Modeling Human Decision Making Using Extended Behavior Networks

Klaus Dorer

Hochschule Offenburg, Elektrotechnik-Informationstechnik, Germany
klaus.dorer@fh-offenburg.de

Abstract. In their famous work on prospect theory Kahneman and Tversky have presented a couple of examples where human decision making deviates from rational decision making as defined by decision theory. This paper describes the use of extended behavior networks to model human decision making in the sense of prospect theory. We show that the experimental findings of non-rational decision making described by Kahneman and Tversky can be reproduced using a slight variation of extended behavior networks.

1 Introduction

Looking forward to the goal of RoboCup to win against the human world champion team of soccer in 2050 one could state the question whether the decision making of the robots should be human like or rational with respect to decision theory. No matter what the answer to this question is, there should be no doubt that the robots should be able to model their opponents to understand and predict their decision making. Since the opponent team will be humans we therefore have to be able to model human decision making.

In their famous work on prospect theory Kahneman and Tversky [1979] have shown that human decision making does violate the tenets of decision theory. In a series of experiments they have shown a couple of deviations to the predictions of decision theory among those are that humans overestimate low probabilities and underestimate high probabilities and that subjective utility can differ from objective utility. Daniel Kahneman was awarded the Nobel prize in economic sciences in response to this work.

Behavior Networks [Maes, 1989] were introduced as a means to combine reactive and deliberative decision making using a mechanism of activation spreading to determine the best behavior. With Extended Behavior Networks (EBNs) [Dorer, 1999a] the mechanism of activation spreading was changed so that activation is a measure of expected utility of a behavior. In this paper we show that the mechanism of activation spreading in EBNs only needs slight modifications to reproduce human decision making reported by prospect theory.

Section 2 presents a number of experiments described by Kahneman and Tversky for their work on prospect theory. Section 3 introduces EBNs and the mechanism of activation spreading. Section 4 describes how EBNs can be used to

J. Baltes et al. (Eds.): RoboCup 2009, LNAI 5949, pp. 81–91, 2010.

model human decision making and reports on experimental results achieved applying EBNs to the same experiments conducted by Kahneman and Tversky
with humans. Section 5 concludes the paper indicating possible future work.

2   Prospect Theory

Decision theory is based on the principle of maximum expected utility. If an
agent is faced with a decision between actions (prospects) of uncertain outcome
it should choose the prospect that has the highest expected utility. More formal:
the expected utility of a prospect $P(u_1, p_1; \ldots u_n, p_n)$ is calculated as

$$eu_P = \sum_{i=1\ldots n} p_i \times u_i \qquad (1)$$

where p_i is the probability of outcome u_i and $\sum p_i = 1$. Choosing the prospect
with highest expected utility will maximize the agent's utility in the long term
and is therefore considered as rational.

On the other side experiments with humans show that human decision making
deviates from the above. Kahneman and Tversky [1979], for example, describe
a series of experiments that led to the formulation of prospect theory, a theory
of human decision making under risk. In this section we describe a selection of
their experiments that are used in section 4 to be repeated by extended behavior
networks.

2.1   Weighting Function

In their first experiment students had the choice between winning 2500 Israeli
pounds[1] with a probability of 0.33, winning 2400 with probability 0.66 and
nothing with probability 0.01 (A) compared to winning 2400 for sure (B). The
expected utility of decision theory for A and B are $eu_A = 0.33 \times 2500 + 0.66 \times$
$2400 + 0.01 \times 0 = 2409$ and $eu_B = 1.0 \times 2400 = 2400$. So a rational agent should
prefer A over B. However, in the experiment 82% (significant*) of the students
chose B, the certain outcome.

The second experiment repeats the first but eliminates a chance of winning
2400 with probability 0.66 from both prospects. Table 2 shows the results of the
experiments. Now with both prospects being uncertain the majority of students
prefer C over D. 62% of the students took combination B and C.

This and more experiments showed that humans overestimate low probabilities (except the impossible outcome) and underestimate high probabilities
(except the certain outcome). Prospect theory therefore introduces a non-linear
weighting function mapping probabilities to decision weights. Figure 1 shows a
qualitative sketch of the weighting function [Kahnemann and Tversky, 1979].

---

[1] The average monthly income of a family was 3000 Israeli pounds at that time.

Table 1. Problem 1: One uncertain one certain prospect

Name	Prospect	Expected Utility	Human choices
A	(2500,0.33; 2400,0.66; 0,0.01)	2409	18
B	(2400, 1.0)	2400	82*

Table 2. Problem 2: Two uncertain prospects

Name	Prospect	Expected Utility	Human choices
C	(2500,0.33; 0,0.67)	825	83*
D	(2400, 0.34; 0,0.66)	816	17

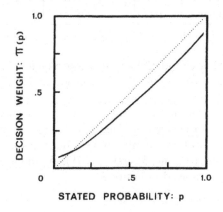

Fig. 1. A hypothetical weighting function as proposed by Kahneman and Tversky

2.2   Value Function

In another experiment Tversky and Kahneman [1981] have shown that students preferred a certain win of 240$ (A) compared to a 25% chance of winning 1000$ (B) despite the fact that expected utility of (A) is less. In the same experiment a 75% chance for a loss of 1000$ (D) was prefered over a certain loss of 750$ (C) despite the fact that both have the same expected utility (see table 3). 73% of the students chose the combination of (A) and (C), 3% chose the combination (B) and (D). In another experiment students had the choice between a 25% chance of winning 240$ and a 75% chance of loosing 760$ (E) or a 25% chance of winning 250$ and a 75% chance of loosing 750$ (F). Not surprisingly all students choose option (F) (see table 4).

This experiment is particularly interesting since the combination (A) and (C) chosen by most students in the first experiment is equivalent with respect to

Table 3. Problem 3: decision under gains and losses

Name	Prospect	Expected Utility	Human choices
A	(240,1.0)	240	84
B	(1000, 0.25; 0,0.75)	250	16
C	(-750, 1.0)	-750	13
D	(-1000,0.75; 0,0.25)	-750	87

Table 4. Problem 4: intransitive decision with respect to problem 3

Name	Prospect	Expected Utility	Human choices
E	(240,0.25; -760,0.75)	-510	0
F	(250,0.25; -750,0.75)	-500	100

decision theory to option (E) of the second experiment while the combination (B) and (D) is equivalent to option (F). So decision making of the majority was intransitive.

Prospect theory suggests that gains and losses are not linearly mapped to the subjective value of human decision makers. The value function is rather "generally concave for gains and commonly convex for losses and steeper for losses than for gains" [Kahnemann and Tversky, 1979]. A qualitative value function with this properties is displayed in figure 2.

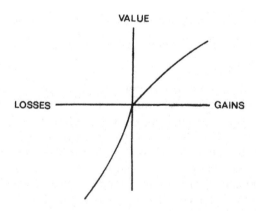

Fig. 2. A hypothetical value function as proposed by Kahneman and Tversky

3   Extended Behavior Networks

Behavior Networks [Maes, 1989] use a mechanism of activation spreading to decide between a couple of executable behaviors combining reactive and deliberative decision making. Extended Behavior Networks (EBNs) [Dorer, 2004; Dorer, 1999a] changed the mechanism of activation spreading so that activation is a measure of expected utility of a behavior according to decision theory. They have been successfully used, for example, as decision mechanism for the magmaFreiburg team scoring 2nd in RoboCup 1999 simulation league competition [Dorer, 1999b]. In this section we give a short overview of the relevant activation spreading mechanism in EBNs before we describe how this mechanism needs to be changed to model human decision making according to prospect theory.

3.1   Network Definition

Extended behavior networks consist of goals, resource nodes and so called competence modules that are linked into a network.

Definition 1. *A* goal *consists of a tuple (*GCon, ι, RCon*) with*

- GCon *the* goal condition *(conjunction of propositions, i.e. possibly negated atoms), the situation in which the goal is satisfied,*
- $\iota \in [0..1]$ *the (static)* importance *of the goal,*
- RCon *the* relevance condition *(conjunction and disjunction of propositions), i.e. the situation-dependent (dynamic) importance of the goal.*

Definition 2. *A* resource node *is a tuple (res, g, θ_{Res}) with*

- $res \in \mathcal{R}$ *the* resource *represented by the node,*
- $g \in \mathbb{R}^+$ *the* amount of *bound resource units, i.e. units that are bound by a currently executing competence module and*
- $\theta_{Res} \in \,]0..\theta]$ *the resource specific* activation threshold *(where θ is the global activation threshold of the network).*

Definition 3. *A* competence module *k consists of a tuple (*Pre, b, Post, Res, a*) with*

- Pre *the* precondition *and $e = \tau_P(Pre, s)$ the* executability *of the competence module in situation s where $\tau_P(Pre, s)$ is the (fuzzy) truth value of the precondition in situation s;*
- b *the* behavior *that is performed once the module is selected for execution;*
- Post *a set of tuples (Eff, ex), where Eff is an expected effect (a proposition) and $ex = P(Eff|Pre)$ is the probability of Eff getting true after execution of behavior b,*
- a *the* activation $\in \mathbb{R}$, *representing a notion of the expected utility of the behavior (see below).*

- Res *is a set of resources res* $\in \mathcal{R}$ *used by behavior b.* $\tau_U(k, res, s)$ *is the situation-dependent amount of resource units expected to be used by behavior b.*

Definition 4. *An* extended behavior network *EBN consists of a tuple* $(\mathcal{G}, \mathcal{M}, \mathcal{U}, \Pi)$, *where* $\mathcal{G}$ *is a set of goals,* $\mathcal{M}$ *a set of competence modules,* $\mathcal{U}$ *a set of resource nodes and* Π *is a set of parameters that control activation spreading (see below)*

- $\gamma \in [0..1[$ *controls the influence of activation of modules,*
- $\delta \in [0..1[$ *controls the influence of inhibition of modules,*
- $\beta \in [0..1[$ *the inertia of activation across activation cycles,*
- $\theta \in [0..\hat{a}]$ *the activation threshold that a module has to exceed to be selected for execution, with* $\hat{a}$ *the upper bound for a module's activation,*
- $\Delta\theta \in]0..\theta]$ *the threshold decay.*

3.2   Activation Spreading

The decision of which behavior to adopt should be based on the the expected utility of executing such behavior. In EBNs, the expected utility of a behavior is approximated by a mechanism called *activation spreading*. The competence modules are connected to the goals and other competence modules of the network. Across those links activation is spread from the goals to the competence modules and among competence modules.

A competence module receives *activation* directly from a goal if the module has an effect that is equal to a proposition of the goal condition of that goal.

$$a_{kg_i}^{t\,\prime} = \gamma \cdot u(\iota_{g_i}, r_{g_i}^t) \cdot \nu_\gamma(p_j) \cdot ex_j \ , \tag{2}$$

$u(\iota_{g_i}, r_{g_i}^t)$ is the utility function mapping importance ι_{g_i} and relevance $r_{g_i}^t$ to a utility value. In section 4 we will show how this utility function has to be changed to reproduce human decision making described in section 2.2. ν_γ determines how activation is distributed to multiple propositions of the goal condition. ex_j is the probability of the corresponding effect to come true. In section 4 we will introduce a weighting function for this probability corresponding to section 2.1. The amount of activation depends on the probability of an effect to come true and the utility of the proposition in the goal condition as described in equation 1.

A competence module is *inhibited* by a goal if it has an effect proposition that is equal to a proposition of the goal condition and one of the two propositions is negated. Inhibition represents negative expected utility and is used to avoid the execution of behaviors that would lead to undesired effects.

$$a_{kg_i}^{t\,\prime\prime} = -\delta \cdot u(\iota_{g_i}, r_{g_i}^t) \cdot \nu_\delta(p_j) \cdot ex_j \ , \tag{3}$$

A competence module x is linked to another competence module y if x has an effect that is equal to a proposition of the precondition of y. y is called a *successor* module of x. Module x gets activation from the successor the amount of which depends on the utility of the precondition and the probability of the effect to come true. The utility of propositions that are not part of a goal condition is not available directly. It can be determined indirectly using the activation of the containing module and the truth value of the proposition. In this way, unsatisfied preconditions get implicit sub-goals of the network. Their utility directly depends on the utility of the competence module itself.

Finally a competence module x is linked to another competence module y if it has an effect that is equal to a proposition of the precondition of y and one of the two propositions is negated. y is called a *conflictor* of x, because it has an effect that destroys an already satisfied precondition of x. Again, a conflictor link from x to y is inhibiting (negative activation) to avoid undesired effects.

The activation of a module k at time t is then the sum of all incoming activation and the previous activation of the module decayed by β (defined in the set of parameters Π):

$$a_k^t = \beta a_k^{t-1} + \sum_i a_{kg_i}^t, \tag{4}$$

where $a_{kg_i}^t$ is the maximal activation module k receives at time t from goal g_i to which the module is linked directly or indirectly across incoming successor and conflictor links of other competence modules. For more details on activation spreading see [Dorer, 1999a].

Behavior selection is done locally in each competence module in a cycle containing the following steps. The details of behavior selection are not relevant in this context.

1. Calculate the activation a of the module.
2. Calculate the executability e of the module.
3. Calculate the execution-value $h(a, e)$ as the product of both.
4. Choose those competence modules for execution that have an execution-value above that of each resource node linked to. For each resource there have to be enough units available.
5. Reduce θ of each resource node not used $\Delta\theta$ and go to 1.

4 EBNs and Prospect Theory

In this section we show how the calculation of activation in EBNs has to be changed to correspond to findings reported in section 2. Experiments with both versions of EBNs reproduce the discrepancy between decision theoretic and human decision making reported in that section.

4.1 Theory

As described in section 2, prospect theory introduces non-linear value and weighting functions to explain results of experiments on human decision making.

The very same can be done for EBNs. The already existing value function u (see equation 2, 3) needs to be changed according to prospect theory. The utility function was chosen to correspond to the measure of risk taking defined by Arrow and Pratt [Eisenführ and Weber, 1999]:

$$r(x) = \frac{u''(x)}{u'(x)} \tag{5}$$

Using US-American tax data, Friend and Blume [1975] showed that investors exhibited decreasing absolute and constant relative risk taking behavior $x \cdot r(x)$. A utility function

$$u(x) = \begin{cases} x^{2\rho} & : & x \geq 0 \\ -x^{2\rho} & : & x < 0 \end{cases} \tag{6}$$

corresponds to this observation if x is a normed value and $\rho \in]0..1]$ is used as risk parameter. Using $\rho = \frac{1}{2}$ the utility function is linear and corresponds to decision theory exhibiting risk neutral behavior. $\rho < \frac{1}{2}$ corresponds to a risk-aversive behavior in case of gains ($x \geq 0$) and risk-taking behavior in case of losses ($x < 0$) as was observed in section 2. For $\rho > \frac{1}{2}$ it is vice versa.

A weighting function was not envisaged in original EBNs, but can easily be introduced. In equations 2 and 3 the probability ex_j is replaced with $\pi(ex_j)$ where the weighting function $\pi(x)$ is defined as follows:

$$\pi(x) = \begin{cases} 0 & : & x = 0 \\ 1 & : & x = 1 \\ e^{x-1} - \frac{1}{4} & : & 0 < x < 1 \end{cases} \tag{7}$$

This weighting function shows the properties described in section 2.1. For the impossible ($x = 0$) and certain ($x = 1$) outcome weighting function and probability match ($\pi(x) = x$). Low probabilities are overestimated ($\pi(x) > x$) while high probabilities are underestimated ($\pi(x) < x$).

The simplicity with which EBNs can be adjusted to model human decision making and prospect theory is also underlined by the amount of code changes necessary for implementation. Changing the existing value function required adding 4 lines of code. Adding the weighting function required changes in 4 lines of code (the calculation of activation for each type of connection) and adding another 6 lines of code.

4.2   Experiments

The changes above have been applied to the problems described in section 2. In the following figures, the upper level nodes of the networks are the goals with corresponding value. The lower level nodes are the competence modules representing the alternatives to choose from. The set of connections from a competence module represent the prospects with corresponding probabilities. The parameters for all the networks were chosen in order to have activation values matching expected utility of decision theory. As values we used $\gamma = 1.0, \delta = 1.0^2$

---

[2] A value of 1.0 for γ and δ is outside the definition area that guarantees convergence of activation. In our case this is no problem since no activation spreading between competence modules is done

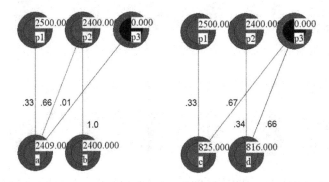

Fig. 3. Decision Theoretic EBN for problem 1 (left) and problem 2 (right) of section 2

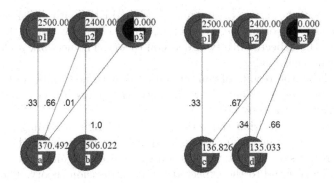

Fig. 4. EBN for problem 1 (left) and problem 2 (right) of section 2

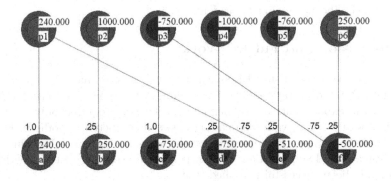

Fig. 5. EBN for problem 3 and problem 4 using decision theoretic activation spreading

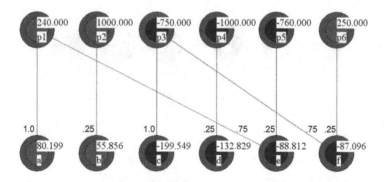

Fig. 6. EBN for problem 3 and 4 using activation spreading based on prospect theory

and $\beta = 0.0$. In the prospect theoretic cases we used as risk parameter $\rho = 0.4$.

Figure 3 shows the results of running an original EBN with decision theoretic activation calculation on problem 1 and problem 2. The activation of the modules correspond to the expected utility of decision theory. The network prefers a over b and c over d accordingly.

Figure 4 shows the results of running an EBN with new value and weighting function on the same problems. Now the network prefers b over a and c over d as the majority of students did.

In the same way, decision theoretic EBNs and EBNs according to prospect theory have been applied to problems 3 and 4. Figure 5 shows the results of an EBN with decision theoretic activation calculation. Again the activation of the modules correspond to the expected utility of decision theory. The network prefers b over a, is indifferent with respect to c and d and prefers f over e.

Figure 6 shows the results of an EBN according to prospect theory on the same problems. Here the network prefers a over b and d over c as the majority of students did. Also it prefers f over e showing the same intransitive decision taken by a significant amount of students.

5   Discussion and Future Work

In this paper we showed that EBNs can be used to reproduce human decision making deviating from rational decision making with respect to decision theory. However, the experiments of Kahneman and Tversky required only relatively small EBNs and no activation spreading between competence modules. Future work should investigate if the results can be used for bigger EBNs using activation spreading. The RoboCup domain is particularly interesting since EBNs have already been successfully applied to it.

The next steps should then be:

1. Play a number of soccer games against agents of an opponent team using EBNs based on prospect theory for decision making. This is already possible

and first results indicate that such a team shows different behavior. Investigate if and how the first team of agents can improve their performance by modeling their opponents using EBNs based on prospect theory compared to a team of agents using EBNs for opponent modeling based on decision theory.

2. Replace the opponent team by a real soccer team and see if it shows improved performance when switching between EBNs using decision theory or prospect theory. If step 1 is successful then the results of this paper indicate that also this step should be.

References

[Dorer, 1999a] Dorer, K.: Behavior networks for continuous domains using situation-dependent motivations. In: Proceedings of the Sixteenth International Conference of Artificial Intelligence, pp. 1233–1238 (1999)

[Dorer, 1999b] Dorer, K.: Extended behavior networks for the magmafreiburg team. In: Coradeschi, S., Balch, T., Kraetzschmar, G., Stone, P. (eds.) Team Descriptions Simulation League, pp. 79–83. Linköping University Electronic Press, Stockholm (1999)

[Dorer, 2004] Dorer, K.: Extended behavior networks for behavior selection in dynamic and continuous domains. In: Visser, U., Burkhard, H.-D., Doherty, P., Lakemeyer, G. (eds.) Proceedings of the ECAI workshop Agents in dynamic domains, ECAI, Valenzia (2004)

[Eisenführ and Weber, 1999] Eisenführ, F., Weber, M.: Rationales Entscheiden. Springer, Heidelberg (1999)

[Friend and Blume, 1975] Friend, I., Blume, M.: The demand for risky assets. American Economic Review 64, 900–922 (1975)

[Kahnemann and Tversky, 1979] Kahnemann, D., Tversky, A.: Prospect theory: An analysis of decision under risk. Econometrica 47, 263–291 (1979)

[Maes, 1989] Maes, P.: The dynamics of action selection. In: Proceedings of the International Joint Conference on Artificial Intelligence, pp. 991–997 (1989)

[Tversky and Kahneman, 1981] Tversky, A., Kahneman, D.: The framing of decisions and the psychology of choice. Science 22, 453–458 (1981)

Motion Synthesis through Randomized Exploration on Submanifolds of Configuration Space

Ioannis Havoutis and Subramanian Ramamoorthy

Institute of Perception, Action and Behaviour,
School of Informatics,
University of Edinburgh,
Edinburgh, EH8 9AB, UK
I.Havoutis@sms.ed.ac.uk,
S.Ramamoorthy@ed.ac.uk

Abstract. Motion synthesis for humanoid robot behaviours is made difficult by the combination of task space, joint space and kinodynamic constraints that define realisability. Solving these problems by general purpose methods such as sampling based motion planning has involved significant computational complexity, and has also required specialised heuristics to handle constraints. In this paper we propose an approach to incorporate specifications and constraints as a bias in the exploration process of such planning algorithms. We present a general approach to solving this problem wherein a subspace, of the configuration space and consisting of poses involved in a specific task, is identified in the form of a nonlinear manifold, which is in turn used to focus the exploration of a sampling based motion planning algorithm. This allows us to solve the motion planning problem so that we synthesize previously unseen paths for novel goals in a way that is strongly biased by known good or feasible paths, e.g., from human demonstration. We demonstrate this result with a simulated humanoid robot performing a number of bipedal tasks.

1 Introduction

One of the most significant recent trends in robotics is the push towards robust autonomy with complex robots such as humanoids. In principle, humanoid robots and other related architectures are highly versatile and capable of performing an unprecedented variety of tasks in applications ranging from service at home to rescue in rugged terrains. However, due to the inherent complexity of these systems, robotics researchers have struggled to realise this promise of robust and flexible operation in a multitude of environments. From the point of view of motion synthesis, i.e., the generation of feasible trajectories for all the joints of a robot given a *family* of task level goals such as, say, foot placement points, one of the big difficulties has been that of reconciling the need for efficient exploration of all possible ways to perform a family of tasks with the need for understanding of the intrinsic constraints that define realisability of the

J. Baltes et al. (Eds.): RoboCup 2009, LNAI 5949, pp. 92–103, 2010.

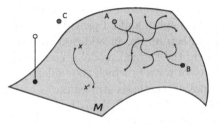

Fig. 1. *Left:* Schematic representation of a low-dimensional manifold as used in our algorithm. Combining an RRT planner with manifold learning focuses random sampling to a task relevant subspace. An RRT is grown by searching this manifold M, a nonlinear subspace of the configuration space. *A, B and C:* Taking a step with the humanoid is a process of following a path connecting configuration A (*green*) with configuration B (*blue*). The point C (*red*) represents a configuration that is not reached in any realisation of the task. Furthermore task specific geodesic distances can be computed, e.g. distance (*dotted line*) between x and x' and samples (*white*) lying close to the manifold can be projected onto it (*black*). The geodesic distance is used to calculate nearest neighbours and step sizes while projection guides sampling onto the task-relevant subspace.

task. Machine learning methods are efficient in capturing intrinsic task-specific constrains within restricted domains, i.e., focussing on properly interpolating between observed examples, while sampling-based motion planning methods are more focussed towards large-scale exploration of the global structure of configuration spaces. In the absence of specialised knowledge of task constraints, this can involve significant computational complexity.

In fact there are a number of tasks, e.g., locomotion in *RoboCup* domain where it is possible to get some human demonstration data but it is hard to explicitly characterize the implicit constraints that define the task. With this in mind, in this paper, we present an approach to motion synthesis that brings together two related but distinct algorithmic threads: sampling-based motion planning and manifold learning. We begin with a small set of example trajectories that are representative of the intrinsic constraints that define a task, e.g., bipedal walking. These trajectories are really just samples drawn from a set of possible trajectories that define a sub-manifold in the configuration space of the robot - indirectly defined by task space, joint space and kinodynamic constraints. We use a manifold learning algorithm to approximate this sub-manifold. In particular, our construction enables us to specify projections onto the manifold and also to compute geodesics. Then, as the robot is presented with different goals that appear in the course of its operation, we use a sampling based motion planning algorithm – Rapidly-exploring Random Trees (RRT) [1] – to synthesise novel trajectories that are restricted to lie on this sub-manifold.

A primary benefit of focussing exploration in this way is that it enables us to bring into the planning process constraints that are only known in terms of observed data from known good behaviours (i.e., not explicitly modelled). This makes our approach a data–driven one, wherein the constraint is inferred from observed data and used in the planning process in the form of a sub–manifold onto which exploration is restricted. The learned sub-manifold provides a basis for higher level deliberation in a layered architecture. So, in addition to computational savings arising from focussed search, the learned model serves as an abstraction that succinctly encodes the variety of ways in which the underlying task may be performed - enabling different motion synthesis strategies.

The main contribution of this paper is the manifold-RRT algorithm, a novel extension to the RRT, which incorporates the focussed sampling idea mentioned above through a data-driven manifold learning algorithm. This enables us to synthesise high quality trajectories for bipedal robotic tasks such that the exploration is focussed to the neighbourhood of demonstrated behaviours. We first provide an overview of the motion planning and manifold learning algorithms as they relate to this work. Then, we describe the mRRT algorithm which combines the benefits of these two approaches. We demonstrate the applicability of this idea through experimental results with a simulated version of the KHR-2HV humanoid robot. Finally, we conclude with a brief discussion of how this specific result may be applied in more general settings involving humanoid and other robot behaviours.

2 Related Work

In the context of biological behaviours, it has been argued [2] and observed [2,3] that the curse of dimensionality is best overcome by utilising synergies and coordination strategies that enforce a restriction of the synthesised motions to low-dimensional spaces. Robotics [4,5,6] and graphics [7] researchers have utilised this fact to devise efficient motion synthesis strategies. Our interest is in incorporating this feature directly into sampling based motion planning. Some recent work [8,9,10] comes close to this issue by considering how task space constraints, e.g., end-effector constraints, can be used to structure search in configuration space with local Jacobian mappings. In other related work, e.g. references [11,12], the goal is to edit a statically stable trajectory, discovered by a sampling based motion planner, in a post-processing step to make the resulting trajectory dynamically realisable. However, the low-dimensional structure of the task is not directly leveraged in on-line planning. Computer animation researchers have arrived at closely related insights in developing structures such as motion-motif graphs [13] which try to abstract families of related trajectories into symbolic nodes so that on-line search is made efficient. However, in that work, the issue of task constraints is not given as much importance as in robotics and the focus is really on efficiently compressing a motion capture database.

3 Background

3.1 Rapidly Exploring Random Trees

Sampling-based motion planning algorithms are based on the idea of approximating the *free* portion of the configuration space by a suitable random structure

that enables efficient computation and fast exploration. The RRT [14] is a re-
markably simple yet effective algorithm for planning a path between two points
in configuration space.

In the algorithm, one adopts a simple set characterisation of the configuration
space, which is the union of the free space, Q_{free} and the obstacle space Q_{obs}.
Sampled configurations, q, are drawn from Q_{full}, $Q_{full} = Q_{free} \cup Q_{obs}$. Q_{full}
can be the configuration space or the phase space for the system, or even just
any composition of state variables within $q \in \mathbb{R}^D$, D being the dimensionality
of the problem space.

We root a tree, T, at the given starting point, q_{init} and grow it by iterating the
following process. Pick a random point $q_{rand} \in Q_{full}$ and calculate its distance
from each point already in T. Select the closest point, q_{near}, from T and grow the
tree toward q_{rand} by a step size Δx. Then evaluate if the resulting configuration
$q_{new} = q_{near} + \Delta x_{q_{rand}}$ belongs to Q_{free} or Q_{obs}. If the former is true q_{new} is
added to T, else the sample is discarded. The procedure is repeated until the goal
configuration q_{goal} is reached, within some tolerance or number of iterations. The
shortest path is then computed on T using a tree search algorithm. Algorithm
2 includes these core RRT (cRRT[1]) steps and is augmented with the LSML
procedure, to be described.

RRTs quickly branch into unexplored regions of the space and when such
regions become small the algorithm begins to fill in gaps with increasing resolution. This ensures that the planner is probabilistically complete, thus it will
find a path if one exists as the number of samples grows to infinity. However,
when considering complex problems involving humanoids, many finer points need
consideration, including convergence to the goal, stability and realisability constrains, space coverage and resolution. For example, as a rule of thumb, in spaces
with $D \geq 8$ convergence is typically slow. It has been shown that including a
bias favouring the goal greatly increases the convergence speed as it steers the
exploration [1].

In general, success of RRTs depends on the metric that is defined over the
space to be explored. Traditionally a metric of the form:

$$d(q, q') = \sum_{i=1}^{n} w_i \, \|q_i - q'_i\| \,,$$

is used where the weights w_i denote the importance of each Degree of Freedom
(DoF). These weights are often empirically chosen based on trial and error but
as the dimensionality grows, and in nonlinear systems, this becomes difficult
from intuition alone, so, there is a need for other ways to arrive such metrics.
We argue that learning such a metric in a data-driven fashion is a desirable and
scalable approach.

The second, related, issue that determines success of RRT-based planning is
coverage. Random sampling in high dimensions can be excessively wasteful when
the underlying task has special structure. The key issue is that sampling a high

---

[1] We term cRRT the classic RRT algorithm as described in [1].

Algorithm 1. Learn Manifold

1: **Lsml**(tr_data, d)
2: INPUT: kinematic task-relevant data tr_data, dimensionality of manifold d
3: OUTPUT: manifold M
4: $NN \leftarrow$ NN_GRAPH(tr_data)
5: $\theta \leftarrow$ OPTIMISE_PARAMETERS(tr_data, NN) {Model Parameters}
6: $M \leftarrow$ MINIMISE_MODEL_ERROR(θ) {Fit the manifold}
7: RETURN M

---

dimensional space densely enough is computationally infeasible. Knowing that many interesting robotic behaviours are restricted to low-dimensional subspaces [15,16,2,4,17], due to a variety of reasons including stability and energy constraints, joint limits and self-collision constraints, it is desirable to leverage this to achieve better coverage where it matters.

3.2 Manifold Learning

The machine learning literature includes many examples of dimensionality reduction methods used to abstract and/or make problem spaces manageable [18,19,16]. One of the big benefits of these methods is that they are data-driven and can be used in a scalable way in novel domains.

In the usual formulation, manifold learning is about finding an embedding or 'unrolling' of a nonlinear manifold onto a lower dimensional space while preserving metric properties such as inter-point distances. Popular examples include MDS [20], LLE [21] and ISOMAP [22]. However, much of this work has been focused on summarisation, visualisation or analysis that explains some aspect of the observed data. Instead, we are more interested in methods that provide a direct representation of a nonlinear subspace in a way that enables standard geometric operations needed in motion planning. Such methods should work with demonstrated motions and provide good interpolation and extrapolation on the learnt manifold. For this, we choose a recently developed method – Locally Smooth Manifold Learning [23,24]. LSML explicitly focuses on generalising to unseen portions of the manifold, which is crucial for use with an exploration algorithm. The learnt manifold can be used to compute geodesic distances, to find projections of points on the manifold and to generate novel sample points. A detailed description of LSML, from [23], follows.

LSML. Given that our D-dimensional data lies on a locally smooth d-dimensional manifold, where $d < D$, there exists a continuous bijective mapping M that converts low dimensional points, $y \in \mathbb{R}^d$, to points, $x \in \mathbb{R}^D$, in the original high dimensional space. The goal is to learn a warping function W that can take a point on the manifold and compute its neighbouring points on the manifold, capturing the modes of variation of the data. Thus we can approximate W by M locally by defining $W(x, \epsilon) = M(y + \epsilon)$ where $y = M^{-1}(x)$ and $\epsilon \in \mathbb{R}^d$. The first order approximation of the above is $W(x, \epsilon) \approx x + \mathcal{H}(x)\epsilon$ where each column $\mathcal{H}_{.k}$ of $\mathcal{H}(x)$ is the partial derivative of M with respect to y_k, i.e. $\mathcal{H}_{.k}(x) = \partial/\partial y_k M(y)$, valid given ϵ is small enough.

The objective then is to learn the unknown parameterised function $\mathcal{H}_\theta$: $\mathbb{R}^D \to \mathbb{R}^{D \times d}$, parameterised by a variable θ (e.g. parameters of an RBF-linear model). For that we first calculate the set of nearest neighbours N^i, for each point x^i of the training data. This way, if x^j is a neighbour of x^i, then there exists an unknown ϵ^{ij} such that $W(x^i, \epsilon^{ij}) = x^j$, or to a good approximation $\mathcal{H}_\theta(\bar{x}^{ij})\epsilon^{ij} \approx \Delta^i_{.j}$, where $\Delta^i_{.j}$ can be regarded as the centred estimate of the directional derivative at $\bar{x}^{ij}$.

To solve for $\mathcal{H}_\theta$ we define the error:

$$err(\theta) = min_{\{\epsilon^{ij}\}} \sum_{i,j \in N^i} \left\| \mathcal{H}_\theta(\bar{x}^{ij})\epsilon^{ij} - \Delta^i_{.j} \right\|^2_2,$$

and minimise for θ with the addition of a regularisation term:

$$\lambda \sum \left\| \epsilon^{ij} \right\|^2_2 + \lambda \sum \left\| \mathcal{H}_\theta(\bar{x}^{ij}) - \mathcal{H}_\theta(\bar{x}^{ij'}) \right\|^2_F,$$

where $\bar{x}^{ij}$ and $\bar{x}^{ij'}$ are two neighboring locations, ϵ^{ij} and λ are regularisation terms that enforce the smoothness of the mapping. To solve this, a radial basis function(RBF)-based linear parametrisation is used, along with an alternating minimisation procedure (with random restarts to avoid local minima). Pseudocode for the method is available in Algorithm 1.

Projection. The projection of a point x on a learnt manifold M cannot be computed in closed form. Instead a gradient descent approach is utilised in finding a new point x' on M that minimises the distance $\|x - x'\|^2_2$. Since $\mathcal{H}_\theta$ is defined over the whole $\mathbb{R}^D$ we calculate the orthonormalised tangent space at x', $H' \equiv orth(\mathcal{H}_\theta(x'))$, and $H'H'^T$ the corresponding projection matrix. We follow the gradient to the local minima on the manifold, using the update rule for x': $x' \leftarrow x' + \alpha H'H'^T(x - x')$, with α being the step size. To find the closest projection we initially set x' to be the nearest point in the training data.

Geodesic distance. To compute the geodesic distance between two points, x and x', on a manifold we use an active contour model, also known as a *snake* [25]. A *snake* defines a discretised path between x and x' and its length is being minimised by gradient descent. The error reflecting the length of the path is given by: $err_{len}(\chi) = \sum_{i=2}^m \left\| \chi^i - \chi^{i-1} \right\|^2_2$, where the χ's are the linearly interpolated -manifold respecting- points between the fixed start and end points. The update rule for each χ^i is very similar to the update rule used for projection.

4   The Manifold-RRT Algorithm

Our algorithm is a variant of the conventional RRT, augmented with the manifold learning operation (Algorithm 1). This hybrid procedure is described in Algorithm 2. We use the learnt manifold, M, to compute distances between points

---

Algorithm 2. Manifold Path Planning

1: **mRrt**(q_{init}, q_{goal}, M)
2: INPUT: start point q_{init}, goal point q_{goal}, learnt manifold M
3: OUTPUT: path in configuration space p
4: T.add(q_{init}) {Initialize tree T}
5: **for** $i = 0$ to k **do**
6:     $q_{rand} \leftarrow$ RANDOM_POINT
7:     $q_{proj} \leftarrow$ PROJECT(q_{rand}, M)
8:     $q_{near} \leftarrow$ GEODESIC_D(q_{proj}, T, M)
9:     $q_{new} \leftarrow$ STEP(q_{near}, q_{proj}, dx) {Construct Snake}
10:    $valid \leftarrow$ EVALUATE(q_{new})
11:    **if** $valid == true$ **then**
12:        T.add(q_{new})
13:        $dist \leftarrow$ GEODESIC_D(q_{new}, q_{goal}, M)
14:        **if** $dist \leq tolerance$ **then**
15:            **break**
16:        **end if**
17:    **end if**
18: **end for**
19: $p \leftarrow$ SHORTEST_PATH(T.first,T.last)
20: RETURN p

---

in configuration space. The metric is the geodesic distance directly learnt from the training data. We utilise the geodesic distance to evaluate nearest neighbour relations and find q_{near}. This is used to decide which node of the tree will be subsequently grown. Moreover we use the learnt manifold to project uniform random samples in configuration space, q_{rand}, onto the manifold - focusing the planner to explore a task-relevant subspace.

Growing the tree T involves this projection, q_{proj}, of the random sample and the computation of a *snake* (Section 3.2) from the nearest neighbour on the graph to the new point, q_{new}. The interpolated points on the manifold that compose the *snake* are then examined and the geodesic distance, $dist$, from the starting point is computed. When the geodesic distance reaches the desired step size dx we set the via-point as the end of the step and evaluate the resulting path in simulation.

Next, the geodesic distance from the new vertex to the goal-point is computed. If the distance is lower than a tolerance threshold, the exploration stops. A shortest path from the start-point to the last vertex added is computed using a standard tree search algorithm. The resulting path, p, is the motion plan.

5   Experimental Setup

We present experiments with a simulated humanoid robot, KHR-2HV (Figure 2). This involves no explicit analytic model of the humanoid robot dynamics. Instead, we treat the simulated robot as an incrementally evaluable black-box.

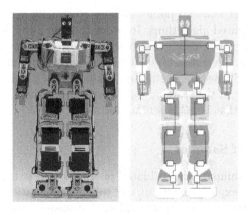

Fig. 2. The KHR-2HV Humanoid robot and the corresponding 17 degrees of freedom

So, although we present the results from simulation, the procedure can be identically applied to a physical robot as well. In particular, even though we search a region in configuration space, the intrinsic dynamics of the nonlinear high dimensional system are taken into consideration implicitly (of course, subject to the restriction of what is expressible in the configuration space).

Our simulation is in Webots [26], a commercial physically realistic modelling and simulation ODE-based environment. In Webots we use an accurate model of the Kondo KHR-2HV humanoid robot, where motion is performed using P-controllers that closely simulate the characteristics of the real robot servos. A controller has been implemented in C that handles the communication between Webots and Matlab and exposes the full functionality of the robot model. Both cRRT and mRRT algorithms are implemented in Matlab and communicate directly with the simulator for the evaluation of configurations. LSML is implemented in Matlab, using Piotr Dollar's LSML code [2].

5.1 Task

We have experimented with a number of bipedal tasks. However, we include nothing in our experiment that is specific to these particular tasks, thus the same procedure is applicable to other bipedal tasks as well. We compare the classical RRT and mRRT on the same tasks of forward and backward stepping and kicking. For the purposes of planning, we consider all the leg and hip DoFs of the humanoid, resulting in a 10-dimensional configuration space.

We begin with a single example – a hand-crafted trajectory, from stance to double support for stepping and to midair reach for kicking. We sample the training data in simulation from the KHR-2HV humanoid and we use a 5 millisecond sampling rate that is equal to the physical simulation time step. The resulting motions are 116, 98 and 96 points long for step, kick and backstep

---

[2] Available at: http://vision.ucsd.edu/~pdollar/

accordingly. The start and end points are chosen equally as the initial and goal points of both compared algorithms. In order to add variability to the learning step of mRRT we further use 2 perturbed instances of the previous motions. These trajectories are generated by random normal sampling in the vicinity of the training data and subsequent stability evaluation in simulation. Furthermore the sequential nature of the training data ensures that the learnt sub–space is a single manifold and that is not disconnected.

5.2   Evaluation of Samples

In both cases, new samples are evaluated in simulation. It is worth noting that we do not have an explicit model of the robot's kinematics or dynamics. So, samples are evaluated in a dynamic fashion that ensures their suitability. Both in cRRT and mRRT, the nearest neighbour of every new sample is computed on each iteration of the exploration cycle. The humanoid's servos are set at the appropriate positions and its global position and rotation is set accordingly. The robot is then commanded to perform the motion that reaches the new configuration point according to the servos' P-controllers. We utilise feedback from the gyroscope and the accelerometers to evaluate the stability and stance. Furthermore we employ two foot force sensors to distinguish between single and double support configurations.

5.3   Other Parameters

We have used the average geodesic distance between data points in the training set to set the step length in both algorithms. Such a choice is well suited to the task at hand and was made for comparability in evaluation. Note that this choice greatly *favours* cRRT as the metric used is now '*informed*' in a systematic manner, in contrast to the often ad-hoc RRT setting. We have set a bias of 0.1 towards the goal point in order to boost convergence. The default LSML parameterisation has been used with no effort at special optimisation as errors have been adequately small. The actual time for learning a manifold depends on the amount of training data and for our experiments required less than a minute in all cases.

6   Results

We compare the performance of cRRT and mRRT with a number of different metrics. Each trial has been repeated 10 times for both algorithms and all reported results are averaged over the number of trials. Examples of a resulting paths discovered by mRRT are depicted in Figure 3.

The evaluation metrics are quite intuitive. In particular, we note the following. Average path length corresponds to the number of points that are traversed from the initial configuration in order to reach the goal configuration. Number of samples denote the total explorative samples needed until the goal is reached.

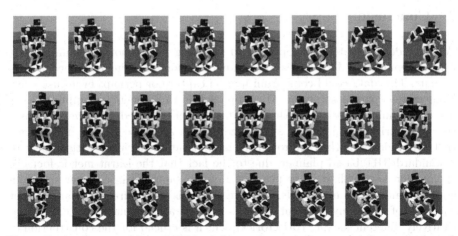

Fig. 3. Paths discovered by mRRT for left step forward *(top row)*, left step backward *(middle row)* and left leg kick *(bottom row)*

Table 1. Results averaged over 10 trials for step forward, kick and step backward

Task	step		kick		backstep	
	cRRT	*mRRT*	*cRRT*	*mRRT*	*cRRT*	*mRRT*
Average path length	40.9	38	52.5	49.4	47.2	37.5
Average number of samples	268.63	199.2	291	249.3	293.4	189.8
Average tree size	127.7	127	140.3	137.7	120.7	108.4
Average number of invalid samples	140.6	74.4	150.7	111.6	172.7	81.4
Smoothness {nRMSE}	0.0051	0.0049	0.0055	0.0041	0.0046	0.0043

Tree size is the number of vertices that the resulting tree consists of. Invalid samples are evaluated points that do not satisfy dynamic stability or collision constraints. Smoothness is defined as the average normalised Root Mean Square Error (nRMSE) with respect to a fitted cubic polynomial at each joint motion for each resulting plan path.

Our experimental results show that mRRT, in all trials, discovers a solution with *much fewer* invalid samples. These are the random configurations that fail in the evaluation step. On average mRRT explores only half as many 'bad' samples as cRRT (57.6%). This translates to an average decrease of 25.2% in overall planning steps for the specific tasks. More importantly, as tasks become more complex in terms of dynamical constraints, this ensures that exploration is accordingly useful.

On average, mRRT discovers shorter paths than cRRT and requires much fewer samples. For all tasks, the average size of trees is approximately equal and both algorithms find smooth paths. The results are summarised in Table 1. In general, we expect that the above mentioned differences would be more pronounced as the tasks are more spatio-temporally extended and dynamical realisability constraints become more severe.

7 Conclusion

We have demonstrated an approach to sampling based motion planning that utilizes demonstrated examples to glean information regarding task-specific constraints. This data could come from motion capture or perhaps even just a few hand crafted partial solutions. Our approach is to augment a sampling-based motion planning algorithm with a manifold learning procedure to provide task-specific metrics and a way to synthesize motion as geodesics. We have shown that this yields a marked improvement in exploration efficiency with respect to a standard RRT based planner, due to the fact that the learnt metric focusses exploration better than other forms of random sampling in a larger space. In addition, and very importantly, we note that this procedure yields an efficient encoding of the many different ways to perform a particular task (e.g., quantitatively different kicking trajectories), which is crucial for the construction of multi-level motion synthesis strategies.

We have shown examples of bipedal tasks in simulation and our current work involves porting the algorithm to a physical robot platform. Also, in future work, we would like to augment the space with velocity and acceleration information which will be required to encode many challenging dynamic behaviours including jumping and running. Finally, our long term goal is to use this procedure as a way to seed the learning of a layered architecture for control, planning and reasoning.

References

1. LaValle, S.M.: Planning Algorithms. Cambridge University Press, Cambridge (2006)
2. Full, R., Koditschek, D.: Templates and anchors: neuromechanical hypotheses of legged locomotion on land. J. Exp. Biol. 202(23), 3325–3332 (1999)
3. Gutfreund, Y., Flash, T., Yarom, Y., Fiorito, G., Segev, I., Hochner, B.: Organization of Octopus Arm Movements: A Model System for Studying the Control of Flexible Arms. J. Neurosci. 16(22), 7297–7307 (1996)
4. Ramamoorthy, S., Kuipers, B.J.: Qualitative hybrid control of dynamic bipedal walking. In: Proceedings of Robotics: Science and Systems, Philadelphia, USA (August 2006)
5. Ramamoorthy, S., Kuipers, B.J.: Trajectory generation for dynamic bipedal walking through qualitative model based manifold learning. In: IEEE International Conference on Robotics and Automation (ICRA), May 2008, pp. 359–366 (2008)
6. Isto, P., Saha, M.: A slicing connection strategy for constructing prms in high-dimensional cspaces. In: Proceedings 2006 IEEE International Conference on Robotics and Automation, ICRA 2006, May 2006, pp. 1249–1254 (2006)
7. Safonova, A., Hodgins, J.K., Pollard, N.S.: Synthesizing physically realistic human motion in low-dimensional, behavior-specific spaces. ACM Trans. Graph. 23(3), 514–521 (2004)
8. Stilman, M.: Task constrained motion planning in robot joint space. In: IEEE/RSJ International Conference on Intelligent Robots and Systems, IROS 2007, 29 November 2 (2007)

9. Yao, Z., Gupta, K.: Path planning with general end-effector constraints: using task space to guide configuration space search. In: 2005 IEEE/RSJ International Conference on Intelligent Robots and Systems (IROS 2005), August 2005, pp. 1875–1880 (2005)
10. Bretl, T., Lall, S., Latombe, J.C., Rock, S.: Multi-step motion planning for free-climbing robots. In: WAFR, pp. 1–16 (2004)
11. James, J., Kuffner, J., Kagami, S., Nishiwaki, K., Inaba, M., Inoue, H.: Dynamically-stable motion planning for humanoid robots. Auton. Robots 12(1), 105–118 (2002)
12. Nakamura, Y., Yamane, K.: Interactive motion generation of humanoid robots via dynamics filter. In: Proc. of First IEEE-RAS Int. Conf. on Humanoid Robots (2000)
13. Beaudoin, P., van de Panne, M., Poulin, P., Coros, S.: Motion-motif graphs. In: Symposium on Computer Animation 2008 (July 2008)
14. LaValle, S.M., Kuffner, J.J.: Randomized kinodynamic planning. The International Journal of Robotics Research 20(5), 378–400 (2001)
15. Vijayakumar, S., D'souza, A., Schaal, S.: Incremental online learning in high dimensions. Neural Comput. 17(12), 2602–2634 (2005)
16. Bitzer, S., Havoutis, I., Vijayakumar, S.: Synthesising novel movements through latent space modulation of scalable control policies. In: Asada, M., Hallam, J.C.T., Meyer, J.-A., Tani, J. (eds.) SAB 2008. LNCS (LNAI), vol. 5040, pp. 199–209. Springer, Heidelberg (2008)
17. Jenkins, O.C., Mataric, M.J.: A spatio-temporal extension to isomap nonlinear dimension reduction. In: International Conference on Machine Learning (ICML), pp. 441–448 (2004)
18. Wang, J.M., Fleet, D.J., Hertzmann, A.: Gaussian process dynamical models for human motion. IEEE Trans. Pattern Anal. Mach. Intell. 30(2), 283–298 (2008)
19. Urtasun, R., Fleet, D.J., Geiger, A., Popovic, J., Darrell, T.J., Lawrence, N.D.: Topologically-constrained latent variable models. In: Proceedings of the 25th international Conference on Machine Learning, Helsinki, Finland, July 5-9, pp. 1080–1087. ACM, New York (2008)
20. Hastie, T., Tibshirani, R., Friedman, J.H.: The Elements of Statistical Learning. Springer, Heidelberg (2001)
21. Roweis, S.T., Saul, L.K.: Nonlinear dimensionality reduction by locally linear embedding. Science 290(5500), 2323–2326 (2000)
22. Tenenbaum, J.B., de Silva, V., Langford, J.C.: A global geometric framework for nonlinear dimensionality reduction. Science 290(5500), 2319–2323 (2000)
23. Dollár, P., Rabaud, V., Belongie, S.: Non-isometric manifold learning: Analysis and an algorithm. In: ICML (June 2007)
24. Dollár, P., Rabaud, V., Belongie, S.: Learning to traverse image manifolds. In: NIPS (December 2006)
25. Blake, A., Isard, M.: Active Contours. Springer, Heidelberg (1998)
26. Michel, O.: Webots: Professional mobile robot simulation. Journal of Advanced Robotics Systems 1(1), 39–42 (2004)

Robust and Computationally Efficient Navigation in Domestic Environments

Dirk Holz[1,2], Gerhard K. Kraetzschmar[1], and Erich Rome[2]

[1] Bonn-Rhein-Sieg University of Applied Sciences, Computer Science Department
dirk.holz@ieee.org, gerhard.kraetzschmar@h-brs.de
[2] Fraunhofer Institute for Intelligent Analysis and Information Systems (IAIS)
erich.rome@iais.fraunhofer.de

Abstract. Presented in this paper is a complete system for robust autonomous navigation in cluttered and dynamic environments. It consists of computationally efficient approaches to the problems of simultaneous localization and mapping, path planning, and motion control, all based on a memory-efficient environment representation. These components have been implemented and integrated with additional components for human-robot interaction and object manipulation on a mobile manipulation platform for service robot applications. The resulting system performed very successfully in the 2008 RoboCup@Home competition.

1 Introduction

Autonomous service robots that assist in housekeeping, serve as butlers, guide visitors through exhibitions in museums and trade fairs, or provide care to elderly and disabled people could substantially ease everyday life for many people and present an enormous economic potential [7,17,19]. Robots for all these applications face, however, the challenging task of operating in real-world indoor and domestic environments, such as those addressed by the RoboCup@Home league. Domestic environments tend to be cluttered, dynamic, and are populated by humans and domestic animals. In order to adequately react to sudden dynamic changes and avoid collisions, these robots need to be able to constantly acquire and process in real-time information about their environment. Furthermore, in order to act in a goal-directed manner, plan actions and navigate effectively, a robot needs an internal representation of its environment. Nature and complexity of these representations highly depend on the robot's task and application space.

For a more concrete example, consider a domestic service robot that is given the task to serve a cold drink from the refrigerator to a guest in the living room. Aside of the activities like interacting with the host and the guest, grasping objects like a can of soft drink, or other manipulation tasks, the robot needs to solve several problems related to navigation: If the environment is initially unknown, the robot must i) *explore the environment* and ii) *build a map*. Both during this exploration and map building phase and during everyday operation later on, the robot needs to iii) *localize itself* and iv) *localize task-relevant objects* (such as the

J. Baltes et al. (Eds.): RoboCup 2009, LNAI 5949, pp. 104–115, 2010.
© Springer-Verlag Berlin Heidelberg 2010

refrigerator) within its environment representation. As self-localization requires a map of the environment, while mapping requires the ability to self-localize, these two problems need to be considered jointly as simultaneous localization and mapping (SLAM). SLAM has not only been a substantial research focus in the robotics community over the last decades but is also regarded as a major precondition of truly autonomous robots [20]. For actually moving to certain locations in the environment, the robot needs to iv) *plan obstacle-free paths* and v) *follow planned paths*. Due to the fact that it operates in a dynamic environment, the robot must also constantly acquire information about the environment during navigation, and use it to vi) *update the map* and vii) *avoid collisions*.

All of the above problems have been well researched in robotics, at least in isolation. For each of these problems a large variety of sophisticated algorithms have been proposed. They coexist legitimately, since they are designed or especially appropriate for a specific purpose. However, despite the huge body of literature available, the problem of robust and computationally efficient navigation in domestic environments cannot be considered solved yet. The first issue is *robustness*. Especially in RoboCup@Home, there is only a short preparation time and only five to ten minutes to solve a complex task. Hence, algorithms need to be robust and the overall system has to act reliably. Advancing robustness, however, often comes with increasing complexity that affects the *real-time applicability* of the algorithm and the overall system which is the second issue. *Scalability* is another issue since the computational complexity of many sophisticated approaches e.g. in SLAM either directly results in prohibitive memory and runtime requirements if applied to realistically-sized or large real-world environments, or at least cannot be used online in a reasonable fast cycle time. The forth issue is *integration*. The aforementioned problems are strongly interwoven as, for example, the choice of the environment representation affects the choice of localization and path-planning algorithms. Identified best-in-class solutions may have different underlying assumptions hindering integration or necessitating possibly complex transformations from one representation into another. Efficiency problems may occur especially if such transformations cannot be done once and offline, but need to be done constantly or in regular intervals due to environmental dynamics. Furthermore, if published implementations are available at all, they are often not modular and easily re-usable as they depend on a specific architecture, development framework or inter-module communication.

Instead of proposing yet another toolkit for navigational purposes, the goal of our work is to design and implement a (complete) set of algorithms for autonomously performing SLAM, planning paths, and controlling the motion of a mobile service robot, i.e. an approach addressing the aforementioned problems ii) to vi) which is robust, efficient and scalable. Exploration and collision avoidance are not addressed in the context of this paper. The algorithms are implemented in a modular and reusable way. Dependencies on external libraries are kept at a minimum. Primary design goals are robustness, simplicity and real-time applicability of the algorithms and the overall system.

The remainder of this paper is organized as follows: Section 2 provides a brief overview on the robot platform used for implementation and in the RoboCup@Home competitions. Section 3 introduces *sparse point maps* as a space efficient environment representation together with the proposed SLAM algorithm. Path-planning based on this representation and the used motion controllers are presented in Section 4 and Section 5, respectively. Finally, Section 6 contains some concluding remarks and an outlook on future work.

2   Base System

For evaluating the performance and robustness of the algorithms presented in this paper, the mobile service robot *Johnny Jackanapes* was used (see Figure 1), which is based on a modular mobile robot platform called *VolksBot* [21]. VolksBot has been designed specifically for rapid prototyping and robot applications in education, research and industry. The customized variant used has an integrated manipulator, a Neuronics Katana 6M180 robot arm equipped with six motors providing five degrees of freedom w.r.t. the gripper's position and orientation in its reachable workspace. It is mounted in a way to provide good reachability and maneuverability. The overall platform size is (51×51×120)cm (W×L×H) and its weight is 60 kg. The drive unit used for locomotion uses a differential drive with two actively driven wheels, powered by two 150 W motors, and two caster wheels to enhance rotating and stability under load. The robot's maximum velocity is 2 m/s.

For perceiving environmental structures, a SICK S300 2D laser scanner is used. The size of the apex angle limiting the scan plane is 270°, with an angular resolution of 0.5°. For accessing other sensors and robot platforms as well

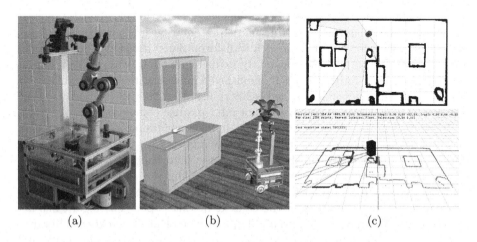

(a)   (b)   (c)

Fig. 1. (a) Robot platform "Johnny Jackanapes". (b) Simulation in Microsoft Robotics Studio. (c) Simulation using Player/Stage (top) and view on the remote inspection/debug application (bottom).

(a)   (b)

Fig. 2. (a) Single laser scan taken from the data set used in [23]. (b) Map constructed from all laser scans in the same data set. The couch table did not intersect the scan plane and is neither perceived nor modeled in the map.

as simulation environments, wrappers and interfaces have been implemented to interact with Microsoft Robotics Studio, Player/Stage and CARMEN (the Carnagie Mellon Navigation Toolkit). However, the drawback of 2D laser range finders for the purposes of collision avoidance and mapping is that objects not intersecting the 2D scan plane cannot be perceived by the robot. See e.g. the couch table that does not cause reflections in the 2D laser scan in Figure 2(a) and is thus not modeled in the point map in Figure 2(b). While in some indoor robot applications this drawback can be neglected, it plays an important role in a human's everyday environment, where typically many objects do not intersect the measurement plane, but still pose a threat to the robot. Examples include open drawers or small objects lying on the ground. In such environments, 3D information becomes crucial. Although we currently do not use a 3D sensor on the robot platform, like e.g. a 3D laser scanner or a time-of-flight camera, the proposed SLAM algorithm is already applicable to both 2D and 3D information.

3   Simultaneous Localization and Mapping

Performing SLAM to build maps and localizing in preliminary built maps are major preconditions for the autonomous operation of mobile robots in changing or preliminary unknown environments. Approaches addressing mapping and localization differ, amongst others, in formulating the problem, the means to cope with the addressed problem and in representing the environment. Occupancy grids [14] are a popular metric map representation for navigation which can be built from various kinds of simple range sensors like sonars and laser range finders. These sensors deliver information that there is some kind of obstacle in a certain distance. Occupancy grids provide a discretized representation of this kind of occupancy information. Furthermore, they distinguish unoccupied and not yet visited areas compared to feature-based representations that only store certain features perceived in the environment or geometric primitives modeling environmental structures. However, occupancy grid maps typically require a large amount of memory and can be computationally expensive to handle. On the other hand, feature-based approaches require robust feature extraction mechanisms which may be computationally expensive.

When addressing SLAM in terms of range image registration, raw measurements (i.e. point clouds) acquired with a laser scanner can directly be used to model environmental structures and to localize a mobile robot by using a matching algorithm. Hence, there is no need for applying additional feature extraction mechanisms. The problem of registering point clouds can be formulated as follows. Given two point clouds M, called model set, and D, called data set, find a transformation $\mathbf{T}$ that minimizes the alignment error between the two sets and correctly maps D onto M. The essential problems derived from this formulation are a) how to define the error function and b) how to minimize this error.

3.1   The ICP Algorithm

A widely used solution to the registration problem is the Iterative Closest Point (ICP) algorithm by Besl and McKay [1], which determines $\mathbf{T}$ in an iterative way. In each iteration step, the ICP algorithm determines pairs of corresponding points from D and M using a nearest-neighbor search. These correspondences are used to quantify and minimize the alignment error:

$$E(\mathbf{R},\mathbf{t}) = \sum_{i=1}^{|M|}\sum_{j=1}^{|D|} w_{i,j}\|\mathbf{m}_i - (\mathbf{R}\mathbf{d}_j + \mathbf{t})\|^2, \ w_{i,j} = \begin{cases} 1, & \mathbf{m}_i \text{ corresponds to } \mathbf{d}_j \\ 0, & \text{otherwise.} \end{cases} \tag{1}$$

$$\mathbf{T} = \begin{pmatrix} \mathbf{R}_{ICP} \ \mathbf{t}_{ICP} \\ 0\ 0\ 0 \quad 1 \end{pmatrix} \text{ with } (\mathbf{R}_{ICP}, \mathbf{t}_{ICP}) = \underset{\mathbf{R},\mathbf{t}}{\arg\min} \ E(\mathbf{R},\mathbf{t}) \tag{2}$$

Finding the nearest neighbors and determining the correspondences is the computationally most expensive step in the ICP algorithm ($O(|D|\,|M|)$) for a brute-force implementation), since for every point $\mathbf{d}_j \in D$ the closest point $\mathbf{m}_i \in M$ needs to be determined. Here, we use an approximate kd-tree search [15], which reduces the complexity of the algorithm to $O(|D|\log|M|)$.

To estimate the rigid transformation $\mathbf{T}$, consisting of a rotation $\mathbf{R}$ and a translation $\mathbf{t}$, that minimizes Eq. (1) there are closed form solutions in both the two- and three-dimensional case (see [13] for a comparison). Extensions to the ICP algorithm for e.g. dealing with partial overlap of D and M or false correspondences as well as weighting and rejecting correspondence pairs can be found in [18]. The primary extension used here is to reject pairs for which the point-to-point distance exceeds a certain threshold. This threshold exponentially decays during the registration process. While initially permitting larger distances between corresponding points guarantees fast convergence of $E(\mathbf{R},\mathbf{t})$, smaller distances in later iteration steps allow fine-tuning the registration result.

3.2   Incremental Registration Using the ICP Algorithm

For registering multiple range scans and constructing a consistent map that models environmental surfaces, an incremental registration procedure is used. The first laser scan D_0 is used as the initial environment model M_0. Thus, the local coordinate frame of D_0 forms the coordinate frame for the overall map. All subsequent scans $D_i, i > 0$ are matched against M_{i-1}. The resulting transformation $\mathbf{T}_i$ is used to correct the position of all points contained in D_i, yielding the transformed point set $\check{D}_i = \{\check{\mathbf{d}}_{i,j} | \check{\mathbf{d}}_{i,j} = \mathbf{R}\mathbf{d}_{i,j} + \mathbf{t}\}$. As an initial estimate $\hat{\mathbf{T}}_i$ for $\mathbf{T}_i$ in this incremental registration we use the transformation from the last registration, i.e. $\hat{\mathbf{T}}_i = \mathbf{T}_{i-1}$. This speeds up the convergence in the ICP algorithm and drastically reduces the probability of converging to a local minimum possibly resulting in an incorrect registration result. If odometry information is available, the estimate $\hat{\mathbf{T}}_i$ is further corrected taking into account the estimated pose shift between the acquisition of D_{i-1} and D_i. Furthermore, we only register a new range scan D_i if the robot traversed more than e.g. 50 cm or turned more than e.g. 25° – a practice being quite common in recent SLAM algorithms.

To account for possibly new information in D_i, the transformed points are than added to M_{i-1}. That is, after matching range image D_i, the model set M_{i-1} computed so far is updated in step i to:

$$M_i = M_{i-1} \cup \{\check{\mathbf{d}}_{i,j} \mid \check{\mathbf{d}}_{i,j} \in \check{D}_i\}. \tag{3}$$

Thus, a model M_N, constructed by incrementally registering N range images, contains all points measured in the environment, i.e.

$$M_N = \bigcup_{i=[0,N]} \{\check{\mathbf{d}}_{i,j} \mid \check{\mathbf{d}}_{i,j} \in \check{D}_i\}. \tag{4}$$

3.3   Sparse Point Maps

The main problem of the incremental registration approach is its scalability with respect to the size of the environment and the number of range images taken. To fully cover a large environment, a lot of range images might be needed. When registering and adding all acquired range images, the model set M can get quite large, e.g. several million points for 3D scans taken in a large outdoor environment [16,22]. However, when acquiring range images in parts of the environment which are already mapped, lots of points would be added to M without providing new information about the environment. This is exploited by the following improvement to our SLAM approach, which makes the point clouds sparse.

The key idea of *sparse point maps* is to avoid duplicate storage of points, and thereby minimize the amount of memory used by the map, by conducting an additional correspondence search. That is, to neglect points that correspond to the same point in the real physical environment as a point already stored in the map. Correspondence is, thereby, defined just like in the ICP algorithm, i.e. a point $\check{\mathbf{d}}_{i,j} \in \check{D}_i$ is not added to M_{i-1}, if the point-to-point distance to its closest point $\mathbf{m}_{i-1,k} \in M_{i-1}$ is smaller than a minimum allowable distance ϵ_D.

$$M_i = M_{i-1} \cup \{\check{\mathbf{d}}_{i,j} \mid \check{\mathbf{d}}_{i,j} \in D_i, \nexists \mathbf{m}_{i-1,k} \in M_{i-1} : \|\check{\mathbf{d}}_{i,j} - \mathbf{m}_{i-1,k}\| < \epsilon_D\} \tag{5}$$

The threshold ϵ_D spans regions in the model in which the number of points is limited to 1, thereby providing an upper bound on the point density in a sparse point map M. Choosing a value of ϵ_D according to the accuracy of the range sensor used will exactly neglect duplicate storage of one and the same point assuming correct alignment of range images. Choosing, however, a larger value allows to reduce the number of points stored in the map. Although some details of the environment might not get modeled, a map constructed in this manner still provides a coarse-grained model of the environment as can be seen in Figure 2(b). In the actual implementation, the additional correspondence search is carried out on the kd-tree built for the ICP algorithm using ϵ_D as the distance threshold in the pair rejection step. However, here the rejected pairs are used to determine the points in $\check{D}_i$ that need to be added to M_{i-1}.

3.4 Examples and Results

The proposed incremental registration approach is so computationally efficient that it can be applied continuously during robot operation, thereby quickly reflecting changes in the environment. The runtime of the algorithm for registering a 2D laser scan lies in the range of milliseconds and increases only slightly $(\log|M|)$ for growing map sizes. Figure 3 illustrates that the maps and trajectories resulting from the application of the proposed SLAM procedure are not inferior compared to those resulting from other state-of-the-art SLAM algorithms, like e.g. Rao-Blackwellized Particle Filters [6].

Having larger loops in the robot's trajectory, however, would require postprocessing such as global relaxation using e.g. Lu-Milios-like approaches [2] or graph-based optimization [5]. Still, for the kind of environments addressed here,

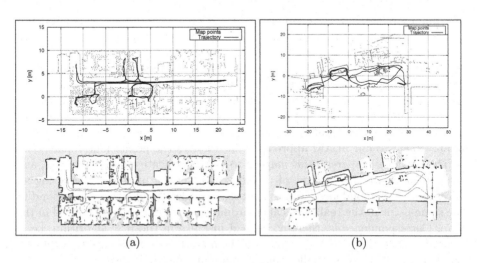

(a)   (b)

Fig. 3. Example on applying the proposed SLAM procedure on typical robot data sets (here two data sets from Cyrill Stachniss and Giorgio Grisetti). The resulting maps (a: 3092 out of 1 123 560 points, b: 2364 out of 1 975 680 points, $\epsilon_D = 15$ cm) and trajectories are shown in the upper plot. Maps provided with the data sets are shown at the bottom.

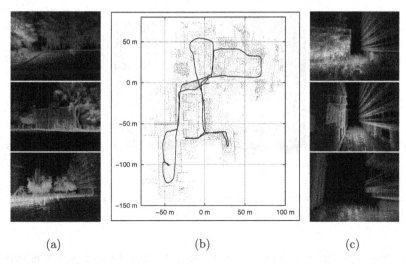

(a)   (b)   (c)

Fig. 4. Example of applying the proposed SLAM procedure on 3D data sets (here a data set from Oliver Wulf). Shown are a topview (b) with $\epsilon_D = 2\,\text{m}$ and detail views (a+c) of a model with $\epsilon_D = 20\,\text{cm}$. Map sizes are 10 060 points ($\epsilon_D = 2\,\text{m}$) and 550 979 points ($D_{min} = 20\,\text{cm}$) out of approx. 10^7 points.

e.g. apartments, the proposed stand-alone single-hypothesis approach seems sufficient. Furthermore, the approach can be integrated into a particle filter framework for multihypotheses SLAM.

An example of matching 3D laser scans to construct a 3D model of the environment and to localize the robot with all six degrees of freedom in space is shown in Figure 4.

4   Path Planning

Grid maps already have an internal structure that can be used directly for path-planning purposes. The sparse point maps used here, however, lack this ability. Instead, path-planning is addressed as a graph-search problem in the Voronoi diagram of the map points. Planning on the Voronoi diagram may not result in the shortest path, but when traveling along a path planned, the robot will always maintain a maximum distance to the obstacles represented in the map.

The Voronoi diagram is constructed using Fortune's Sweep-Line Algorithm [3]. The runtime complexity of this algorithm is $O(n \log n)$ with space complexity $O(n)$ for n points. A typical result of applying this algorithm to sparse point maps is shown in Figure 5. However, as shown in Figure 5(b), the Voronoi diagram constructed contains edges that lie outside of the modeled environment. Other edges cannot be traversed by the robot as the distance to the nearest obstacles is too short. Therefore, we prune the Voronoi diagram, first by removing all edges lying outside of or intersecting the convex hull for the map points, and

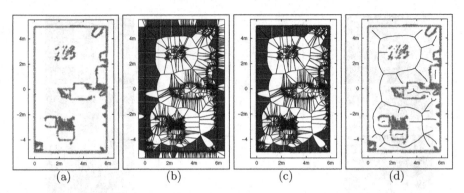

Fig. 5. Voronoi diagram construction and pruning based on a sparse point map acquired during the RoboCup GermanOpen in Hannover

second by removing all edges whose distance to neighboring points is smaller than half of the robot's width plus some safety distance (e.g. 5 cm). The latter pruning step can directly be integrated into Fortune's algorithm not affecting its complexity. The convex hull is computed by Graham's Scan Algorithm [4], which has a runtime complexity of $O(n \log n)$ for n map points. The results of both pruning steps are shown in Figure 5(c) and Figure 5(d).

Path planning is performed on the resulting graphs using $A^\star$ search [8]. The Euclidean distance to the target position ($\mathbf{x}_{goal}$) is used as an admissible heuristic in the cost function. Therefore, $A^\star$ is optimal and guaranteed to find the shortest path, if a solution exists. The overall cost function for a path from the start position through a node n to the goal is thereby defined as:

$$f(n) = g(n) + h(n) = \left(\sum_{i=1}^{n} \|\mathbf{x}_i - \mathbf{x}_{i-1}\| \right) + \|\mathbf{x}_{\text{goal}} - \mathbf{x}_n\| \qquad (6)$$

where $\mathbf{x}_0 = \mathbf{x}_{\text{start}}$ and the sequence $< \mathbf{x}_0, \mathbf{x}_1, \ldots, \mathbf{x}_n >$ represents the shortest path between $\mathbf{x}_{\text{start}}$ and $\mathbf{x}_n$. As $A^\star$ can only plan paths between nodes in the graph, representatives for the true start and goal poses need to be found. The algorithm simply chooses the closest nodes in the graph and in cases where multiple nodes have similar distances to the true poses, we prefer the nodes in the direction of the other true pose. A result from applying this path-planning procedure is shown in Figure 6(a). Also shown in the map is a part of a topological layer on top of the map, storing a vector of learned and predefined objects with positions, orientations, shapes and names used to communicate with a human user.

5   Motion Control

To actually follow a planned path and reach a target location and orientation, we subsequently apply two non-linear motion controllers which have been especially

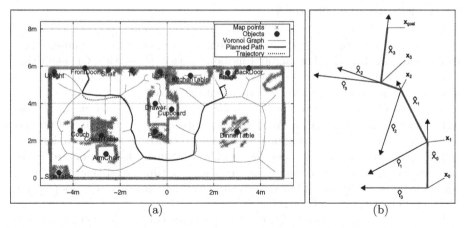

Fig. 6. (a) Path planning and following in a map of the RoboCup@Home arena at the GermanOpen 2008. Shown are the path graph (thin red lines), the planned path (thicker black lines) and the trajectory of the robot (dotted blue lines). (b) Local coordinate frames (thin black axes) in the path-tracking controller for an example path (thicker red lines).

designed for motion control of non-holonomic vehicles. The first motion controller by Indiveri and Corradini [11] is for tracking linear paths and is applied until the robot reaches the immediate vicinity of the target location. Then a second controller is used, which controls both linear velocity v and angular velocity ω of the robot, to reach the target pose while traversing a smooth trajectory [10]. The latter motion controller has previously been successfully used in an affordance-based robot control architecture in the EU FP6 project MACS [12].

For the application of the path following controller, planned paths are represented as a chain of local coordinate frames, as shown in Figure 6(b). Transforming the robot's pose into the local coordinate frame of the currently traversed graph edge allows for directly applying Indiveri's steering control law

$$\omega = -hvy\frac{\sin\theta}{\theta} - \gamma\theta \quad : \quad h, \gamma > 0 \tag{7}$$

where the controller gains h and γ are calculated depending on the current situation, i.e. whether the robot is regaining or maintaining the currently tracked path segment. Furthermore, transforming the latest laser scan into the local frame allows for checking whether the current segment is obstacle-free and can be traversed. If the segment is block, the corresponding edge is marked as being not traversable and the path is re-planned. In addition, the robot performs reactive collision avoidance [9]. The x-axes $\hat{X}_i$ of the local frames are formed by the path segments, whereas y, θ form the error signal of the controller (position and orientation deviation). The linear velocity v can be freely chosen and adapted. For details on both motion controllers it is referred to [10] and [11].

6   Concluding Remarks

We presented a complete system for autonomous navigation, including algorithms for SLAM, path planning and motion control. Using the ICP algorithm in an incremental registration procedure and sparse point maps, simulated and real robots were able to construct memory-efficient environment representations online. Path-planning on the resulting point maps has been done using A^* and fast algorithms for computing Voronoi diagrams and convex hulls for obtaining a pruned path graph. Using non-linear motion controllers for non-holonomic systems, simulated and real robots were able to robustly follow planned paths and reach target poses while localizing in and updating the sparse point map. All algorithms are highly efficient and run within the main control loop of the mobile robot platform (50-100 Hz).

Future work will focus on the development of efficient exploration and inspection strategies based on and consistent with the proposed algorithms. Extensions to this system for 3D collision avoidance and filtering out dynamics from raw range data can be found in [9]. The proposed algorithms as well as further details will be made publicly available through the RoboCup@Home Wiki[1]. Videos showing the proposed system in action and the performace of the presented SLAM algorithm are available at http://www.b-it-bots.de/media.

Acknowledgments

The continued support by the b-it Foundation is gratefully acknowledged. Our thanks go to Cyrill Stachniss, Giorgio Grisetti and Oliver Wulf for providing the data sets and Zoran Zivkovic for providing the sketch of the apartment environment used in Figures 2(a) and 2(b) as well as the corresponding data set.

References

1. Besl, P.J., McKay, N.D.: A Method for Registration of 3-D Shapes. IEEE Transactions on Pattern Analysis and Machine Intelligence 14(2), 239–256 (1992)
2. Borrmann, D., Elseberg, J., Lingemann, K., Nüchter, A., Hertzberg, J.: Globally Consistent 3D Mapping with Scan Matching. Journal of Robotics and Autonomous Systems 56(2), 130–142 (2008)
3. Fortune, S.J.: A Sweepline Algorithm for Voronoi Diagrams. Algorithmica, 153–174 (1987)
4. Graham, R.L.: An Efficient Algorithm for Determining the Convex Hull of a Finite Planar Set. Information Processing Letters 1(4), 132–133 (1972)
5. Grisetti, G., Lordi Rizzini, D., Stachniss, C., Olson, E., Burgard, W.: Online Constraint Network Optimization for Efficient Maximum Likelihood Map Learning. In: Proceedings of the IEEE International Conference on Robotics and Automation (2008)

---

[1] RoboCup@Home wiki site: http://robocup.rwth-aachen.de/athomewiki

6. Grisetti, G., Stachniss, C., Burgard, W.: Improved Techniques for Grid Mapping with Rao-Blackwellized Particle Filters. IEEE Transactions on Robotics 23(1), 34–46 (2007)
7. Haegele, M., Neugebauer, J., Schraft, R.: From Robots to Robot Assistants. In: Proceedings of the International Symposium on Robotics (2001)
8. Hart, P.E., Nilsson, N.J., Raphael, B.: A Formal Basis for the Heuristic Determination of Minimum Cost Paths. IEEE Transactions on Systems Science and Cybernetics 4(2), 100–107 (1968)
9. Holz, D., Lörken, C., Surmann, H.: Continuous 3D Sensing for Navigation and SLAM in Cluttered and Dynamic Environments. In: Proceedings of the International Conference on Information Fusion, FUSION (2008)
10. Indiveri, G.: Kinematic Time-invariant Control of a $2d$ Nonholonomic Vehicle. In: Proceedings of the 38th Conference on Decision and Control (1999)
11. Indiveri, G., Corradini, M.L.: Switching linear path following for bounded curvature car-like vehicles. In: Proceedings of the IFAC Symposium on Intelligent Autonomous Vehicles (2004)
12. Lörken, C.: Introducing Affordances into Robot Task Execution. In: Kühnberger, K.-U., König, P., Ludewig, P. (eds.) Publications of the Institute of Cognitive Science (PICS), vol. 2-2007. University of Osnabrück, Osnabrück, Germany (2007)
13. Lorusso, A., Eggert, D.W., Fisher, R.B.: A Comparison of Four Algorithms for estimating 3-D Rigid Transformations. In: Proceedings of the British conference on Machine vision, BMVC, pp. 237–246 (1995)
14. Moravec, H.P.: Sensor fusion in certainty grids for mobile robots. Sensor Devices and Systems for Robotics, 253–276 (1989)
15. Mount, D., Arya, S.: ANN: A library for approximate nearest neighbor searching. In: Proceedings of the 2nd Annual Fall Workshop on Computational Geometry (1997)
16. Nüchter, A., Lingemann, K., Hertzberg, J., Surmann, H.: 6D SLAM – 3D Mapping Outdoor Environments. Journal of Field Robotics 24(8-9), 699–722 (2007)
17. Pollack, M.E., Engberg, S., Matthews, J.T., Thrun, S., Brown, L., Colbry, D., Orosz, C., Peintner, B., Ramakrishnan, S., Dunbar-Jacob, J., McCarthy, C., Pineau, J., Montemerlo, M., Roy, N.: Pearl: A Mobile Robotic Assistant for the Elderly. In: Proceedings of the AAAI Workshop on Automation as Eldercare (August 2002)
18. Rusinkiewicz, S., Levoy, M.: Efficient Variants of the ICP Algorithm. In: Proceedings of the International Conference on 3D Digital Imaging and Modeling (2001)
19. Siegwart, R., Arras, K.O., Bouabdallah, S., Burnier, D., Froidevaux, G., Greppin, X., Jensen, B., Lorotte, A., Mayor, L., Meisser, M., Philippsen, R., Piguet, R., Ramel, G., Terrien, G., Tomatis, N.: Robox at Expo.02: A large-scale installation of personal robots. Robotics and Autonomous Systems 42(3-4), 203–222 (2003)
20. Wang, C.-C.: Simultaneous localization, mapping and moving object tracking. PhD Thesis CMU-RI-TR-04-23, Robotics Institute, Carnegie Mellon University, Pittsburgh, PA (April 2004)
21. Wisspeintner, T., Bose, A.: The VolksBot Concept – Rapid Prototyping for real-life Applications in Mobile Robotics. it – Information Technology 47(5), 274–281 (2005)
22. Wulf, O., Nüchter, A., Hertzberg, J., Wagner, B.: Benchmarking Urban Six-Degree-of-Freedom Simultaneous Localization and Mapping. Journal of Field Robotics 25(3), 148–163 (2008)
23. Zivkovic, Z., Booij, O., Kröse, B.: From images to rooms. Robotics and Autonomous System 55(5), 411–418 (2007)

Robust Collision Avoidance in Unknown Domestic Environments

Stefan Jacobs[1], Alexander Ferrein[1,2], Stefan Schiffer[1],
Daniel Beck[1], and Gerhard Lakemeyer[1]

[1] Knowledge-Based Systems Group
RWTH Aachen University
Aachen, Germany
stefan_j@gmx.de,
{schiffer,dbeck,gerhard}@cs.rwth-aachen.de
[2] Robotics and Agents Research Lab
University of Cape Town
Cape Town, South Africa
alexander.ferrein@uct.ac.za

Abstract. Service robots operating in domestic indoor environments must be endowed with a safe collision avoidance and navigation method that is reactive enough to avoid contacts with the furniture of the apartment and humans that suddenly appear in front of the robot. Moreover, the method should be local, i.e. should not need a predefined map of the environment. In this paper we describe a navigation and collision avoidance method which is all of that: safe, fast, and local. Based on a geometric grid representation which is derived from the laser range finder of our domestic robot, a path to the next target point is found by employing A^*. The obstacles which are used in the local map of the robot are extended depending on the speed the robot travels at. We compute a triangular area in front of the robot which is guaranteed to be free of obstacles. This triangle serves as the space of feasible solutions when searching for the next drive commands. With this triangle, we are able to decouple the path search from the search for drive commands, which tremendously decreases the complexity. We used the proposed method for several years in RoboCup@Home where it was a key factor to our success in the competitions.

1 Introduction

One of the most important and most basic tasks for a mobile robot in domestic domains is safe navigation with reliable collision avoidance. In this paper we present our navigation and collision avoidance algorithm which we successfully deployed over many years in the domestic robot domain. The navigation scheme presented here was one of the key features for our success in the domestic robot competition RoboCup@Home. Our method relies on a distance measurement sensor from which a local map of the surrounding is constructed. In our case, we make use of a laser range finder. In this local map, which is in fact a grid representation, we search for a path to a target point. We employ A^* for this.

J. Baltes et al. (Eds.): RoboCup 2009, LNAI 5949, pp. 116–127, 2010.

The such calculated path serves as an initial solution to the navigation task. For computing the path the robot's kinematic constraints have not been taken into account. This was decoupled in order to decrease the size of the search space. We integrate it in a second step where we construct the so-called *collision-free triangle*, where the path as it is calculated by A* serves as the leg of this triangle. In particular, the current setting of the robot's parameters like speed and orientation are taken into account. By this, we explicitly take care of the robot's kinematic constraints. In the sequel, we prove that this triangle is obstacle-free and can be traversed safely.

The success of our method is founded on two ideas. (1) The first one is to represent the size of the surrounding obstacles depending on the speed of the robot, i.e. the faster the robot drives the larger the obstacles will become, since the robot needs more time to break in front of them. This way we can represent the robot as a mass point. (2) The second idea lies in decoupling the search for a path from its realization. In particular, we propose to construct a collision-free triangle which the robot can traverse safely.

In the past, many different approaches for this fundamental problem have been proposed. So why does this paper go beyond proposing yet another collision avoidance approach? The answer is three-fold:

1. We believe that extending the size of the obstacles depending on the speed of the robot is innovative and worth to be mentioned; with this the robot drives only as fast as possible not to collide with any obstacles. In narrow passages it reduces iteratively its speed until it can safely travel through, while in broad areas it will try to reach its maximal speed.
2. With the collision-free triangle we have an area in the motion area of the robot which is guaranteed to be collision-free.
3. Finally, we deployed this approach for many years in robotics competitions and it was a key to succeed in the domestic robot competition RoboCup@Home for the past three years.

The rest of this paper is organized as follows: In the next section we will present some of the huge body of related articles. In Sect. 3 we introduce our robot platform. In Sect. 4 we present how obstacles as perceived by the sensors of the robot are integrated into its local map, and how a path is calculated. We prove that the collision-free triangle is in fact without obstacles, and show how this triangle bounds the kinematic parameters of the robot. Sect. 5 discusses some implementation details and experimental results. We conclude with Sect. 6

2   Related Work

Approaches to mobile robot navigation can be categorized along several criteria. Some approaches make use of randomized planning techniques, e.g. [1,2], other approaches use reactive schemes, for instance [3,4,5], and/or make use of a navigation function to follow a path like [6] or plan directly in the velocity space like [7,8,9,10]. Yet other approaches employ a search, some in physical space, some in configuration space.

In their early approach, Borenstein and Koren [7] proposed to use vector field histograms. The target point exerts an attractive force to the robot while obstacles impose repulsive forces. The trajectory of the robot is then formed by the sum of both these forces. They propose a special wall-following mode to avoid getting stuck in local minima which could otherwise cause oscillating behavior. The method was tested with robots equipped with sonar sensors driving with an average speed of 0.58 m/s.

In [3], Fox et al. proposed the dynamic window approach. It is directly derived from the motion equations. In the velocity space circular collision-free trajectories are searched. To handle the state space they define a *dynamic window* around the robot to only consider those velocities that the robot can reach in the next time interval. Finally, a trajectory is found by maximizing over the minimal target heading, maximal clearance around the robot, and maximal speed. The method was tested on an RWI B21 robot with a maximum speed of 0.95 m/s. A similar approach except for the dynamic window was proposed in [11].

Seraji and Howard describe a behavior-based navigation scheme in [12]. They distinguish between different terrain types such as roughness, slope, and discontinuity. A fuzzy controller selects between traverse-terrain, avoid-obstacles, and seek-goal behaviors. While their method aims at outdoor navigation, it is an example for a local reactive navigation scheme. Another reactive approach is presented in [13]. The difference to our work is that we use an optimal path as an initial solution to avoid nearby obstacles.

Besides range sensors, imaging sensors are commonly used to navigate a mobile robot. In [14] an approach to build an occupancy map from a stereo camera on an RWI B14 robot is presented. The map is used to navigate through previously unknown environments. We want to note that our method is different in the sense that we here present a reactive local method while the focus of [14] is on vision-based exploration. A large number other papers deals with navigation approaches using several sensors. Among those, the fusion of imaging and proximity sensors (cameras and LRFs) is popular, see for example [15,16,17].

Koenig and Likhachev [18] present a search heuristic called Lifelong Planning A^*. They propose an incremental search heuristic where only the relevant parts of a path are recalculated. While Lifelong Planning A^* is an interesting extension to the basic A^* we use in this paper, we here focus on the obstacle representation and the decoupling of path and velocity planning to decrease the dimensionality of the search problem.

The most related research to our approach is the method proposed by Stachniss and Burgard [10]. They use a laser range finder as sensor and employ A^* to find a shortest trajectory to a given target. The state space used here is a five-dimensional pose-velocity space consisting of the pose x, y, θ and the translational and rotational velocities v, ω. At first, a trajectory to the target is calculated using A^* in the $\langle x, y \rangle$-space. This trajectory serves as the starting point for the search in the pose-velocity space. With a value iteration approach a 70 cm broad channel around the calculated trajectory is calculated which restricts the state space for the five dimensional search. Stachniss and Burgard tested their approach on a Pioneer I and an RWI B21 both having a maximum speed below 1 m/s. By restricting the search for velocities to the collision-free triangle which

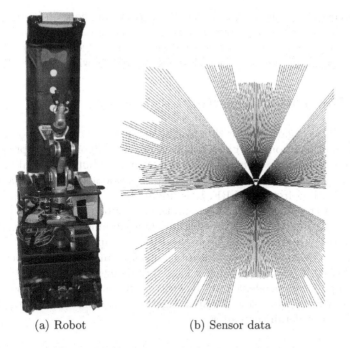

(a) Robot   (b) Sensor data

Fig. 1. The robot "Caesar" and its sensor data. The four blind regions are due to the rods supporting the platforms above the laser range finder.

is known to be obstacle-free, we are able to avoid the search in the five dimensional pose-velocity space which leads to a much more reactive navigation and collision avoidance scheme. A similar method applied to holonomic small-size league robots in the robotic soccer domain can be found in [19].

3   Service Robot Platform

Our hardware platform has a size of 40 cm × 40 cm × 160 cm. It is driven by a differential drive, the motors have a total power of 2.4 kW and are originally developed for electric wheel chairs. With those motors the robot reaches a top speed of 3 m/s and a maximal rotational velocity of 1000°/s at a total weight of approximately 90 kg. On-board we have two Pentium III PCs at 933 MHz running Linux. Only one of these PCs is used for the navigation and collision avoidance approach presented in this paper. A 360° laser range finder (LRF) with a resolution of 1 degree provides new distance readings for each direction at a frequency of 10 Hz. Fig. 1(a) shows our robot.

We assume accelerated motion with our approach. Thus, the connection between pose, velocity and acceleration is given by $x(t) = \frac{1}{2} \cdot \ddot{x}(t) \cdot t^2 + \dot{x}(t) \cdot t + x_0$, $\dot{x}(t) = \ddot{x}(t) \cdot t + v_0$, and $\ddot{x}(t) = const.$ As usual, $\ddot{x}(t)$ refers to acceleration, $\dot{x}(t)$ to the velocity, and $x(t)$ to the displacement at any given time t. In the algorithm we present in the next section, we assume that at each time instance the robot is accelerated only in the first time instance and from there on is driving with constant velocity till the next acceleration command is settled. We need this relation because, for efficiency reasons, we decouple the search in the pose-velocity space into a pose space and a velocity space in the following.

4 The Navigation Algorithm

The task of the navigation algorithm is to steer the robot on a collision-free path from its current location to a given target point. The algorithm we present does not rely on a global map as many other algorithms do but on a local map of its environment. The dimension of the local map corresponds to the area covered by the current reading from the LRF. It is updated every time new laser-readings are received. Although, in our implementation, we integrate new sensor readings into the previous local map if possible, we here assume that the local map is set up from scratch every cycle. In juxtaposition to approaches which rely on a global map a local map has the advantage that it allows to easily account for dynamic obstacles. Moreover, using a local map makes the successful execution of a planned path independent from the localization of the robot. In the following we will give a rough overview of our navigation algorithm and discuss the key aspects in greater detail thereafter.

Input: $\Delta x, \Delta y$ the target point in relative coordinates
while *not reached target* **do**
    $d_1, \ldots, d_n \leftarrow$ getLaserReadings() ;
    $v_t^{cur}, v_r^{cur} \leftarrow$ getCurrentVelocities() ;
    $map \leftarrow$ extendObstacles($d_1, \ldots, d_n, v_t^{cur}, v_r^{cur}$) ;
    $path \leftarrow$ findInitialPath(map) ;
    if $path.isEmpty()$ **then**
        sendMotorCommands(0, 0); break;
    end
    $v_t, v_r \leftarrow$ findVelocities($path, map$) ;
    if *no* v_t, v_r **then**
        sendMotorCommands(0, 0);
        break;
    end
    sendMotorCommands(v_t, v_r) ;
    $\Delta x, \Delta y \leftarrow$ updateTarget(v_t, v_r) ;
end
sendMotorCommands(0, 0) ;

Algorithm 1. The navigation algorithm in pseudo-code.

If the robot is in close proximity to the given target, i.e., the target is reached, it stops. Otherwise the current distance readings from the LRF and the current translational and rotational velocities are obtained. This data is then used to build a (local) map. For this we employ a technique we refer to as *dynamic obstacle extension* (cf. Sect 4.1) which yields a (local) grid-based map of the robot's environment. The cells of the grid map may either be occupied or free.

With this representation of the surrounding of the robot we search for an initial path to the target, first. In a second step, we search for an approximation of the initial path which takes into account the kinematic constraints of the robot. In both steps we employ the A^* search algorithm. Splitting up the path-planning problem into two independent search problems reduces the original problem of dimensionality four to two search problems over two-dimensional search spaces. Namely, finding a path in the xy-plane and appropriate velocities v_t, v_r. This is only possible since the search for an appropriate approximation of the initial path is restricted to a certain area which is guaranteed to be free of obstacles. We refer to this area as the *collision-free triangle*. More details on this are given in Sect. 4.2.

4.1   Dynamic Obstacle Extension

A technique often referred to as *obstacle growing* [20] extends the obstacles by the dimensions of the robot. This alleviates the problem of collision detection in the path-planning problem since the robot can now be treated as a mass point. We leapfrog on this idea and additionally extend the obstacles in dependence on their respective imminence of collision which takes into account the current speed of the robot as well as the position of the obstacle relative to the current trajectory of the robot. The intuition behind this is to mark potentially dangerous areas as occupied in the local map and thereby force the search to not consider paths leading through those areas.

The most threatening obstacle for the next step is the obstacle lying on the trajectory defined by the current and the projected position of the robot in the next step. We assume to recompute a new path every iteration and, consequently, it is not necessary to project further into the future then the next step. The next-step trajectory is computed from the current translational velocity v_r, the current rotational velocity v_r and the time between two iterations Δt:

$$v_x = \frac{x(t+1) - x(t)}{\Delta t} \qquad v_y = \frac{y(t+1) - y(t)}{\Delta t} \qquad \alpha = \tan \frac{v_y}{v_x}$$

For each detected obstacle we place an ellipse centered at the reflection point of the laser beam in the local map and align the axes such that the semi-major axis is parallel to the laser beam. The radius for the semi-major axis r_1 and the radius for the semi-minor axis r_2 are computed as:

$$r_1 = l + l_{sec} + |\cos(\theta - \alpha)| \cdot d \cdot n$$
$$r_2 = l + l_{sec}$$

where l is the radial extension of the robot[1], l_{sec} is an additional security distance, θ is the angle of the laser beam that hits the obstacle and d is the euclidean distance between the current position and the position projected for the next step

$$d = \sqrt{(v_x \cdot \Delta t)^2 + (v_y \cdot \Delta t)^2}$$

---

[1] The formula could be further detailed to account for a rectangular shape of the robot, but this is omitted here for clarity reasons.

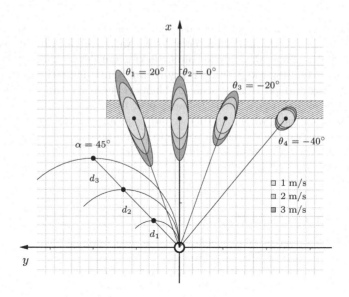

Fig. 2. In this illustrating example a wall is located in front of the robot. The extension of the obstacles is shown for the obstacles detected at 20°, 0°, -20°, and -40°. The translational velocities are 1 m/s, 2 m/s, and 3 m/s; the rotational velocity remains fixed at 1 rad/s. For illustrating purposes we chose $\Delta t = 0.25$ s and $n = 1$.

Then, the obstacles are extended in such a way that the robot will stay out of the "dangerous area" for the next n steps.

By means of extending the obstacles in such a way we capture the current obstacle configuration as well as the current configuration of the robot (in terms of translational and rotational velocity) in the local map. Fig. 2 illustrates the extension of the obstacles for different configurations of the robot: the rotational velocity remains fixed whereas the translational velocity is altered between 1 m/s and 3 m/s.

4.2   The Collision-Free Triangle

As we pointed out in the introduction, within the search for an initial path we ignore the kinematic constraints of the robots as well as its current configuration in terms of velocity and orientation. With this, we are in good company with several other search-based methods like [10]. However, for successfully realizing a path on a real robot, the kinematic constraints need to be taken into account, of course. In our algorithm, we do so by performing an A^* search on the possible accelerations in translation and rotation making use of the standard motion equations for accelerated motion. The kinematic constraints of the robot are only one part of the velocity planning problem of the robot. The other part is to take the robot's current state into account, i.e. its current translational and rotational velocities. In the following, we therefor present the collision-free triangle. We prove that, by construction, each grid cell inside this triangle is free

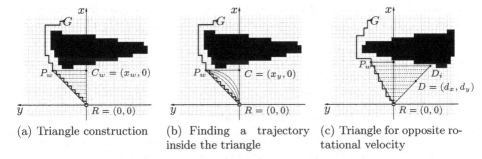

(a) Triangle construction

(b) Finding a trajectory inside the triangle

(c) Triangle for opposite rotational velocity

Fig. 3. Construction of the Collision-free Triangle

of obstacles. Poses inside this triangle are thus safe and therefore it can be used to restrict the search for the optimal trajectory the robot should follow.

Definition 1 (Collision-Free Triangle). *The robot is located in the origin of a Cartesian coordinate system, such that $R = (0,0)$ is the position of the robot facing the positive x-axis. Let $C_k = (x_k, 0)$ be a grid cell in front of the first obstacle along the x axis. In case there is no obstacle along the x axis, the last cell within the perception range of the robot is taken.*

Let $p = \langle (0,0), (x_1, y_1), \dots, (x_{g-1}, y_{g-1}), (x_g, y_g) \rangle$ be the path to the target point G. The path is given as a sequence of grid cells. For each path point $P_i = (x_i, y_i)$, $1 \le i \le g$ we ray-trace to the point $C_i = (x_i, 0)$. A path point (x_i, y_i) is called safe *iff the ray $\overline{P_i C_i}$ does not hit any obstacle, otherwise it is unsafe. The collision-free triangle is now given by the points R, P_w, and C_w with w being the index of the last safe path point.*

Fig. 3 depicts the construction of the collision-free triangle as described in the definition. Note that the x axis always points into the direction of the robot's orientation as the map is given by a Cartesian coordinate system with the robot in its origin. Hence, the point C denotes the last free cell before the robot would collide with an obstacle if, from now on, it would drive only with translational velocities. Now, for each path point it is checked if the orthogonal projection of a path point onto the segment $|RC|$ will hit an obstacle. The robot's position R, the last safe path point P_w and the point C_w, the projection point of P_w onto $|RC|$, yields the corner points of the triangle.

Fig. 3(c) illustrates the situation in which the robot is turning away from the path. In that case, we span an additional triangle by projecting the robot's position according to its current velocity for the next step (cf. point D). We put a straight line from the robot's position through D until we hit an obstacle (cf. point D_i). Then we ray-trace analogous to the original triangle to ensure the additional triangle is also obstacle-free.

Theorem 1. *Each cell inside the triangle is obstacle-free.*

Proof. Suppose, P_i is the next path point to be considered in the construction of the collision-free triangle as described in the definition. Hence, the triangle given by $\triangle R, P_{i-1}, C_{i-1}$ is collision-free. Now we ray-trace from P_i orthogonal to the segment given by $|RC|$. If the ray hits an obstacle, the collision-free triangle is given by $\triangle R, P_{i-1}, C_{i-1}$, otherwise we conclude that the $\triangle R, P_i, C_i$ is collision-free. $\square$

P_w is the next point for the robot to reach (safely). P_w closes in on G and the robot will thus eventually reach the target point.

4.3   Computing the Drive Commands

Lastly, a suitable sequence of velocities to reach the intermediate target point P_w needs to be determined. Again we employ A^* search to find such a sequence. We restrict the search space by only considering velocity sequences which steer the robot to locations within the collision-free triangle—as soon as the robot leaves the collision-free triangle the search is aborted.

The initial state in the search space is $\langle 0, 0, 0, v_t, v_r \rangle$, i.e., the robot is located at the origin of the coordinate system, its orientation is 0°, and the current translational and rotational velocities are v_t and v_r, respectively. Possible successor states in each step are $\langle x', y', \theta', v_t', v_r' \rangle$ where $v_t' = v_t + c_t \cdot a_t^{max} \cdot \Delta t$ with $c_t \in \{-1, -\frac{2}{3}, \ldots, 1\}$ and a_t^{max} being the maximal translational acceleration. Analogously for v_r'. (x', y', θ') is the projected pose at time $t + \Delta t$ when sending the drive commands $\langle v_t', v_r' \rangle$ to the motor controller and the robot is located at $\langle x, y, \theta \rangle$ at time t. The change in position and orientation is computed according to the standard equations for differentially driven robots The heuristic value for each state is computed as the straight-line distance to the intermediate target; the costs are uniform. A goal state is reached if the distance between the projected position and the intermediate goal is smaller than a certain threshold.

The velocities returned by the function "findVelocities(·)" in Alg. 1 are the first translational and rotational velocities in the sequence of velocities that was determined by the search.

5   Implementation and Evaluation

Occupancy Grid. Although the LRF has a far longer range we limited the local map to a size of 6×6 m^2 for practical reasons. The local map is subdivided into grid-cells with a size of 5×5 cm^2. Consequently, the complete local map is made up of 14400 cells. Recomputing the ellipses that result from extending the obstacles and their respective rasterizations with every update is quite costly. Therefore, we pre-computed a library of rasterized ellipses of various sizes and at various angles. For a given obstacle we look-up the ellipse matching the current velocities of the robot and the angle at which the obstacle is detected and integrate it into the local map.

Searching for the Path. In order to avoid that the changed perception of the environment which is due to the movement of the robot leads to an oscillating behavior we accumulate the sensor readings for a short duration, i.e., we do not only consider the current distance readings but also a number of readings from the past. Each obstacle in this accumulated sensor reading is then extended

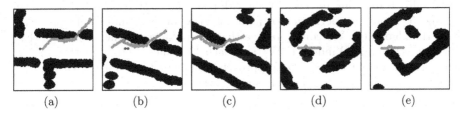

Fig. 4. Example traces

in the same way described in Sect. 4.1. Further, for reasons of efficiency, we try to avoid re-computing a path and proceed with the remainder of the path computed previously, instead. Of course, this is only possible if the "old" path is still a valid path. This means we have to check whether the robot strayed too far from the projected path and whether the collision-free triangle computed for the "old" path is still free of any obstacles. Thus, we still maintain a high degree of reactivity. The implementation of the A^* search algorithm calculates a path of a length up to 300 grid cells (i.e. a path length of 15 m) in less than 10 ms on the Pentium-III 933 machine on the robot. Given that the frequency of the laser range finder is 10 Hz and the navigation module runs at 20 Hz (not to lose any update from the laser range finder) there are about 40 ms left for the other steps of the algorithm.

Evaluation. The method proposed in this paper was extensively tested during several RoboCup tournaments as well as with indoor demonstrations where the robot had to safely navigate through crowds of people. Fig. 4 shows the path visualization of a run of a robot in our department hallway. The red dot represents the robot, the green line represents the planned path, the black objects are the walls of the hallway. The robot should navigate from the hallway into a room. Note that the path is calculated in such a way that the shortest possible connection between robot and target is chosen. The second picture in the series shows the robot a few moments later. The calculation of the drive commands and the realization of these commands on the robot have the effect that the robot slightly deviates from the path. We remark that the position of the robot is still inside the collision-free triangle (which is not shown in the figure). In the fourth picture the robot entered the room.

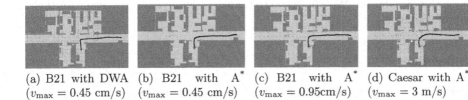

(a) B21 with DWA (b) B21 with A^* (c) B21 with A^* (d) Caesar with A^*
($v_{max} = 0.45$ cm/s) ($v_{max} = 0.45$ cm/s) ($v_{max} = 0.95$cm/s) ($v_{max} = 3$ m/s)

Fig. 5. Comparison of the DWA with our method

We also tested our navigation algorithm on our B21 robot Carl as shown in Fig. 5. From the performance point of view we encountered no problems with the algorithm. We could observe that Carl reached higher velocities than with the Dynamic Window (DW) approach [3] which is part of the original control software of Carl. The DW approach has inherently problems with narrow doorways as well as with relatively sharp turns.

6   Conclusion

In this paper we presented a navigation and collision avoidance method for service robots operating in domestic indoor environments. Particularly in these environments, a domestic robot must navigate carefully and be able to drive around obstacles that suddenly cross its path as it is interacting with humans. The core of our method is a grid representation which is generated from the sensory input of a laser range finder. Depending on the speed of the robot and several other security parameters, the detected obstacles are extended in the grid representation. This is done to speed up the collision detection when searching for a path to the target. For finding a path, we employ A^* on the grid. Next, we construct a so-called collision-free triangle from the obstacle configuration, the current parameters of the robot (its actual speed) and the desired path. For each grid cell inside this triangle we can guarantee that it is collision-free. In a second step, we use this triangle to calculate the drive parameter for the next time step. Again, we employ A^* for this task, this time we search for accelerations which keep the robot inside the triangle. The collision-free triangle relates the search for a path with the search for drive commands and allows to decouple both. Positions inside the triangle are safe and therefore feasible. This decreases the complexity of the search problem tremendously. This method allows for fast and adaptive navigation and was deployed for RoboCup@Home competitions over several years without ever colliding with the furniture or humans. For example, we were able to solve the Lost&Found task in 25 seconds while driving through a large part of the apartment looking for an object.

Acknowledgments

This work was supported by the German National Science Foundation (DFG) in the Priority Program 1125, Cooperating Teams of Mobile Robots in Dynamic Environments. A. Ferrein is currently funded by a grant of the Alexander von Humboldt foundation. We would like to thank the anonymous reviewers for their helpful comments.

References

1. Kavraki, L., Svestka, P., Latombe, J.C., Overmars, M.H.: Probabilistic roadmaps for path planning in high-dimensional configuration space. IEEE Transaction on Robotics and Automation 12(4), 566–580 (1996)

2. Petti, S., Fraichard, T.: Safe motion panning in dynamic environemts. In: Proc. of the IEEE/RSJ International Conference on Intelligent Robots and System (IROS 2005), pp. 2210–2215 (2005)
3. Fox, D., Burgard, W., Thrun, S.: The dynamic window approach to collision avoidance. IEEE Robotics & Automation Magazine 4(1), 23–33 (1997)
4. Khatib, M., Bouilly, B., Simeon, T., Chatila, R.: Indoor navigation with uncertainty using sensor-based motions. In: Proc. of the IEEE/RSJ International Conference on Robotics and Automation (ICRA 1997), pp. 3379–3384 (1997)
5. Fiorini, P., Shiller, Z.: Motion planning in dynamic environments using velocity obstacles. The International Journal of Robotics Research 17, 760–772 (1998)
6. Brock, O., Khatib, O.: High-speed navigation using the global dynamic window approach. In: Proc. of the 1999 IEEE International Conference on Robotics and Automation, vol. 1, pp. 341–346 (1999)
7. Borenstein, J., Koren, Y.: The vector field histogram - fast obstacle avoidance for mobile robots. IEEE Transactions on Robotics and Automation 3(7), 278–288 (1991)
8. Ulrich, I., Borenstein, J.: Vfh*: Local obstacle avoidance with look-ahead verification. In: Proc. of the 2000 IEEE/RSJ International Conference on Robotics and Automation, pp. 2505–2511 (2000)
9. Large, F., Sekhavat, S., Shiller, Z., Laugier, C.: Towards real-time global motion planning in a dynamic environment using the nlvo concept. In: Proc. of the 2002 IEEE/RSJ International Conference on Robots and Systems, pp. 607–612 (2002)
10. Stachniss, C., Burgard, W.: An integrated approach to goal-directed obstacle avoidance under dynamic constraints for dynamic environments. In: Proc. of the 2001 IEEE/RSJ International Conference on Intelligent Robots and Systems (IROS 2001), pp. 508–513 (2002)
11. Simmons, R.: The curvature-velocity method for local obstacle avoidance. In: Proc. of the 1996 IEEE International Conference on Robotics & Automation (ICRA 1996), vol. 1, pp. 1615–1621 (1996)
12. Seraji, H., Howard, A.: Behavior-based robot navigation on challenging terrain: A fuzzy logic approach. In: Proc. of the IEEE Transactions on Robotics and Automation (ICRA 2002), pp. 308–321 (2002)
13. Minguez, J., Montano, L.: nearness diagram navigation (nd): Collision avoidance in troublesome scenarios. IEEE Transactions on Robotics and Automation 20(1), 45–59 (2004)
14. Murray, D., Little, J.: Using real-time stereo vision for mobile robot navigation. Autonomous Robots 8(2), 161–171 (2000)
15. Asensio, J.R., Montiel, J.M.M., Montano, L.: Goal directed reactive robot navigation with relocation using laser and vision. In: ICRA, pp. 2905–2910 (1999)
16. Dedieu, D., Cadenat, V., Soueres, P.: Mixed camera-laser based control for mobile robot navigation. In: Proc. of the 2000 IEEE/RSJ International Conference on Intelligent Robots and Systems (IROS 2000), vol. 2, pp. 1081–1086 (2000)
17. Wijesoma, W., Kodagoda, K., Balasuriya, A.: A laser and a camera for mobile robot navigation. In: 7th International Conference on Control, Automation, Robotics and Vision (ICARCV 2002), vol. 2, pp. 740–745 (2002)
18. Koenig, S., Likhachev, M.: Improved fast replanning for robot navigation in unknown terrain. In: Proc. of the 2002 IEEE International Conference on Robotics and Automation (ICRA 2002), pp. 968–975 (2002)
19. Bruce, J., Veloso, M.: Safe multi-robot navigation within dynamics constraints. Proc. of the IEEE, Special Issue on Multi-Robot Systems 94(7), 1398–1411 (2006)
20. Meystel, A., Guez, A., Hillel, G.: Minimum time path planning for a robot. In: Proc. of the 1986 IEEE International Conference on Robotics and Automation (ICRA 1986), April 1986, pp. 1678–1687 (1986)

Real-Time Ball Tracking in a Semi-automated Foosball Table

Rob Janssen, Jeroen de Best, and René van de Molengraft

Eindhoven University of Technology
Department of Mechanical Engineering
Den Dolech 2, 5600MB Eindhoven, Netherlands
r.j.m.janssen@tue.nl

Abstract. In this article a method is proposed for ball tracking using 100 Hz computer vision in a semi-automated foosball table. In this application the behavior of the ball is highly dynamic with speeds up to 10 m/s and frequent bounces occur against the sides of the table and the puppets. Moreover, in the overhead camera view of the field the ball is often fully or partially occluded and there are other objects present that resemble the ball. The table is semi-automated to enable single user game play. This article shows that it is possible to perform fast and robust ball tracking by combining efficient image processing algorithms with a priori knowledge of the stationary environment and position information of the automated rods.

Keywords: computer vision, automated foosball table, visual servoing, ball segmentation, object tracking, perspective projection, Kalman observer, real-time systems.

1 Introduction

Tracking a ball in a highly dynamic and non predictive environment by use of computer vision becomes difficult when the ball is occluded or surrounded by similar looking objects. An example can be found in the Robocup environment, where autonomous robots have to track a ball and grab it [1]. Another example can be found in the Hawkeye computer vision system [2], where cameras are positioned alongside a tennis court to track the ball. In this article the considered environment is the soccer field used in a foosball table, and the object to be tracked is a white ball of approximately 3.5 cm in diameter. The occluding objects are the puppets and the rods on which these puppets are mounted. In the captured images the ball can cross a white field line, or it can resemble one of the white dots on the field. On one side of the table the rods are electro-mechanically controlled. These puppets need to intercept and return the ball, and therefore ball tracking must be fast and accurate. Other project groups that have been working on the development of an automated foosball table

J. Baltes et al. (Eds.): RoboCup 2009, LNAI 5949, pp. 128–139, 2010.
© Springer-Verlag Berlin Heidelberg 2010

used various techniques to track the ball. The Danish University of Technology has used different colors for the ball and the environment [3]. In this setup, color segmentation [4] can be used to segment the ball from its environment. The University of Victoria developed a laser grid to acquire the coordinates of the ball on the field [5]. They positioned the laser grid such that the lasers would go underneath the puppets and therefore only the ball could cross it. An automated foosball table that eventually became commercially available comes from the University of Freiburg and is named KiRo [6]. This table was developed to a commercial version named StarKick [7]. This group mounted a camera underneath their table and replaced the standard field with transparant glass so that the camera could see the ball from below.

Although all of these automated tables worked and some were able to defeat professional players, the methods they used to track the ball were not flawless. Color segmentation only works on a table where the ball has a different color than its environment. With a laser grid that is mounted beneath the feet of the puppets, the ball can be found by determining where the grid has been broken. However, performance lacks because in a real game the ball bounces such that robust tracking becomes cumbersome. These bounces also make ball tracking difficult in the KiRo and StarKick projects.

In this article a different approach is used to track the ball. A priori knowledge of the environment and perspective projection algorithms are used to segment the ball from its environment. An observer is used to obtain estimates for the position and the velocity. As the table is meant for professional foosball play, after any alterations it should still comply to the official USTSA regulations defined in [8]. There are two more restrictions made on the setup.

− the used camera is mounted above the table,
− the camera can only capture monochrome images,

In this article first the hardware will be explained briefly. In the second part the segmentation of the ball will be explained, and the observer that was used. In the last part the results are discussed, which will focus on the robustness of the image processing algorithms and their real-time performance. Finally, a small summary and a prospect on the overall performance on the table will be given in section 5.

2   The Hardware

To have a platform that complies with the USTSA regulations a professional foosball table is acquired. A steel frame is placed on top of the table, on which a camera [9] is mounted. One side of the table is equipped with electro-mechanically controlled rods. This way the puppets can be moved towards a certain position, and they can also lift their legs or perform a kicking movement. A schematic overview of the whole setup is depicted in Fig. 1.

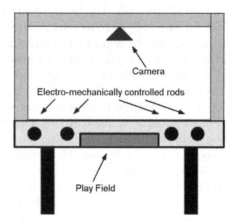

Fig. 1. Schematic overview of the table

The EtherCAT module [12] that is used for the data acquisition allows to control the rods at 1 kHz. The desktop PC that processes the images and interfaces with the EtherCAT module is an Intel Dual Core 2GHz processor, running a Linux low latency kernel. A complete overview of the connections between the hardware components and their corresponding dataflow is depicted in Fig. 2.

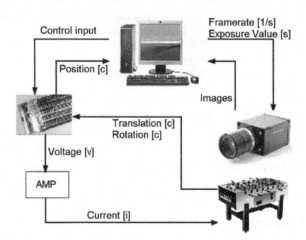

Fig. 2. The different components and their dataflow

3 Ball Segmentation

To obtain the coordinates of the ball, the ball has to be segmented from its environment. Possible methods to segment the ball from its environment are the Circular Hough Transform [10] or more advanced methods such as gradient based circle detection algorithms [11]. There are three reasons why these methods are not applicable in this setup.

- creating a 3D parameter space requires a lot of storage and is therefore not preferable in real-time applications,
- in an occluded situation the ball often does not resemble a circle,
- there are other objects present that resemble the ball such as the white dots and the heads of the puppets,
- the presence of nonconstant light intensities make:
  - the radius criterion used in the CHT difficult to define
  - the intensity gradients differ throughout the field

Therefore in this article a different method is chosen to segment the ball. The coordinate frame that is used throughout this article is depicted in Fig. 3.

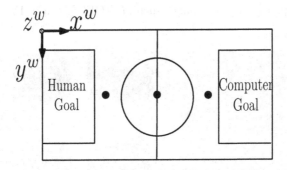

Fig. 3. The field and the coordinate system

3.1 Defining the Region of Interest

The first step in the segmentation algorithm is to crop the captured images to a region of interest, or ROI, in order to remove redundant image data. The largest

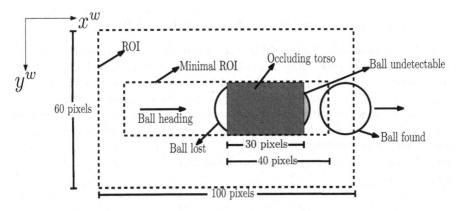

Fig. 4. Worst case loss of ball

object that can occlude the ball is the upper torso of a puppet. This upper torso is 30x20 pixels. When the ball of 20 pixels in diameter rolls underneath, it can become completely occluded. The position of the ROI is programmed to remain static when the ball is lost. When the ball is partially occluded the camera needs approximately 1/4th of the diameter of the ball to detect it. Worst case scenario could be, that the ball rolls underneath a puppet while the ROI stays at the position where it lost the ball. This is schematically depicted in Fig. 4.

" A length of 40 pixels allows the ROI to find the ball when it reappears. Because there can be made no distinction between ball movement in the positive x^w and in the negative x^w direction, the minimum size for the ROI in the x^w direction (as defined in Fig. 3) should be 80 pixels. The minimum size for the ROI in the y^w direction should be 20 pixels as it is the diameter of the ball, but in this direction also the rods have an influence in occluding the ball. Therefore some margins need to be taken into account. A good estimate for the size of the ROI is 100x60 pixels. An example of such an ROI is given in Fig. 5.

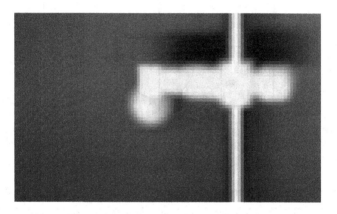

Fig. 5. Region of interest 100x60 pixels

3.2   Determining the Minimum Required Frame Rate

With the minimum size for the ROI determined, also the minimum frame rate at which the image processing algorithms have to run can be calculated. Tests have shown that the maximum ball speed V_{max} can reach up to 10 m/s. It can be assumed that the ball experiences this maximum velocity only in the x^w direction. Because the ROI was chosen such that the ball cannot be occluded twice at its maximum velocity, the radius of the ball r_{ball} has to be added to half the length of the ROI in x^w to determine the maximum distance over which the ball is allowed to travel. The minimum frame rate FPS_{min} can then be calculated as

$$FPS_{min} = \frac{V_{max}P_{res}}{r_{ball} + \frac{1}{2}ROI_{x^w}}, \tag{1}$$

where P_{res} is the pixel resolution of 584 pixels/m. Therefore the minimum frame rate is determined to be 97.3 Hz. In the following steps a frame rate of 100 Hz will be assumed.

3.3 Undistorting the Image

To find the correct coordinates of the ball on the field, the images that come from the camera have to be undistorted. To solve for this distortion, a camera calibration is carried out using the Camera Calibration Toolbox for Matlab [13]. The resulting distortion coefficients are then used to restore the images as shown in Fig. 6.

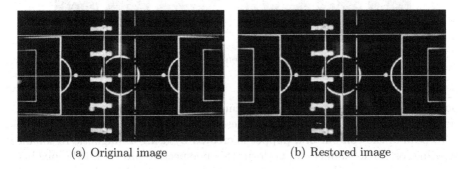

(a) Original image   (b) Restored image

Fig. 6. The original and the restored image. Straight reference lines in white.

3.4 Removing Static Objects by Creating a Mask

In the foosball table there are objects present that do not move over time. These static objects include the field lines, the field dots and the rods that hold the puppets (but not the puppets themselves). These objects look similar to the ball in shape or intensity value, and should therefore be disregarded when searching for the ball. This can be done be creating a static mask. This mask can then be subtracted from the captured image.

The center of the field is somewhat brighter than the rest of the field. This is due to the fact that the field is not uniformly illuminated. This illumination difference also has to be included in the mask. The mask that contains all the static objects and the illumination differences is depicted in Fig. 7.

3.5 Masking the Yellow Puppets

As can be seen in Fig. 6 the puppets on the field have two different colors. The bright puppets have to be masked, because the intensity values of these puppets are close to those of the ball. Therefore the bright puppets are chosen to be the electro-mechanically controlled puppets, so that during a game their position and orientation will be available for processing. With this information the 3D

Fig. 7. The mask that removes the static objects

position and orientation of the puppet can be calculated and through a perspective projection a 2D mask of each of these puppets can be determined. For this a 3D scan of the contour of a puppet is created by using a Magnetic Resonance Imaging, or MRI, scanner. To perform this perspective projection the pixel coordinates D_i^p in the 3D scan have to be transformed to 3D pixel coordinates in the camera frame D_i^c. The formula that describes this transformation is given below.

$$\underline{D}_i^c = \mathbf{R}_s^r(\underline{O}_p^s + \underline{D}_i^p) + \underline{O}_r^c \tag{2}$$

A corresponding graphical interpretation is given in Fig. 8.

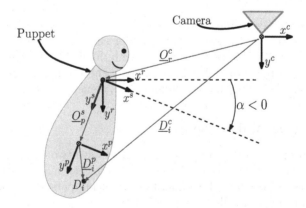

Fig. 8. The transformation of the scan coordinates D_i^p to camera coordinates D_i^c

The vector $\underline{O}_r^c$ describes the translation from camera frame to rod frame and consists of $[x_r^c \ y_r^c \ z_r^c]$ where z_r^c is the variable translation of the rod that can be controlled by the computer. The rotation matrix $\mathbf{R}_s^r$ holds the other computer controlled variable α.

$$\mathbf{R}_s^r = \begin{bmatrix} \cos(\alpha) & \sin(\alpha) & 0 \\ -\sin(\alpha) & \cos(\alpha) & 0 \\ 0 & 0 & 1 \end{bmatrix}.$$

After the pixel transformation the pixels can be placed into the image plane $[b_x^I , b_y^I]$ as depicted in Fig. 9

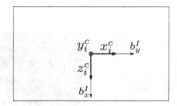

Fig. 9. 3D camera pixel coordinates to pixels in the image plane

according to

$$b_x^I = \frac{\mathbf{f}}{P_s} \frac{z_i^c}{y_i^c} \tag{3}$$

$$b_y^I = \frac{\mathbf{f}}{P_s} \frac{x_i^c}{y_i^c} \tag{4}$$

where $\mathbf{f}$ is the focal length parameter of 6 mm and P_s is the camera pixel size and is 9.9 μm.

A result of the perspective projection is given in Fig. 10. This way the mask is thus fully constructed from the known geometry of the puppet and the measured position and orientation.

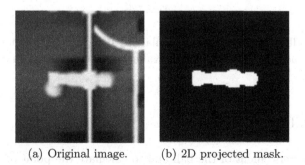

(a) Original image.   (b) 2D projected mask.

Fig. 10. The original image of the puppet and the 2D projected mask

3.6   Thresholding the Dark Puppets

The dark puppets are removed by simply thresholding all that is left over in the image.

3.7   What Is Left over

What is left over in the captured image now can only be the ball itself. The ball will show its full shape, or when occluded by any of the above objects, a part of it. An example of how the ball looks like when the masking and thresholding has been done, is shown in Fig. 11.

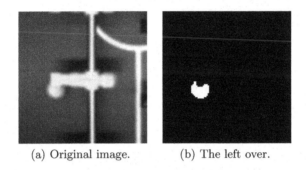

(a) Original image.   (b) The left over.

Fig. 11. The original image and what is left over after masking and thresholding

Finding the center of the ball is done by take the average of the minimum and maximum pixel coordinates x^w_{min} and x^w_{max} for the x^w direction, and the minimum and maximum pixel coordinates y^w_{min} and y^w_{max} for the y^w direction. Therefore the center of the ball $[x^w_c , y^w_c]$ becomes

$$x^w_c = \frac{x^w_{min} + x^w_{max}}{2} \tag{5}$$

$$y^w_c = \frac{y^w_{min} + y^w_{max}}{2} \tag{6}$$

In case the ball is occluded this formula will not give the exact center of the ball. This will be solved for by using a Kalman observer as explained in the next section.

4   Controlling the Rods

To predict the movement of the ball and to calculate the point of interception, the velocity of the ball is determined. For this a linear Kalman observer [14] is implemented.

4.1   The Kalman Observer

To let the Kalman observer converge as fast as possible to an accurate state estimate, the estimates for the process noise covariance $\tilde{Q}$ and measurement noise covariance $\tilde{R}$ have to be accurate. Estimating the measurement noise covariance $\tilde{R}$ can be done by looking at the error between a static ball position and the corresponding measurements. In case the ball is almost fully occluded, the actual measurement of the bal will be completely on the outer edge of the ball. Therefore the maximum value that the measurement noise covariance $\tilde{R}$ can have is the radius of the ball squared. This value has to be given in pixels and is determined to be

$$\tilde{R} = \begin{bmatrix} r_b{}^2 \\ & r_b{}^2 \end{bmatrix} \tag{7}$$

where r_b is the radius of the ball and is approximately 10 pixels. The estimate for the process noise $\tilde{Q}$ is more difficult to determine. The easiest way is to perform an off-line analysis of the logged measurements. The distance between the human controlled attacker and the electro-mechanically controlled keeper is 30 cm, and the maximum velocity of the ball can be 10 m/s. Therefore tuning of the process noise covariance should lead to a maximum convergence time of 0.03 s. In this period the error between the measurement and the actual position estimate should converge to less then the width of the puppet's feet, which is 3 cm. Because the ball bounces and the acceleration was not estimated it can be assumed that $\tilde{Q}$ depends on a deviation of the acceleration of the ball. Movement of the ball in x^w and y^w is decoupled, and therefore $\tilde{Q}$ will only have values on its diagonal.

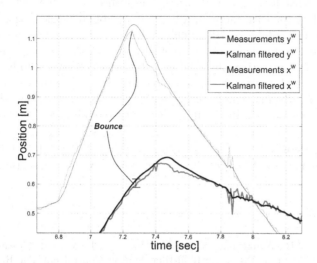

Fig. 12. Measurement data and the Kalman filtered estimate in x^w and y^w direction

When the position data is evaluated a good estimate for $\tilde{Q}$ becomes

$$\tilde{Q}_k = \begin{bmatrix} \frac{1}{2}\tilde{a}\delta t^2 & 0 & 0 & 0 \\ 0 & \tilde{a}\delta t & 0 & 0 \\ 0 & 0 & \frac{1}{2}\tilde{a}\delta t^2 & 0 \\ 0 & 0 & 0 & \tilde{a}\delta t \end{bmatrix} \tag{8}$$

where $\tilde{a}$ is about 20.000 pixels/s^2. With this value the outcome for the Kalman filtered estimate in x^w and y^w (as defined in Fig. 3) is given in Fig. 12. When a bounce occurs in the x^w direction at t=7.26 s it can be seen that the error in the y^w direction converges to less than 3 cm in 0.03 s, which is accurate enough for the mechanically controlled keeper to intercept the ball. An animated illustration where the points of interception are determined can be found at [15].

5   Summary and Prospect

In this article a method was proposed for ball tracking using 100 Hz computer vision in a semi-automated foosball table. A region of interest was defined to reduce the dataflow. Undistortion and static masking were used to remove the static objects. Perspective projection and dynamic masking were applied to mask the bright puppets that resemble the ball in intensity values. Finally an observer was implemented to direct the rods smoothly to their point of interception with the ball. This observer was fast and accurate enough for the keeper to prevent a goal from an opponent that shoots from its most forward attacker at the maximum assumed velocity of 10 m/s. Currently the table is being tested and the results are promising. At this point the image processing algorithm has a 100% found-ratio of the ball. This means that nobody was able to play the ball in such a way that it went out of the region of interest as defined in 3.1. Even a professional player with tactical skills was not able to mislead the algorithm. A video is made available at [16] which indicates the region of interest and the mask of the electro-mechanically controlled puppets. More videos will be made available during the testing phase. When the table has successfully undergone this phase, it can also serve as a test bed for high level strategic control schemes and learning algorithms. The results from these tests can then be used in other research areas such as the Robocup robot soccer domain.

Acknowledgements. The author wants to thank G.J.Strijkers for his assistance in creating the 3D scan of the puppets that was used in section 3.5.

References

1. Bruijnen, D.J.H., Aangenent, W.H.T.M., Helvoort, J.J.M., v.d. Molengraft, J.G., Steinbuch, M.: From Vision To Real-time Motion Control For The Robocup Domain. In: IEEE Conference on Control Applications, Singapore (2007)

2. Owens, N., Harris, C., Stennet, C.: Hawk-Eye Tennis System. In: International Conference on Visual Information Engineering, pp. 182–185 (2004) ISBN 0-85296-757-8

3. Danish University of Technology: Automated Foosball Table, http://foospmp.myl.dk/

4. Gonzalez, R.C., Woods, R.E.: Digital Image Processing, 3rd edn., pp. 445–450. Pearson Prentice Hall, London (2008)

5. Extreme Automation Foosball Table, http://web.uvic.ca/~tiran/

6. University of Freiburg: KiRo - The Table Soccer Robot, http://www.informatik.uni-freiburg.de/~kiro/

7. Gauselmann StarKick, http://www.merkur-starkick.de/

8. USTSA Foosball Rules Of Play, http://www.foosball.com/learn/rules/ustsa/

9. Prosilica GC640 Gigabit Ethernet Camera, http://www.prosilica.com/products/gc640.html

10. Ballard, D.H.: Generalizing The Hough Transform To Detect Arbitrary Shapes. Pattern Recognition 13(2), 111–122 (1981)

11. Rad, A.A., Faez, K., Qaragozlou, N.: Fast Circle Detection Using Gradient Pair Vectors. In: Proc. 8th Digital Image Computing: Techniques and Applications, Syndey (2003)

12. Beckhoff Ethernet for Control Automation Technology, http://www.beckhoff.com/english.asp?ethercat/

13. California Institute of Technology: Camera Calibration Toolbox For Matlab, http://www.vision.caltech.edu/bouguetj/calib_doc/

14. Kalman, R.E., Bucy, R.S.: New Results In Linear Filtering And Prediction Theory. Transactions of the ASME. Series D: Journal of Basic Engineering 83, 95–108 (1961)

15. Eindhoven University Automated Foosball table, animated video 1, http://nl.youtube.com/watch?v=d7xgL5mEyeY

16. Eindhoven University Automated Foosball table, animated video 2, http://nl.youtube.com/watch?v=TN2ENdw3Gac

Three Humanoid Soccer Platforms: Comparison and Synthesis

Shivaram Kalyanakrishnan, Todd Hester, Michael Quinlan,
Yinon Bentor, and Peter Stone

Department of Computer Sciences, The University of Texas at Austin
{shivaram,todd,mquinlan,yinon,pstone}@cs.utexas.edu

Abstract. In this article, we provide an overview of three humanoid soccer platforms currently in use at RoboCup: 3D simulation, the humanoid Standard Platform League (SPL), and the Webots-based simulator released with the SPL. Although these platforms trace different historical roots, today they share the same robot model, the Aldebaran Nao. Consequently, they face a similar set of challenges, primary among which is the need to develop reliable and robust bipedal locomotion. In this paper, we compare and contrast these platforms, drawing on the experiences of our team, UT Austin Villa, in developing agents for each of them. We identify specific roles for these three platforms in advancing the overarching goals of RoboCup.

1   Introduction

The long-term goal of RoboCup is to field a team of humanoid soccer players that can compete with the best human teams on a regulation soccer field by the year 2050 [11]. This goal is still very far away, in part because we do not yet have "human-level" humanoid robots. Since the start of RoboCup in 1997, however, steady progress has been made on all aspects of the challenge, which has been achieved by planning, prioritizing efforts, and by imagining the technology of a few years ahead. We attribute the ability of RoboCup to maintain its momentum in advancing the frontiers of technology to two main characteristics:

1. Its competitive structure, which fosters enthusiasm in diverse groups from all over the world, and at the same time, encourages collaboration. As Behnke [1] notes, competitions promote the evaluation of complete systems, providing a standardized testbed on which comparisons are fair. This is supplemented by community-based development, brought about by the sharing of ideas, solutions, and organizational responsibilities.
2. The division of effort into well-defined leagues and challenge problems, each with self-contained, specific focus areas that are challenging by themselves. Currently at RoboCup there are five main leagues: Simulation League, Small Size League, Middle Size League, Standard Platform League, and Humanoid League. Within each of these, there are multiple sub-leagues.

J. Baltes et al. (Eds.): RoboCup 2009, LNAI 5949, pp. 140–152, 2010.

The focus of this paper is in the context of the second characteristic: the leagues that constitute RoboCup. Ideally, these various leagues, while addressing separate challenges, should also be carefully connected so as to address complementary and synergistic research challenges. Based on the extensive firsthand experiences gathered by our team, UT Austin Villa, we have the opportunity to closely examine three platforms (two sub-leagues and one supplement) that are very new, are spawned from different threads of history, and yet have much in common. These are the 3D simulation sub-league within the Simulation league (Sim-3D), the humanoid-based Standard Platform League (SPL), and its accompanying Webots-based simulator (SPL-Sim).

These three platforms trace different origins with different research foci, but they are now converging in the need to tackle bipedal humanoid locomotion. Incidentally, these three platforms share the same robot model, the Aldebaran Nao.[1] At this point, it seems that in any platform that has bipedal locomotion, the dominant challenge is that of stable, fast movement: the team that is able to walk the fastest and kick the most accurately is favored to win, with minimal need for sophisticated higher-level reasoning. However, the goal is to reach the point where locomotion can be taken as mainly a skill to be fine-tuned within the larger context of individual and team decision-making in strategic situations.

An important ongoing effort within the RoboCup community over the past few years has been the inclusion of "road-map" discussions at the symposia, which are the culmination of discussions within the individual leagues. However, there are only a limited (though significant) number of participants bringing experiences from multiple leagues to these discussions. This paper offers a focused road-map proposal pertaining to three closely related platforms within RoboCup Sim-3D, SPL and SPL-Sim are all 1-3 years old, with much potential yet to be realized. We compare and contrast the current state of progress in these three platforms and identify specific roles for them in advancing the overarching goals of RoboCup. We present these recommendations from the point of view of keeping the platforms complementary, each with its important role to fill. We recognize that individual communities may have perspectives that are not entirely consistent with our proposals. Yet, we believe that some level of *inter-league* planning is necessary for the future.

This paper is organized as follows. In Section 2 we survey the current state of the three platforms, following which, in Section 3 we describe our experiences in developing agents for each platform. Section 4 lays down some of the long-term challenges facing humanoid robotics, and Section 5 earmarks specific roles to each platform in moving forward. We summarize the paper in Section 6.

2   Overview of Platforms: Past and Present

In this section, we present the history and the current challenges facing each platform. Figure 1 shows snapshots of the Nao robot from the three platforms.

---

[1] See: http://www.aldebaran-robotics.com/eng/

Fig. 1. Pictures of the Nao robot in the various leagues. From left to right: Sim-3D, SPL, and SPL-Sim.

2.1   Sim-3D

Sim-3D has evolved from the 2D Simulation sub-league, one of the earliest competitions in RoboCup [3], having been in existence since 1997. The 2D Simulation sub-league simulates a 2D world, with cylindrical robots that have access to abstract actions such as Turn, Kick, Dash, and Catch. It has models of noisy, asynchronous sensation and actuation, real-time decision making, restricted vision and hearing, and player stamina. In short, it incorporates an extensive range of realistic considerations in simulation. Stable "11 versus 11" simulations can be run on current desktop hardware, with agents able to communicate with the simulation server through a network interface following a well-defined protocol. While the first few years of the 2D simulation sub-league witnessed an emphasis on developing agent skills (such as reliable passing and ball interception), its focus soon progressed to high-level strategy. Not only do player formations and communication play an important role in games today, teams even employ strategic reasoning for determining whether to play offensively or defensively depending on the goal difference and time left in the game.

Despite all the realism modeled in the 2D simulator, there exists one significant omission: the third dimension. To address this issue, the 3D Simulation sub-league (Sim-3D) was introduced into RoboCup in 2004. The migration to a 3D world called for physical simulation engines to replace the simple physics models of the 2D simulation. With the intent of staying ahead of the hardware leagues and to encourage high-level reasoning, Sim-3D adopted agents in the form of spheres, with access to abstract commands for kicking, turning and dashing. However, it soon became apparent that while interesting in itself, this version of Sim-3D was lacking direct relevance to the long-term goal of playing soccer with humanoid robots. Indeed in their 2007 paper, Mayer *et al.* [13] argued that the simulation league should embrace humanoid robots as early as possible. To this end, in 2007, Sim-3D transitioned to a more realistic and challenging humanoid model: the Soccerbot, based on the Fujitsu HOAP-2 robot[2]. This step marked a defining moment in the history of the Simulation League: for

---

[2] See: http://jp.fujitsu.com/group/automation/en/services/humanoid-robot/hoap2/

the first time, robots had to be programmed through the low-level interface of controlling motor torques. While this has caused inevitable backtracking in terms of the performance levels of games, its long-term benefits will be significant.

In the initial days of development, the Soccerbot-based simulation encountered a string of systems-related problems, such as unstable physics simulation and performance issues. To correct this, the Soccerbot model was constantly changed, and by the 2007 RoboCup competitions, it was 5 meters tall! However, subsequent contributions from the community towards developing the server have increased its reliability, as has the change of the robot model from Soccerbot to the Aldebaran Nao for the 2008 competitions. In the 2008 competitions, complete, noise-free world information was provided to the agents. In addition, there was no actuator noise. As a result, there has been a quick development of locomotion skills. Both in 2007 and in 2008, skills (in particular, walking speed) have been a major factor determining team performance. It must be noted, however, that some passing behavior began to emerge during RoboCup 2008.

The 3D simulator is under active development. Earlier the platform used the SPADES timer [21], and it still uses the SPARK simulation engine [17]: both of these were developed by the RoboCup community. The simulation server code is fully open source. Apart from the annual RoboCup competitions, Sim-3D is now also popular at the regional open competitions in Iran, Germany, and China.

2.2   SPL

The Standard Platform League is unique in that all the teams use the same standardized hardware, making it essentially a *software* competition. SPL allows teams to work with an affordable humanoid robot platform without having to invest as much time or money as the other humanoid leagues [2]. The current SPL sub-league (using humanoids) emerged from the Sony Aibo "Four-legged" league. Areas of research in SPL include vision [9,22], localization [6,7], and motion and skill development [12]. Since the robots are fully autonomous and all processing is performed on-board, CPU cycles need to be used efficiently.

In 2006, Sony stopped manufacturing the quadrupedal Aibo robot; consequently, the SPL switched to the Aldebaran Nao for 2008. The Nao is a two-legged humanoid robot, on which balance, walking and other soccer skills are more difficult to implement than on the Aibo. Teams competing in the SPL served as beta testers for the Nao; the original robots were very fragile and motors would frequently overheat or break. The robot also was unable to see its own feet with its camera without bending over, which made teams spend significant amounts of time searching for and lining up to the ball.

The Nao has 25 degrees of freedom, compared to the 20 on the Aibo. The Nao robot, a biped, stands 57cm tall, while the Aibo, a quadruped, measures 28cm tall. The V3 version of the Nao has two color cameras in its head, each with a much higher resolution than the Aibo. The Nao has a 500 MHz AMD Geode processor and 256 MB of RAM, which is more sophisticated than what the Aibo uses (a 576 MHz RISC CPU and 64 MB of RAM).

The first competition with the Nao robot was held at RoboCup 2008 in Suzhou, China. Since the robots broke frequently, many teams arrived at the competition with very little code tested on actual robots. In Suzhou, robots were fixed as they broke, giving some teams their first opportunity to test code on the robots. It was a struggle for teams to even "close the loop" on robot behaviors, i.e., reliably score a goal on an empty field. There were only 14 goals scored in total in the 29 games during RoboCup 2008 (including one own goal), even though most teams had very minimal defense strategies. In stark contrast, the Aibo competition the same year witnessed a total of 151 goals scored in 28 competitive games.

2.3 SPL-Sim

Two companies released Nao models in their respective simulators after the SPL humanoid sub-league was introduced. These were Microsoft Robotics Studio[3] and Cyberbotics Webots [15]. Both simulators are more sophisticated than Sim-3D, using commercial physics software and providing useful tools for programming the agent, such as for visualization and debugging. The simulators aid code development for the physical robots in the SPL, proving especially useful in making initial progress in developing skills, as the robot itself is quite fragile.

For code development on the Nao, UT Austin Villa employed the Webots simulator, which we denote SPL-Sim. The simulator was helpful in developing code for SPL, especially since the robots were unavailable or unusable for long durations. While the physics were not very realistic, and therefore the learned motions were not directly transferable to the robot, the simulator was useful in testing vision, localization and behaviors. The simulator was also useful for learning parameters for motions, which made good starting points for motions on the physical robot. Although there is no official competition at RoboCup using Webots, there was an Internet-based competition called "ROBOTSTADIUM"[4], as well as an informal competition at RoboCup 2008. About eight teams played 4 versus 4 soccer games, exhibiting superior skills compared to those from the SPL games. Versions of Webots with the Nao model were made available free to the teams. UT Austin Villa did not take part in this Webots-based competition.

3   Experiences in Agent Development

The agent architectures that UT Austin Villa developed were similar across the three platforms. At the low level, we have PID control for the joints, as well as inverse kinematics for arms and legs. The main thrust of our effort was in developing skills, such as walk, kick, turn, and get-up. These in turn were tied together

---

[3] See: http://msdn.microsoft.com/en-us/robotics/

[4] See: http://www.robotstadium.org/

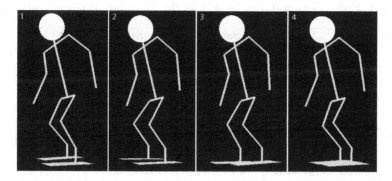

Fig. 2. States from our SPL(-Sim) walk engine. In the state 1, the robot shifts its weight to its left leg and lifts its right leg. In state 2, it re-balances its weight and brings the right leg forward and down. In states 3 and 4, the robot repeats this motion with the legs reversed.

by fairly simple high-level behavior. Here we summarize the salient aspects of our agent behavior, which is described in detail in our technical reports [5,10].

For SPL, we developed a walk that comprises four key-frame states, inspired by Yin *et al.* [24]. Each state is represented by the spatial coordinates of the feet relative to the hips. Joint angles are calculated from these spatial coordinates using inverse kinematics. The motion between states is interpolated and the transition to the next state begins either at a specific time or after getting close to the current target state. The specific coordinates of the key-frames are determined through a set of 10 walk parameters, such as step height and length. When turning, the "HipYawPitch" angle is set to turn the robot's legs. In addition, the robot's shoulder joints are moved to help balance the robot during walking. We used the Downhill Simplex algorithm [19] to learn the best walk parameters through trials in the Webots simulator (SPL-Sim).

While the walk we developed worked well in simulation, it did not work consistently on the physical robots, likely because of lag and jitter introduced by the interface we used to send commands to the robot. Our Sim-3D agent too employed a similar state-based approach for developing skills, although with vastly different parameter settings. Specifically, Sim-3D used higher gains for motor torques in order to realize fast, dynamically stable motions.

We summarize the different walking speeds and kick distances that we achieved on the different platforms in Table 1.[5] The results on the different platforms are not directly comparable, as they were obtained on different surfaces, with different parameters, different physics, different code, etc. We used our own code for all the Sim-3D skills and for the kicks in SPL(-Sim). We used the walk engine provided by Aldebaran with parameters that we tuned for the walking experiments in SPL(-Sim). For comparison, the robot walked at 28.56 ± 1.31 mm/sec in SPL-Sim using

---

[5] The authors acknowledge Hugo Picado from the FC Portugal robot soccer team for suggesting this set of relevant statistics.

Table 1. Performance statistics for skills of the Nao robot on different platforms

Skill	Sim-3D	SPL	SPL-Sim
Forward walking: linear velocity (mm/sec)	144.27 ± 1.22	82.40 ± 1.54	91.62 ± 2.77
Side walking: linear velocity (mm/sec)	62.80 ± 2.00	18.13 ± 2.66	24.48 ± 2.23
Turning: angular velocity (deg/sec)	19.96 ± 3.15	27.20 ± 0.54	18.83 ± 0.21
Kicking: distance reached by ball after kick (mm)	3122.68 ± 14.47	1200.00 ± 312.64	5122.12 ± 756.78
Get-up: time to rise after falling forwards (sec)	10.21 ± 0.94	10.56 ± 0.27	NA
Get-up: time to rise after falling backwards (sec)	23.14 ± 0.81	11.30 ± 0.37	NA

our walk engine. The results in SPL and SPL-Sim were taken over 5 trials each, and the ones from Sim-3D over 10 trials.[6]

There is a significant difference between the results from SPL-Sim and SPL, as we switch from simulation to reality. For example, although the same kick was used on the real robot and in SPL-Sim, the ball traveled an average of about four times as far in the simulator because it has less friction than the real carpet. The main difference between the Sim-3D and SPL-Sim is in the magnitude of the gains. Interestingly, the real robot turns much faster than either simulated robot. The turn is implemented by moving the HipYawPitch joint to turn the legs relative to each other. This is more effective on the real robot than in simulation because the extra friction arrests sliding of the feet. We note that the reported statistics are not representative of all soccer teams; for example, some Sim-3D teams obtained very fast walks through optimization. Yet, these results showcase some key differences in the platforms and the behaviors developed on them.

The key element in our architecture for SPL(-Sim) was to enforce that the environment *interface*, the agent's *memory* and its *logic* were kept distinct (Figure 3). In this case, *logic* encompasses the vision, localization, behavior and motion modules. The main advantages of our architecture are:

Consistency. The *core system* remains identical irrespective of whether the code is run on the robot, in the simulator or inside our debug tool. As a result, we can test and debug code in any of the three environments without fear of code discrepancies. The robot, simulator and tools each have their own *interface* class which is responsible for populating *memory*.

Flexibility. We can "plug & play" modules into our system by allowing each module to maintain its own local memory and communicate to other modules using the common memory area. For example, a Kalman Filter localization module would read the output of the vision module from common

---

[6] Videos of several skills are available from our team website: http://www.cs.utexas. edu/~AustinVilla/

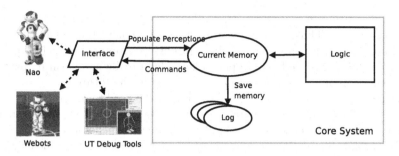

Fig. 3. Agent architecture for SPL(-Sim)

memory, work in its own local memory and then write object locations back to common memory.

Debug-ability. At every time step only the contents of current *memory* are required to make logic decisions. We can therefore save a "snapshot" of the current memory to a log file and then examine the log subsequently in our debug tool. The debug tool not only has the ability to read and display logs, it can also take logs and process them through the logic modules. As a result we can modify code and watch the full impact of the change in our debug tool before testing it on the robot or in the simulator.

4   Shared Challenges

Sim-3D, SPL, and SPL-Sim are all just coming out of their nascency. In this section, we discuss some of the challenges facing humanoid robotics, which should be treated as long-term issues on all these platforms.

The key concern at this stage for all three platforms is also the most basic: reliable, robust bipedal locomotion. Locomotion is a prerequisite for a large number of activities performed by a humanoid robot, and its role is particularly important in soccer. The RoboCup community has successfully addressed the challenge of quadrupedal locomotion [12], but bipedal locomotion requires solutions of a more complex nature and scale. Bipedal locomotion offers great flexibility in developing a wide range of stepping patterns, including climbing stairs and lateral motion. Yet, this flexibility comes at the price of constantly having to balance. While balance is almost like a second nature to human beings, the state-of-the-art with humanoid robotics is yet to provide satisfactory solutions to this problem [23].

Pratt's thesis [18] provides useful insights into the nature of bipedal locomotion. Whereas human beings rely extensively on the natural dynamics of walking (uncontrolled "falling" for some part of the gait, followed by a swift catching step), most algorithms for walking today control every part of the walk cycle. This places a severe limit on the robustness of the developed walking gait: even slight variations in the cycle can potentially cause fall. Another fallout is the energy efficiency of walking. Fully controlled walks spend roughly an order of magnitude more energy for walking than human beings with comparable masses [20]. Thus, developing robust, energy efficient walks is of utmost importance for any humanoid platform in RoboCup. Approaches such as passive dynamic walking [14] are likely to play a key role in realizing "natural" patterns of bipedal locomotion, and need to be considered in the future.

Another challenge facing humanoid robotics is the need to develop human-like models. Currently, the body components of the Nao robot in Sim-3D are implemented as cuboidal elements: this hinders the development of natural gaits as the feet are flat. Simulating complex mesh geometries can be time inefficient; at the same time, they might be *necessary* for developing robust locomotion skills. Another related possibility is to explore the use of more human-like actuators to augment the motors available at the joints. For example, Hosada *et al.* [8] have demonstrated that pneumatic artificial muscles can lead to gaits that require minimal correction. Likewise, it is relevant to note that the Nao (and most humanoid robots) do not have a flexible spine. Flexible spines such as the ones being developed by Mizuuchi *et al.* [16] allow the robot to have more natural movement and enable it to absorb shocks much better. Indeed a major problem in developing walks for SPL is ensuring that the robot's feet touch the ground softly and do not create oscillations in the robot's movement.

In the following section, we argue that in coming to terms with the multiple challenges ahead, the three platforms considered in this paper should concentrate on separate issues, in order that their *combination* will be most effective in advancing towards the goals of RoboCup.

5   Roles for Platforms in Future

As Mayer *et al.* [13] observe, it will become important to start integrating the various leagues in RoboCup; in the year 2050, one "league" will stand to summarize the progress made over the years. At this stage, humanoid robotics still faces a broad range of problems, chief among which is reliable bipedal locomotion. Yet, there is the need to look ahead and preempt problems that are bound to arise in future. In this section, we identify specific roles for Sim-3D, SPL, and SPL-Sim such that they divide their attention in doing so. Encouraging diversity in the platforms is important at this juncture because it will alert us to a wider range of issues, which would remain occluded if our focus remains narrow. However, we note that the desire for diversity needs to be balanced with serving the interests of the members of the individual leagues.

There is the need to study problems in hardware because simulation is invariably subject to unrealistic modeling assumptions, and of course, the ultimate goal of RoboCup needs to be realized in hardware. However, hardware has the disadvantage of being brittle, expensive, and tedious to work with. Thus, it becomes efficient to bypass it and make progress in simulation environments. For example, simulation is far more convenient as a platform for learning and optimization of skills and behaviors. Ultimately, simulation needs to become as close to reality as possible. As long as simplifying assumptions have to be made by simulators, we argue that Sim-3D and SPL-Sim should maintain separate foci: Sim-3D should stay ahead of current possibilities of hardware, and SPL-Sim should become as close to reality as possible. We also believe that each platform should continue to get gradually more difficult as earlier challenges are met.

5.1 Sim-3D

We believe that as with the 2D simulation sub-league that spawned it, Sim-3D should stay ahead of the hardware by a few years and realize what is currently possible *only in simulation*, such as sophisticated coordination strategies among the players and complex soccer skills such as heading, kicks in the air, and the interception of balls in a 3D trajectory. These aspects do not manifest in today's hardware platforms, but it would be inadvisable to wait for the hardware to catch up before devoting research to these issues, which are integral to soccer. Such research might guide the evolution of the hardware itself.

Among the current simulation platforms, Sim-3D has developed the best robotic skills to date, which are evolving at a steady rate. Also Sim-3D has the support of a very active development community. Thus, we propose that Sim-3D consciously make the effort to tackle increasingly complex skills and behavior that humanoid robots will ultimately possess. In so doing, we need to exercise intelligence in choosing parts of reality to approximate and parts to model exactly. For example, we do not need to develop perfectly realistic models of the cameras of today, for it is very likely that in future, robots will be equipped with superior vision systems. However, it would be unwise to persist with very high joint torque limits because these are unrealistic, and will remain so until the year 2050.

5.2 SPL

Unlike Sim-3D, it is not possible to look very deep into the future of SPL, because SPL is inherently restricted by the limits of hardware. At this point, the main focus of the SPL is on important low-level tasks such as vision, localization, and bipedal control. The SPL humanoid sub-league is more challenging than the SPL four-legged (Aibo) sub-league owing to the difficulties involved in developing robust motions on two legs. For the short term, it will be quite an achievement to be able to reproduce the proficiency level of the Aibo robots on the Nao robots. This would entail developing algorithms to for walking stably, robustly, and fast; kicking optimally; ball interception, etc. It might not be worthwhile to plan beyond such a level of proficiency for the Nao robots because they may undergo fundamental changes to the hardware in a few years' time.

5.3 SPL-Sim

Simulators are essential in modern scientific research, as it is time-consuming, expensive, and tedious to conduct several types of useful experiments in the real world. Ultimately, it will be beneficial to have simulators that are as close to reality as possible, for they can function as a more convenient substitute. We propose that at least one thread of research within RoboCup pursue the goal of developing accurate simulators. The simulator being developed need not be specific to the SPL; any hardware league can profit from the use of a realistic physical simulator. But it is especially important to have simulators of humanoid robots, which are going to be principal in the future development of RoboCup.

We believe that SPL-Sim should adopt the aim of becoming more realistic, partly because it possesses superior systems performance at this stage, which make it the more promising alternative for modeling complex mesh geometries (including feet), simulating realistic collisions (which are common in soccer), and modeling surface properties. Currently Webots already has a reliable and accurate physics engine, developed based on the Open Dynamics Engine (ODE).[7]

Gaps between simulation and reality need to be closed; likely, they will get more pronounced when there are more collisions and robots move faster. Recently, Hebbel and Laue [4] have proposed an interesting idea for doing so: by optimizing simulator parameters based on a fitness function that is evaluated in reality. They demonstrate significant results on the Aibo platform. It would be ideal if efforts such as theirs are complemented by code development dedicated to realizing more realistic simulations for RoboCup. It will also be good for the community to develop open source packages. We could start from the current 3D simulator, SPARK, and evolve distinct threads for Sim-3D and SPL-Sim.

6 Summary

UT Austin Villa's participation in Sim-3D and SPL, as well as our use of the SPL-Sim to develop our code gives us a unique perspective on these platforms. There are many similar challenges on these platforms: developing a good code base that is easily debug-able and extendable, developing good robust bipedal motions, and creating good soccer skills. There are differences as well; for instance, SPL(-Sim) has to cope with vision and localization issues, while Sim-3D currently does not.

For RoboCup to continue its progress towards the goal of fielding a human-level team in 2050, we believe that it is important to continue with different leagues that focus on separate aspects of the problem. SPL should continue to deal with problems related to using a real two-legged robot, such as vision, localization, and bipedal motion. We believe that there should be a push towards more realistic simulators to allow more development in simulation instead of on fragile and expensive robots. Finally, the Sim-3D should continue to focus on problems a few years ahead of the hardware leagues, such as coordination, teamwork, intelligence, and developing skills on advanced robot models. All the platforms should continue to gradually increase in difficulty as teams progress.

---

[7] See: http://www.ode.org/

Acknowledgments

The authors thank anonymous reviewers for their comments. This work has taken place in the Learning Agents Research Group (LARG) at the Artificial Intelligence Laboratory, The University of Texas at Austin. LARG research is supported in part by grants from the National Science Foundation (CNS-0615104, EIA-0303609 and IIS-0237699), DARPA (FA8750-05-2-0283, FA-8650-08-C-7812 and HR0011-04-1-0035), General Motors, and the Federal Highway Administration (DTFH61-07-H-00030).

References

1. Behnke, S.: Robot competitions - ideal benchmarks for robotics research. In: Proc. of IROS 2006 Workshop on Benchmarks in Robotics Research (October 2006)
2. Behnke, S., Schreiber, M., Stückler, J., Renner, R., Strasdat, H.: See, walk, and kick: Humanoid robots start to play soccer. In: Proc. of IEEE-RAS Int. Conf. on Humanoid Robots (Humanoids 2006) (December 2006)
3. Chen, M., Foroughi, E., Heintz, F., Huang, Z., Kapetanakis, S., Kostiadis, K., Kummeneje, J., Noda, I., Obst, O., Riley, P., Steffens, T., Wang, Y., Yin, X.: Users manual: RoboCup soccer server — for soccer server version 7.07 and later. The RoboCup Federation (August 2002)
4. Hebbel, M., Laue, T.: Automatic parameter optimization for a dynamic robot simulation. In: Iocchi, L., Matsubara, H., Weitzenfeld, A., Zhou, C. (eds.) RoboCup 2008. LNCS (LNAI), vol. 5399, pp. 121–132. Springer, Heidelberg (2009)
5. Hester, T., Quinlan, M., Stone, P.: UT Austin Villa 2008: Standing on Two Legs. Technical Report UT-AI-TR-08-8, The University of Texas at Austin, Department of Computer Sciences, AI Laboratory (November 2008)
6. Hester, T., Stone, P.: Negative information and line observations for monte carlo localization. In: Proc. of IEEE International Conference on Robotics and Automation (ICRA), May 2008, pp. 2764–2769 (2008)
7. Hoffmann, J., Spranger, M., Göhring, D., Jüngel, M.: Negative information and proprioception in monte carlo self-localization for a 4-legged robots. In: Proc. of Nineteenth International Joint Conference on Artificial Intelligence, Workshop on Agents in Real-Time and Dynamic Environments (2005)
8. Hosada, K., Takuma, T., Nakamoto, A., Hayashi, S.: Biped robot design powered by antagonistic pneumatic actuators for multi-modal locomotion. Robotics and Autonous Systems 56(1), 46–53 (2008)
9. Jüngel, M.: Using layered color precision for a self-calibrating vision system. In: Nardi, D., Riedmiller, M., Sammut, C., Santos-Victor, J. (eds.) RoboCup 2004. LNCS (LNAI), vol. 3276, pp. 209–220. Springer, Heidelberg (2005)
10. Kalyanakrishnan, S., Bentor, Y., Stone, P.: The UT Austin Villa 3D Simulation Soccer Team 2008. Technical Report AI09-01, The University of Texas at Austin, Department of Computer Sciences, AI Laboratory (February 2009)
11. Kitano, H., Asada, M., Kuniyoshi, Y., Noda, I., Osawa, E.: RoboCup: The robot world cup initiative. In: Proceedings of The First International Conference on Autonomous Agents, pp. 340–347. ACM Press, New York (1997)
12. Kohl, N., Stone, P.: Machine learning for fast quadrupedal locomotion. In: The Nineteenth National Conf. on Art. Intelligence, July 2004, pp. 611–616 (2004)

13. Mayer, N.M., Boedecker, J., da Silva Guerra, R., Obst, O., Asada, M.: 3D2Real: Simulation league finals in real robots. In: Lakemeyer, G., Sklar, E., Sorrenti, D.G., Takahashi, T. (eds.) RoboCup 2006. LNCS (LNAI), vol. 4434, pp. 25–34. Springer, Heidelberg (2007)
14. McGeer, T.: Passive dynamic walking. Int. J. Rob. Res. 9(2), 62–82 (1990)
15. Michel, O.: Webots: Professional mobile robot simulation. Journal of Advanced Robotics Systems 1(1), 39–42 (2004)
16. Mizuuchi, I., Yoshiada, S., Inaba, M., Inoue, H.: The development and control of a flexible-spine for a human-form robot. Advanced Robotics 17(2), 179–196 (2003)
17. Obst, O., Rollmann, M.: SPARK – A Generic Simulator for Physical Multiagent Simulations. Computer Systems Science and Engineering 20(5), 347–356 (2005)
18. Pratt, J.: Exploiting Inherent Robustness and Natural Dynamics in the Control of Bipedal Walking Robots. PhD thesis, Massachusetts Institute of Tech. (2000)
19. Press, W., Teukolsky, S., Vetterling, W., Flannery, B.: Numerical Recipes in C, 2nd edn. Cambridge University Press, Cambridge (1992)
20. Ramamoorthy, S., Kuipers, B.: Trajectory generation for dynamic bipedal walking through qualitative model based manifold learning. In: IEEE International Conference on Robotics and Automation, pp. 359–366 (2008)
21. Riley, P., Riley, G.: SPADES — a distributed agent simulation environment with software-in-the-loop execution. In: Chick, S., Sánchez, P.J., Ferrin, D., Morrice, D.J. (eds.) Winter Simulation Conference Proceedings, vol. 1, pp. 817–825 (2003)
22. Seysener, C.J., Murch, C.L., Middleton, R.H.: Extensions to object recognition in the four-legged league. In: Nardi, D., Riedmiller, M., Sammut, C., Santos-Victor, J. (eds.) RoboCup 2004. LNCS (LNAI), vol. 3276, pp. 274–285. Springer, Heidelberg (2005)
23. Wieber, P.-B.: Viability and predictive control for safe locomotion. In: 2008 IEEE/RSJ International Conference on Intelligent Robots and Systems, Acropolis Convention Center, Nice, France, September 22-26, pp. 1103–1108 (2008)
24. Yin, K., Loken, K., van de Panne, M.: Simbicon: Simple biped locomotion control. ACM Trans. Graph. 26(3), Article 105 (2007)

Learning Complementary Multiagent Behaviors: A Case Study

Shivaram Kalyanakrishnan and Peter Stone

Department of Computer Sciences, The University of Texas at Austin
{shivaram,pstone}@cs.utexas.edu

Abstract. As machine learning is applied to increasingly complex tasks, it is likely that the diverse challenges encountered can only be addressed by combining the strengths of different learning algorithms. We examine this aspect of learning through a case study grounded in the robot soccer context. The task we consider is Keepaway, a popular benchmark for multiagent reinforcement learning from the simulation soccer domain. Whereas previous successful results in Keepaway have limited learning to an isolated, infrequent decision that amounts to a turn-taking behavior (passing), we expand the agents' learning capability to include a much more ubiquitous action (moving without the ball, or getting open), such that at any given time, multiple agents are executing learned behaviors simultaneously. We introduce a policy search method for learning "GETOPEN" to complement the temporal difference learning approach employed for learning "PASS". Empirical results indicate that the learned GETOPEN policy matches the best hand-coded policy for this task, and outperforms the best policy found when PASS is learned. We demonstrate that PASS and GETOPEN can be learned simultaneously to realize tightly-coupled soccer team behavior.

1 Introduction

Learning to play soccer can be framed elegantly as a multiagent reinforcement learning (RL) problem. However, the state-of-the-art in multiagent RL is yet to cope with the demands of such a complex problem. In the context of multiagent RL, a number of models have been proposed to exploit task-specific regularities such as coordination of actions [5], state abstraction [4], and information sharing [12]. While such measures all pave the way towards learning increasingly complex behavior, they still assume that the task being considered is simple enough to be learned using a *single* learning algorithm. Yet complex tasks such as soccer comprise multiple overlapping behaviors, whose diverse demands can only be met by combining the strengths of qualitatively different learning approaches. Identifying this as a crucial direction for future research, we present a detailed case study of one such task that is grounded in the RoboCup 2D simulation soccer platform [2].

The task we consider is Keepaway [14], which has become a popular test-bed for multiagent RL [8,9]. Keepaway is a realistic, continuous, high-dimensional,

J. Baltes et al. (Eds.): RoboCup 2009, LNAI 5949, pp. 153–165, 2010.

stochastic task, and is significantly more complex than synthetic, discrete tasks such as Predator-Prey [1] that have been used in the past for studying agent cooperation [7] and games such as Tic-Tac-Toe for studying agent competition [12]. However, all the learning in Keepaway to date has addressed just one aspect of the task, in which the learned decision is made on a turn-taking basis among teammates. These studies have all focused on the "Pass" behavior of the player with possession of the ball in deciding whether (and to which teammate) to pass. They assume that its teammates, when moving to positions on the field likely to induce successful passes, execute fixed, hand-coded "GetOpen" strategies.

In contrast, we formulate GETOPEN as a multiagent learning problem, thereby extending learning in Keepaway from PASS to PASS+GETOPEN. Consequently, Keepaway becomes an instance of a learning problem composed of highly interdependent behaviors executing simultaneously. Each player executes multiple behaviors (PASS and GETOPEN) that affect the outcome of its teammates' behaviors, and in the long run, also interact with one another. Such a scenario poses a significant challenge for designing a credit assignment scheme that both reflects the intended objectives in the underlying task and guides learning in a natural, incremental manner.

We present a novel solution for learning GETOPEN using policy search, which contrasts with the temporal difference learning method used for PASS. Results show that the learned GETOPEN policy matches the best performing hand-coded policy for this task. Further experiments illustrate that learning these complementary behaviors results in a tight coupling between them, and indeed that PASS and GETOPEN can be learned simultaneously. These results demonstrate the effectiveness of applying separate learning algorithms to distinct components of a significantly complex task. As a direct consequence of our formulation of GETOPEN for learning, numerous opportunities arise for conducting research in the Keepaway test-bed.

This paper is organized as follows. In Section 2, we review the standard PASS task and formalize GETOPEN similarly. In Section 3, we describe algorithms for learning PASS and GETOPEN, both individually and together. Experimental results are discussed in Section 4, which is followed by a presentation of related and future work in Section 5. Our conclusions are summarized in Section 6.

2   Keepaway PASS and GETOPEN

The RoboCup 2D simulation soccer domain [2] models several difficulties that agents must cope with in the real world. Soccer is necessarily a multiagent enterprise, in which agents have both teammates and opponents. In the simulation, they are only provided partial and noisy perceptions, and have imperfect actuators. Their sensing and acting routines are not synchronized, and in the interest of keeping real time, do not admit extensive deliberation. The atomic actions available to an agent are Turn, Turn-Neck, Dash, Kick, and Catch; skills such as passing to a teammate or going to a point must be composed of a string of these low-level actions executed sequentially. For all these reasons, simulated RoboCup soccer becomes a challenging domain for machine learning.

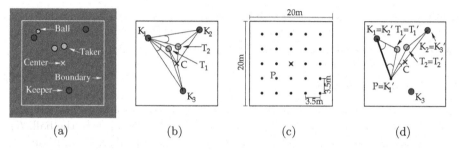

(a)   (b)   (c)   (d)

Fig. 1. (a) A snapshot of Keepaway. (b) Corresponding PASS state variables. (c) Target points for GETOPEN, among them P. (d) Corresponding GETOPEN state variables. $dist(K_1', K_2')$ and $dist(K_1, K_1')$ (darkened) overlap.

Keepaway [14] is a subtask of soccer in which a team of 3 *keepers* aims to keep possession of the ball[1] away from the opposing team of 2 *takers*. The game is played within a square region of side $20m$.[2] Each episode begins with some keeper having the ball, and ends when some taker claims possession or the ball overshoots the region of play. It is the objective of the keepers to maximize the expected length of the episode, referred to as the episodic **hold time**. The keepers must cooperate with each other in order to realize this objective; they compete with the team of takers that seeks to minimize the hold time. Figure 1(a) shows a snapshot of a Keepaway episode in progress.

In order to make the task amenable to learning, it becomes necessary to constrain the scope of decision making by the keepers. Figure 2 outlines the policy followed by *each* keeper in the scheme employed by Stone *et al.* [14]. The keeper closest to the ball intercepts the ball until it has possession. Once it has possession, it must execute the PASS behavior (not to be confused with a pass action), by way of which it may retain ball possession or pass to a teammate. Keepers other than the one closest to the ball move to a position conducive for receiving a pass by executing GETOPEN behavior.

Fig. 2. Policy followed by each keeper

PASS and GETOPEN, by offering a *choice* of high-level actions based on the keeper's state, are candidates for the application of learning. Most prior work assumes GETOPEN, and indeed the behavior followed by the takers, to follow fixed, hand-coded strategies. In other words, the teammates and opponents of the keeper with the ball do not *adapt* to the specific characteristics of that keeper, as they do in real soccer. As a step in the direction of furthering team adaptation, we extend the frontier of learning in Keepaway to include GETOPEN. Thus, we treat Keepaway as a composite of two distinct behaviors to be learned: PASS and

---

[1] A player is deemed to have *possession* of the ball if it is close enough to be kicked.
[2] Keepaway can be generalized to varying numbers of keepers and takers, as well as field sizes [14].

GETOPEN. As in previous work [14], we restrict the takers to the fixed policy of moving towards the ball. In recent work, Iscen and Erogul [8] explore learning taker behavior, which complements the work in our paper (see Section 5).

2.1   Keepaway PASS

Here we revisit the problem of PASS defined by Stone *et al.* [14]. The keepers and takers assume roles that are indexed based on their distances to the ball: K_i is the ith closest keeper to the ball, and T_j the jth closest taker. From Figure 2, we see that the keeper executing PASS must be K_1.

The three high-level actions available to K_1 are HoldBall, which is composed of a series of kicks close to itself, but away from any approaching takers; and PassBall-i, $i \in 2, 3$, a direct pass to K_i. Each player processes its low-level perceptual information to construct a world model, which constitutes a continuous state space. This space is represented through a vector of 13 state variables, comprising distances and angles among the players and the center C of the field. These are marked in Figure 1(b), and enumerated in Table 1.

A policy for PASS maps a 13-dimensional vector representing the state variables to one of the high-level actions: Hold-Ball, PassBall-2, and PassBall-3. An example of such a policy is PASS:HAND-CODED (Algorithm 1), which implements a well-tuned manually programmed strategy [14]. Under this policy, K_1 executes HoldBall until the takers get within a certain range, after which distances and angles

---

Algorithm 1. PASS:HAND-CODED

---

input PASS state variables (13)
output Action $\in$ {HoldBall, PassBall-2, PassBall-3}
  if $dist(K_1, T_1) > C_1$ **then**
    Return HoldBall.
  for $i \in 2, 3$ **do**
    $valAng_i \leftarrow \min_{j \in 1,2} ang(K_i, K_1, T_j)$.
    $valDist_i \leftarrow \min_{j \in 1,2} dist(K_i, T_j)$.
    $val_i \leftarrow C_2 \cdot valAng_i + valDist_i$.
  if $\max_{i \in 2,3} val_i > C_3$ **then**
    $passIndex \leftarrow \text{argmax}_{i \in 2,3} val_i$.
    Return PassBall-passIndex.
  else
    Return HoldBall.
  $\{C_1 = 5.0, C_2 = 0.25, C_3 = 22.5;$ distances are taken to be in meters and angles in degrees.$\}$

---

involving its teammates and opponents are used to decide whether (and to which teammate) to pass. Yet another policy for PASS is PASS:RANDOM, under which K_1 chooses one of the three available actions with equal likelihood. PASS:LEARNED denotes a learned PASS policy, which is described in Section 3.

2.2   Keepaway GETOPEN

Whereas learning the PASS behavior has been studied extensively in the literature [9,10], to the best of our knowledge, all previous work has used the hand-coded GETOPEN policy originally defined by Stone *et al.* [14], which we refer to here as GETOPEN:HAND-CODED. Thus, while previous work on this task has considered multiple agents learning, they have never been executing their learned behaviors concurrently (only one player executes PASS at any given time). This paper introduces a learned GETOPEN behavior, thereby expanding the scope of multiagent learning in Keepaway significantly. Below we describe our formulation of GETOPEN.

In principle, there are infinitely many positions that K_2 and K_3 can occupy on the square playing field, However, they only get a small amount of time to pick a target. Since nearby points are likely to be of similar value, an effective strategy is to evaluate only a small, finite set of points spread out across the field and choose the most promising. Figure 1(c) shows a uniform grid of 25 points overlaid on the field, with a 15% margin on the sides. GETOPEN is implemented by evaluating each grid point P, and moving to the one with the highest value. Indeed, we define the GETOPEN learning problem to be learning an evaluation function that assigns a value to every target point P, given the configuration of the players.

As with PASS, it becomes necessary to define a set of state variables for learning GETOPEN. In Figure 1(d), K_3 is shown seeking to evaluate the point P at some time t. The distances and angles marked correspond to the GETOPEN state variables used for the purpose, which we identify based on informal experimentation. None of the state variables involve K_3, as K_3 is examining a situation at time t' in the future when it would itself be at P. At time t', K_3 expects to have possession of the ball, and re-orders the other players based on their distances to it. Thus K_3 becomes K'_1, and in the state from Figure 1(d), K_1 becomes K'_2, T_1 becomes T'_1, and so on. Conceptually, the evaluation of the target point P should consider both the likelihood of receiving a pass at P, and the value of being at P with the ball afterwards. This leads to two logical groups within the state variables. One group contains 2 variables that influence the success of a pass from K_1 to K'_1, the latter being at P. These are the distance between K_1 and K'_1, and the minimum angle between K_1, K'_1 and any taker. The other group of state variables bear direct correspondences with those used for learning PASS, but computed under the re-ordering at t'. Of the 13 state variables used for PASS, we leave out the 5 distances between the players and the center of the field, as they do not seem to benefit the learning of GETOPEN. This results in a total of 10 state variables for GETOPEN, which are listed in Table 1.

In defining the state variables for GETOPEN, it is implicitly assumed that players other than K'_1 do not change their positions between t and t'. This clearly imperfect assumption does not have too adverse an impact since GETOPEN is executed *every* cycle, always with the *current* positions of all players. Revising the target point every cycle, however, has an interesting effect on a random GETOPEN policy. In order to get from point A to point B, a

Table 1. PASS, GETOPEN state variables

PASS	GETOPEN
$dist(K_1, K_2)$	$dist(K'_1, K'_2)$
$dist(K_1, K_3)$	$dist(K'_1, K'_3)$
$dist(K_1, T_1)$	$dist(K'_1, T'_1)$
$dist(K_2, T_2)$	$dist(K'_2, T'_2)$
$\min_{j \in 1,2} dist(K_2, T_j)$	$\min_{j \in 1,2} dist(K'_2, T'_j)$
$\min_{j \in 1,2} ang(K_2, K_1, T_j)$	$\min_{j \in 1,2} ang(K'_2, K'_1, T'_j)$
$\min_{j \in 1,2} dist(K_3, T_j)$	$\min_{j \in 1,2} dist(K'_3, T'_j)$
$\min_{j \in 1,2} ang(K_3, K_1, T_j)$	$\min_{j \in 1,2} ang(K'_3, K'_1, T'_j)$
$dist(K_1, C)$	$dist(K_1, K'_1)$
$dist(K_2, C)$	$\min_{j \in 1,2} ang(K'_1, K_1, T_j)$
$dist(K_3, C)$	
$dist(T_1, C)$	
$dist(T_2, C)$	

player must first turn towards B, which takes 1-2 cycles. When a random target point is chosen each cycle, K'_1 constantly keeps turning, achieving little or no net displacement. To redress this effect, our implementation of GETOPEN:RANDOM only allows K_1' to revise its target point when it reaches its current target. Such a measure is not necessary when the targets remain reasonably stable, as they do for GETOPEN:LEARNED, the learned policy, and GETOPEN:HAND-CODED [14], which we describe below.

Under GETOPEN:HAND-CODED (Algorithm 2), the value of a point P is inversely related to its *congestion*, a measure of its distances to the keepers and takers. Assuming that K_1 will pass the ball from *predictedBallPos*, P is deemed an inadmissible target (given a value of $-\infty$) if any taker comes within a threshold angle of the line joining *predictedBallPos*

Algorithm 2. GETOPEN:HAND-CODED

input Evaluation point P, World State
output Value at P

$teamCongestion \leftarrow \sum_{i \in 1,2,3, i \neq myIndex} \frac{1}{dist(K_i, P)}$.

$oppCongestion \leftarrow \sum_{j \in 1,2} \frac{1}{dist(T_j, P)}$.

$congestion \leftarrow teamCongestion + oppCongestion$.

$value \leftarrow -congestion$.

$safety \leftarrow \min_{j \in 1,2} ang(P, predictedBallPos, T_j)$.

if $safety < C_1$ **then**

    $value \leftarrow -\infty$.

Return $value$.

{$C_1 = 18.4$; angles are taken to be in degrees.}

---

and P. Thus, GETOPEN:HAND-CODED is a sophisticated policy using complex entities such as congestion and the ball's predicted position, which are not captured by the set of state variables we define for learning GETOPEN. In Section 4, we compare GETOPEN:HAND-CODED with GETOPEN:LEARNED to verify if simple distances and angles indeed suffice for describing competent GETOPEN behavior.

2.3   Keepaway PASS+GETOPEN

PASS and GETOPEN are separate behaviors of the keepers, which together may be viewed as "distinct populations with coupled fitness landscapes" [12]. At any instant, there are two keepers executing GETOPEN; their teammate, if it has intercepted the ball, executes PASS. Specifically, each keeper executes GETOPEN when it assumes the role of K_2 or K_3, and executes PASS when it has possession of the ball, as K_1. The extended sequence of actions that results as a combination each keeper's PASS and GETOPEN policies determines the team's performance. Indeed, the episodic hold time is precisely the temporal length of that sequence. PASS has been the subject of many previous studies, in which it is modeled as a (semi) Markov Decision Problem (MDP) and solved through temporal difference learning (TD learning) [9,10,14]. In PASS, each action (HoldBall, PassBall-1, PassBall-2) is taken by exactly one keeper; hence only the keeper that takes an action needs to get rewarded for it. Indeed, if this reward is the time elapsed until the keeper takes its next action (or the episode ends), the episodic hold time gets maximized if each keeper maximizes its own long-term reward.

Unfortunately, GETOPEN does not admit a similar credit assignment scheme, because at any instant, two keepers (K_2 and K_3) take GETOPEN actions to move to target points. If K_1 executes the HoldBall action, none of them will reeceive a pass; if K_1 passes to K_2 (K_3), it is not clear how K_3 (K_2) should be rewarded. In principle, the sequence of *joint actions* taken by K_2 and K_3 up to the successful pass must be rewarded. Yet, such a joint action is taken *every* cycle (in contrast with PASS actions, which last 4-5 cycles on average), and the large number of atomic GETOPEN actions (25, compared to 3 for PASS) leads to a very large joint action space. In short, GETOPEN induces a far more complex MDP than PASS. An additional obstacle to be surmounted while learning PASS and GETOPEN together is non-stationarity introduced by each into the other's environment. All these reasons, combined with the inherent complexity of RoboCup 2D simulation soccer, make PASS+GETOPEN a demanding problem for machine learning.

3   Learning Framework

Each of the 3 keepers must learn one PASS and one GET-OPEN policy; an array of choices exists in deciding whether the keepers learn separate policies or learn them in common. Thus, the total number of policies learned may range from 2 (1 PASS, 1 GETOPEN) to 6 (3 PASS, 3 GETOPEN). Different configurations have different advantages in terms of the size of the overall search space, constraints for communication, the ability to learn specialized behaviors, etc. It falls beyond the scope of this paper to systematically comb the space of solutions for learning PASS and GETOPEN. As an exploratory study, our emphasis in this work is rather on verifying the *feasibility* of learning these behaviors, guided by intuition, trial and error. In the learning scheme we adopt, each keeper learns a unique PASS policy, while all of them share a common GETOPEN policy. We proceed to describe these. As in Section 2, we furnish pseudo-code and parameter settings to ensure that our presentation is complete and our experiments reproducible.

3.1   Learning PASS

We apply the *same* algorithm and parameter values employed by Stone *et al.* for learning PASS [14], under which each keeper uses Sarsa to make TD learning updates. Owing to space restrictions, we do not repeat the specifications of this method here, which is described in detail in Section 4 of their paper [14].

3.2   Learning GETOPEN

The solution to be learned under GETOPEN is an evaluation function over its 10 state variables, by applying which the keepers maximize the hold time of the episode. Whereas TD learning is a natural choice for learning PASS, the difficulties outlined in Section 2.3 to solve GETOPEN as a sequential decision making problem make direct policy search a more promising alternative. Thus, we represent the evaluation function as a parameterized function and search for parameter values that lead to the highest episodic hold time.

Our learned GETOPEN policy is implicitly represented through a neural network that computes a value for a target location given the 10-dimensional input state. The player executing GETOPEN compares the values at different target points on the field, and moves to the point with the highest value. Note that unlike with PASS, these values do not have the same semantics as action values computed through TD learning; rather, they merely serve as *action preferences*, whose relative order determines which action is chosen. We achieve the best results using a 10-5-5-1 network, with a total of 91 parameters (including biases at each hidden node). The parameters are initialized to random values drawn uniformly from $[-0.5, 0.5]$; each hidden node implements the sigmoid function $f(x) = 1.7159 \cdot tanh(\frac{2}{3}x)$, suggested by Haykin [6].

A variety of policy search methods are applicable for optimizing the 91-dimensional policy. We verify informally that methods such as hill climbing, genetic algorithms, and policy gradient methods all achieve qualitatively similar results. The experiments reported in this paper are conducted using the cross-entropy method [3], which evaluates a population of candidate solutions drawn from a distribution, and progressively refines the distribution based on a selection the fittest candidates. We use a population size of 20 drawn initially from $N(0,1)^{91}$, picking the fittest 5 after each evaluation of the population. Each keeper follows a fixed, stationary PASS policy across all evaluations in a generation; within each evaluation, all keepers share the same GETOPEN policy (the one being evaluated). The fitness function used is the average hold time over 125 episodes, which negates the high stochasticity of Keepaway.

3.3 Learning PASS+GETOPEN

Algorithm 3 outlines our method for learning PASS+GETOPEN. Learning is bootstrapped by optimizing a GETOPEN policy for a random PASS policy. The best GETOPEN policy found after two iterations (a total of $2 \times$

Algorithm 3. Learning PASS+GETOPEN

output Policies π_{PASS} and π_{GETOPEN}
 $\quad \pi_{\text{PASS}} \leftarrow$ PASS:RANDOM.
 $\quad \pi_{\text{GETOPEN}} \leftarrow$ GETOPEN:RANDOM.
repeat
 $\quad \pi_{\text{GETOPEN}} \leftarrow learnGetOpen(\pi_{\text{PASS}}, \pi_{\text{GETOPEN}})$.
 $\quad \pi_{\text{PASS}} \leftarrow learnPass(\pi_{\text{PASS}}, \pi_{\text{GETOPEN}})$.
until convergence
Return $\pi_{\text{PASS}}, \pi_{\text{GETOPEN}}$.

$20 \times 125 = 5000$ episodes) is fixed, and followed while learning PASS using Sarsa for the next 5000 episodes. The PASS policy is now frozen, and GETOPEN is once again improved. Thus, inside the outermost loop, either PASS or GETOPEN is fixed and stationary, while the other is improved, starting from its current value. Note that π_{PASS} and π_{GETOPEN} are still *executed* concurrently during each Keepaway episode as part of *learnPass()* and *learnGetOpen()*.

Whereas Algorithm 3 describes a general learning routine for each keeper to follow, in our specific implementation, the keepers execute it in phase, and indeed share the same π_{GETOPEN}. Also, we obtain slightly better performance in learning PASS+GETOPEN by spending more episodes on learning GETOPEN than on learning PASS, which we report in the next section.

4   Results and Discussion

In this section, we report the results of a systematic study pairing three PASS policies (PASS:RANDOM, PASS:HAND-CODED, and PASS:LEARNED) with three

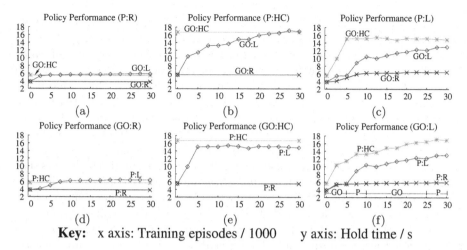

Key: x axis: Training episodes / 1000   y axis: Hold time / s

Fig. 3. Learning curves corresponding to conjunctions of various PASS and GETOPEN policies. Each curve represents an average over at least 20 independent trials. Each reported point corresponds to an evaluation (non-learning) for 500 episodes; points are reported every 2500 episodes. Note that each of the nine experiments appears once in the left column, where experiments are grouped by common PASS policies, and once in the right column, where they are grouped by GETOPEN.

GETOPEN policies (GETOPEN:RANDOM, GETOPEN:HAND-CODED, and GET-OPEN:LEARNED). For the sake of notational convenience, we use abbreviations: thus, PASS:RANDOM is denoted P:R, GETOPEN:LEARNED is denoted GO:L, and their conjunction P:R-GO:L. Nine configurations arise in total. Figure 4 shows the performance of each PASS policy when paired with different GETOPEN policies, and vice versa.[3] Policies in which both PASS and GETOPEN are either random or hand-coded are static, while the others show learning.

Figure 3(c) shows the performance of P:L. P:L-GO:HC corresponds to the experiment conducted by Stone *et al.* [14], and we see similar results. After 30,000 episodes of training, the hold time achieved is about 14.9 seconds, which falls well short of the 16.7 seconds registered by the static P:HC-GO:HC policy (Figure 3(b)). Although P:L-GO:HC is trained in these experiments with a constant learning rate of $\alpha = 0.125$, we posit that annealing α will improve its performance by avoiding the gradual dip in hold time we observe between episodes 12,500 and 30,000. In the absence of any guarantees about convergence to optimality, we consider the well-tuned P:HC-GO:HC to serve as a near-optimal benchmark for the learning methods. Interestingly, under the random GETOPEN policy GO:R (Figure 3(d)), P:HC is overtaken by P:L at 30,000 episodes ($p < 0.0001$). This highlights the ability of learning methods to adapt to different settings, for which hand-coded approaches demand manual attention.

---

[3] Videos of policies are posted on the following web page:
   http://www.cs.utexas.edu/~AustinVilla/sim/keepaway-getopen/.

Figure 3(f) confirms the viability of our policy search method for learning GETOPEN, and its robustness in adapting to different PASS policies. Practical considerations force us to terminate experiments after 30,000 episodes of learning, which corresponds roughly to one day of real training time. After 30,000 episodes, P:HC-GO:L achieves a hold time of 16.9 seconds, which indeed exceeds the hold time of P:HC-GO:HC (Figure 3(b)); yet despite running 20 independent trials of each, this result is not statistically significant. Thus, we only conclude that when coupled with P:HC, learning GETOPEN, a novel contribution of this work, matches the hand-coded GETOPEN policy that has been used in all previous studies on the Keepaway task. This result also highlights that well-crafted state variables such as *congestion* and *predictedBallPos*, which are used by P:HC-GO:HC, are not necessary for describing good GETOPEN behavior. Interestingly, the hold time of P:HC-GO:L is significantly higher than that of P:L-GO:HC ($p < 0.001$). In other words, our GETOPEN learning approach outperforms the previously studied PASS learning when each is paired with a hand-coded counterpart, underscoring the relevance of *learning* GETOPEN.

An important result we observe from Figures 3(c) and 3(f) is that not only can PASS and GETOPEN be learned when paired with static policies, they can indeed be learned in tandem. In our implementation of Algorithm 3, we achieve the best results by first learning GETOPEN using policy search for 5000 episodes, followed by 5000 episodes of learning PASS using Sarsa. Subsequently, we conduct 6 generations of learning GETOPEN (episodes 10,000 to 25,000), followed by another 5000 episodes of Sarsa, as depicted along the x axis in Figure 3(f). The hold time of P-L:GO-L (13.0 seconds after 30,000 episodes) is significantly lower than P:L-GO:HC, P:HC-GO:L, and P:HC-GO:HC ($p < 0.001$), reflecting the additional challenges encountered while learning PASS and GETOPEN simultaneously. Indeed, we notice several *negative* results with other variant methods for learning PASS+GETOPEN. In one approach, we represent both PASS and GETOPEN as parameterized policies and evolve their weights concurrently to maximize hold time. In another approach, GETOPEN uses the value function being learned by PASS as the evaluation function for target points. In both these cases, the performance never rises significantly above random.

We conduct a further experiment in order to ascertain the degree of specialization achieved by learned PASS and GETOPEN policies, i.e., whether it is beneficial to learn PASS specifically for a given GETOPEN policy (and vice versa). In Table 2, we summarize the performances of learned PASS and GETOPEN policies trained and tested with different counterparts. Each column corresponds to a test pairing. We notice that the best performing PASS policy for a given GETOPEN policy is one that was trained with the same GETOPEN policy (and vice versa); the maximal sample mean in each column coincides with the diagonal. It must be noted, however, that despite conducting at least 20 trials of each experiment, some comparisons are not statistically significant. A possible reason for this is the high variance caused by the stochasticity of the domain. Yet, it is predominantly the case that learned behaviors adapt to work best with the counterpart behavior with which they are playing. Thus, although

Table 2. In the table on the left, PASS learned while trained with different GETOPEN policies is tested against different GETOPEN policies. Each entry shows the mean hold time and one standard error of at least 20 independent runs, conducted for 500 episodes. Each *column* corresponds to a test GETOPEN policy. The largest entry in each column is in boldface; entries in the same column are marked with "-" if not significantly lower ($p < 0.05$). The cell GO:L-GO:L shows two entries: when the learned PASS policy is tested against the same ("s") learned GETOPEN policy as used in training, and when tested against a different ("d") learned GETOPEN policy. The table on the right is constructed similarly for GETOPEN, and uses the same experiments as PASS for the cell P:L-P:L.

PASS:LEARNED				GETOPEN:LEARNED			
Train	Test			Train	Test		
	GO:R	GO:HC	GO:L		P:R	P:HC	P:L
GO:R	**6.37**±.05	11.73±.25	10.54±.26	P:R	**5.89**±.05	10.40±.39	11.15±.43
GO:HC	6.34±.06⁻	**15.27**±.26	12.25±.32	P:HC	5.48±.04	**16.89**±.39	12.99±.43⁻
GO:L	5.96±.07	13.39±.35	**13.08**±.26 (s) / 12.32±.32 (d)⁻	P:L	5.57±.06	11.78±.56	**13.08**±.26 (s) / 12.32±.32 (d)⁻

different learning algorithms are applied to PASS and GETOPEN, the behaviors are tightly-coupled in the composite solution learned.

5   Related and Future Work

Multiple learning methods are used in the layered learning architecture developed by Stone [13] for simulated soccer. These include neural networks for learning to intercept the ball, decision trees for evaluating passes, and TPOT-RL, a TD learning method for high-level strategy learning. This work shares our motivation that different sub-problems in a complex multiagent learning problem can benefit from specialized solutions. Yet a key difference is that in Stone's architecture, skills learned using supervised learning are employed in higher-level sequential decision making, to which RL is applied; in our work, the two learning problems we consider are themselves both sequential decision making problems.

The policy search approach we use for GETOPEN is similar to one used by Haynes *et al.* [7] for evolving cooperative behavior among four predators that must collude to catch a prey. The predators share a common policy, represented as a LISP S-expression, in contrast with the neural representation we engage for computing a real-valued evaluation function. The Predator-Prey domain [1], which is discrete and non-stochastic, is much simpler compared to Keepaway.

By decomposing Keepaway into PASS and GETOPEN, our work enriches the multiagent nature of the problem and spawns numerous avenues for future work. For example, a new promising dimension is agent communication. Consider K_1 "yelling" to K_2 where it is about to pass, as is common in real soccer. K_1's PASS and K_2's GETOPEN behaviors could conceivably exploit such information to further team performance.

The Brainstormers team [11] has applied RL for learning attacking team behavior. In their work, the actions available to the player with the ball are several variants of passing and dribbling. Its teammates can move in different directions or head to a home position. Assuming the availability of an environmental model, TD learning is used to estimate a value function over the possible states. The team attack is shown to increase its goal-scoring percentage. Iscen and Erogul [8] consider applying TD learning to the behavior of the takers. The actions available to the takers are ball interception and player marking. Whereas PASS+GETOPEN models cooperation, extending Keepaway to include taker behavior would also incorporate competition.

6   Conclusion

Through a concrete case study, we advance the case for applying different learning algorithms to qualitatively distinct behaviors present in a complex multiagent system. In particular, we introduce Keepaway GETOPEN as a multiagent learning problem that complements Keepaway PASS, the well-studied reinforcement learning test-bed problem from the robot soccer domain. We provide a policy search method for learning GETOPEN, which compares on par with a well-tuned hand-coded GETOPEN policy, and which can also be learned simultaneously with PASS to realize tightly-coupled behaviors. Learning GETOPEN with a hand-coded PASS policy outperforms the earlier result in which PASS is learned and GETOPEN is hand-coded. Our algorithm for learning both PASS and GETOPEN in an interleaved manner confirms the feasibility of learning them together, but also shows significant scope for improvement. This work widens the scope for conducting research on the Keepaway test-bed. It puts together distinct techniques that apply to sequential decision making, which is a crucial element in scaling to more complex multiagent learning problems.

Acknowledgments

The authors thank Ian Fasel and anonymous reviewers of the current and previous versions of this paper for providing useful comments. This work has taken place in the Learning Agents Research Group (LARG) at the Artificial Intelligence Laboratory, The University of Texas at Austin. LARG research is supported in part by grants from the National Science Foundation (CNS-0615104, EIA-0303609 and IIS-0237699), DARPA (FA8750-05-2-0283, FA-8650-08-C-7812 and HR0011-04-1-0035), General Motors, and the Federal Highway Administration (DTFH61-07-H-00030).

References

1. Benda, M., Jagannathan, V., Dodhiawala, R.: On optimal cooperation of knowledge sources - an empirical investigation. Technical Report BCS–G2010–28, Boeing Advanced Technology Center, Boeing Comp. Serv., Seattle, WA (July 1986)

2. Chen, M., Foroughi, E., Heintz, F., Huang, Z., Kapetanakis, S., Kostiadis, K., Kummeneje, J., Noda, I., Obst, O., Riley, P., Steffens, T., Wang, Y., Yin, X.: Users manual: RoboCup soccer server — for soccer server version 7.07 and later. The RoboCup Federation (August 2002)
3. De Boer, P.T., Kroese, D.P., Mannor, S., Rubinstein, R.: A tutorial on the cross-entropy method. Annals of Operations Research 134(1), 19–67 (2005)
4. Ghavamzadeh, M., Mahadevan, S., Makar, R.: Hierarchical multi-agent reinforcement learning. Aut. Agents and Multi-Agent Sys. 13(2), 197–229 (2006)
5. Guestrin, C., Lagoudakis, M.G., Parr, R.: Coordinated reinforcement learning. In: Sammut, C., Hoffmann, A.G. (eds.) Proceedings of the Nineteenth International Conference on Machine Learning, University of New South Wales, Sydney, Australia, July 8-12, pp. 227–234. Morgan Kaufmann, San Francisco (2002)
6. Haykin, S.: Neural Networks: A Comprehensive Foundation. Prentice Hall PTR, Upper Saddle River (1998)
7. Haynes, T., Wainwright, R., Sen, S., Schoenefeld, D.: Strongly typed genetic programming in evolving cooperation strategies. In: Forrest, S. (ed.) Proc. of the 6th Int. Conf. Gen. Alg., San Mateo,CA, pp. 271–278. Morgan Kaufman, San Francisco (1995)
8. Iscen, A., Erogul, U.: A new perpective to the keepaway soccer: The takers. In: Proceedings of the Seventh International Joint Conference on Autonomous Agents and Multi–Agent Systems, pp. 1341–1344. International Foundation for Autonomous Agents and Multiagent Systems, Richland (2008)
9. Jung, T., Polani, D.: Learning Robocup-Keepaway with kernels. In: Gaussian Processes in Practice: JMLR Workshop and Conference Proceedings, vol. 1, pp. 33–57 (2007)
10. Metzen, J.H., Edgington, M., Kassahun, Y., Kirchner, F.: Analysis of an evolutionary reinforcement learning method in a multiagent domain. In: Proceedings of the Seventh International Joint Conference on Autonomous Agents and Multi–Agent Systems, pp. 291–298. International Foundation for Autonomous Agents and Multiagent Systems, Richland (2008)
11. Riedmiller, M., Gabel, T.: On experiences in a complex and competitive gaming domain: Reinforcement learning meets robocup. In: 3rd IEEE Symposium on Computational Intelligence and Games, April 2007, pp. 17–23 (2007)
12. Rosin, C.D., Belew, R.K.: Methods for competitive co-evolution: Finding opponents worth beating. In: Forrest, S. (ed.) Proc. of the 6th Int. Conf. Gen. Alg., pp. 373–380. Morgan Kaufmann, San Mateo (1995)
13. Stone, P.: Layered Learning in Multiagent Systems: A Winning Approach to Robotic Soccer. MIT Press, Cambridge (2000)
14. Stone, P., Sutton, R.S., Kuhlmann, G.: Reinforcement learning for RoboCup-soccer keepaway. Adaptive Behavior 13(3), 165–188 (2005)

Rollover as a Gait in Legged Autonomous Robots: A Systems Analysis

Vadim Kyrylov[1], Mihai Catalina[2], and Henry Ng[2]

[1] Rogers State University; Claremore, OK 74017 USA
vkyrylov@sfu.ca
[2] Simon Fraser University – Surrey; Surrey, British Columbia V3T 0A3 Canada
mscatali@sfu.ca, henry_ng@alumni.sfu.ca

Abstract. Rollover has been normally regarded as an undesirable way of legged robot locomotion. In spite of this, we are looking for the improved robot movement by combining the rollover with regular gaits. By considering gaits on the macro level, we identify the conditions when adding rollover may result in a faster movement of a legged robot. Our method is generic because the number of legs does not matter; we are taking into account only the features and constraints shared by all legged robots. As the outcome of this study, we propose a mathematical model for estimating the efficiency of using rollover as additional gait. This model is illustrated by evaluating the speed gain achieved in 4-legged Sony Aibo ERS-7 robot if it is equipped with a rollover skill. While having in our implementation the same linear speed as walking, rollover in some situations eliminates the need to make turns, thus saving time for the robot to change its pose.

1 Introduction

Legged robots are designed to walk using different gaits, i.e. particular manners of moving on their feet. We want to relax this constraint by allowing the robot actively touching the floor with any part of its body, neck, head, and legs while moving by rolling over. The main hypothesis of this study is that, by combining the rollover with regular gaits, in some conditions a robot can move faster. We determine these conditions and propose a mathematical model for planning robot movements as a combination of walking and rollovers. We also use this model for evaluating speed gains using the 4-legged Sony Aibo robot as an example. Still our method is generic because the number of legs does not matter; we are taking into account only the features and constraints shared by all legged robots.

The term 'rollover of an artificial object' normally implies something negative; rollovers in most cases should be avoided. In particular, motor vehicle developers are doing their best to prevent rollovers. Same attitude appears to exist in the legged robot developer community; by design, rolling over in legged robots is regarded an unsuitable way to move [1, 2]. However, for the appropriately shaped robots rollovers may be deemed suitable. For example, in [3], this 'gait' was studied for the purpose of locomotion of a cylinder-shaped robot. One more recent study of rollover has been conducted for humanoid robots [4]. In this case, this mode of locomotion

J. Baltes et al. (Eds.): RoboCup 2009, LNAI 5949, pp. 166–178, 2010.

was used for the sole purpose of rising of the robot after it falls down. Still in general for the legged robots rollover as a gait has not been systematically studied as yet.

Unlike the works [1-4] that concentrate on the micro level design of the robot locomotion, in this paper we are taking a higher-level view of different gaits; hence 'A Systems Analysis' in the title. Our main objective is to create a mathematical model that allows estimating the time needed to change the robot pose as the function of the ordered set of applied gaits. The sequence of gaits and their parameters are the controlled variables. Then we solve the discrete optimization problem by minimizing time with and without rollover and measure the time gain.

Although this study was started with 4-legged robots in mind, we make assumptions neither about the number of legs nor do we look into the details of leg motions. Rather, we consider the robot gaits on the macro level; they are dashing, pulling, sidestepping, turning, and optionally rolling over. To make our approach applicable to legged robots, we impose some constraints on the way how these gaits can be combined; we believe that these constraints are specific to most, if not all, legged robots. We view the robot motion planning task as deciding on what sequence of available macro level gaits to apply for making the desired change to the robot pose. For the purpose of planning, we are using generalized approximations of the gaits, each such sub model containing very few parameters that are easy to estimate in given robot.

The objective of this study is to show that using the rollover in a legged robot may result in moving faster. In doing so, we identify the main factors and their parameters affecting the time required robot to change its initial state to the desired final one. Then we derive equations that allow calculating this time and determine the best sequence of gaits and parameters thereof that result in the minimal time. Reducing the travelling time is especially critical in robotic soccer.

Since 1998, the RoboCup legged league was exclusively using Sony 4-legged robots; however, these robots have been recently phased out by the manufacturer. Because new types of legged robot may likely be developed for the RoboCup competitions in the future, we believe that it makes sense to investigate the locomotion capabilities of a generic legged robot with rolling over as one of possible gaits. We anticipate that the results of this study will be taken into account by the developers of the new generation RoboCup legged robots.

Section 2 describes the problem addressed in this study and explains the main assumptions. In Section 3 we analyze the minimal-time optimization problem for the robot movement without rollovers. In Section 4 we introduce the rollover as a gait and provide the algorithm for optimizing robot movement with this optional gait. Section 5 describes the experimental results and Section 6 concludes this study.

2 The Optimized Robot Motion Problem: Main Assumptions

For the purpose of this study, we limit our consideration to 2D space. In doing so, we represent the state of the robot (pose) just by three parameters (x, y, δ); x, y being the coordinates of its center on the plane and δ the facing direction. We also assume that the robot has a set of gaits $\{G_0, G_1,..., G_n\}$, each having a set of parameters. To reduce the dimension of the planning problem, we are using very limited set of parameters for each gait.

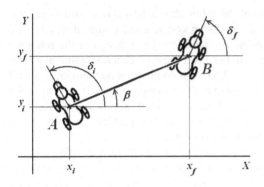
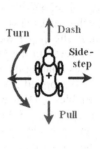

Fig. 1. Robot must change its initial state $A(x_i, y_i, \delta_i)$ to $B(x_f, y_f, \delta_f)$ in the minimal time

Fig. 2. The 'regular' gaits of a legged robot on the macro level

Consider the following optimization problem (Fig.1). Robot located in the initial state $A(x_i, y_i, \delta_i)$ wants to get in the final state $B(x_f, y_f, \delta_f)$ in the minimal time. Thus the robot must select the sequence of gaits and values of their parameters that minimize time to move.

Note that we consider planning ahead the desired robot orientation δ in the final point; in the robotic soccer the success of ball interception highly depends on this parameter.

To make this problem tractable, we make several simplifying assumptions.

1. There are no obstacles on the robot path.

2. We do not consider short-range moves, say, shorter than 1-2 robot length; rather, we are interested in optimizing longer-range movements when rollover may result in some gains. This allows neglecting the robot movement dynamics by assuming that starting/stopping are instant. (This is in particular reasonable for rather slow robots like Sony Aibos.)

3. Walking is split in four atomic macro level gaits: dashing, pulling, sidestepping, and turning while staying in the same place (Fig.2). These gaits can be applied one at a time; at any given instance, robot can either dash, or pull, or side-step, or turn, or roll.

4. Switching from one gait to another can be made instantly at any time (except the rollover in which each revolution must be fully completed before changing to different gait).

5. As time is the only criterion, each of these four gaits is fine tuned to achieve the maximal speed.

Assumptions 3-5 need more explanation. Generally, even with this very limited set of gaits, there is a continuum of walking options (from dashing/pulling/sidestepping straight to dashing-turning on a curve to just turning while staying on the same place). However, in legged mechanical systems, be they animals or robots, the movement along a non-straight trajectory can be achieved by interweaving of the four basic macro gaits shown in Fig.2. Because we assume that gaits could be applied in small discrete time intervals, the order of the three translation gates (dashing, pulling, and sidestepping) does not matter; Fig.3 illustrates this point. With dynamics neglected, the time to translate the robot thus does not depend on the trajectory if turns are not executed. This is similar to the well known *city block distance* [5]; the straight path in Fig.3 is as fasts as the other two. We believe that this assumption is true to most legged mechanical systems. If that was not the case, legged animals could be able to achieve higher speed while running in the diagonal directions; yet this does not happen. So in our model, this assumption implicitly sets the constraints inherent to legged robots.

We believe that further simplification of this model by merging the three translation gaits in just one is unreasonable because most legged robots are asymmetrical. Thus we assume that dashing, pulling, and sidestepping can be executed with different maximal speeds. This is one more macro level feature of legged robots that is reflected in the proposed model.

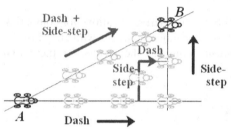
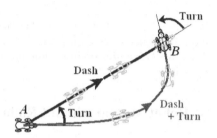

Fig. 3. Three alternative paths for translating the robot. All are taking same time to travel.

Fig. 4. Alternative paths: (top) turn-dash-turn and (bottom) dashes interweaved with turns

With turns included, applying same gaits with different timing may result in the trajectories having different end points. In Fig.4, the two trajectories have same end points and robot direction in it. However, for the bottom trajectory this is achieved by special selection of the time intervals when the robot is making turns. Finding the optimal sequence of gaits thus becomes a non-trivial problem, especially if dynamics is a factor. To simplify the analysis, we are making one more assumption: turns, if any, are only made in the start and end points like in the top trajectory in Fig. 4. With neglected dynamics, time necessary to move the robot in the required end pose by first making a turn followed by a translation followed by a second turn thus cannot be improved by applying turns and translations differently.

Rollovers provide extra choice to the robot how to move. With rollovers, there is an additional constraint: the robot must be always making integer number of revolutions.

The above assumptions imply that:
1. Robot has five alternative actions associated with macro level gaits:
    - G_t= "turn in place by angle α",
    - G_d= "dash distance d",
    - G_p= "pull distance d",

- G_s= "sidestep distance d", and
- G_r= "make N rollovers".

where α, d, and N are robot decision parameters.

2. Robot can execute any one of these actions at any time except the rollover which, once started, must be completed.
3. As the maximal force is always assumed, force is not a gait parameter.
4. Because legged robots are typically asymmetrical, each translation gait has its own speed.
5. The set of the robot actions and their timing is limited to an optional turn in the start point A, translation to the end point B and a second optional turn in this point.
6. Translation may be any combination of gates $\{G_d, G_p, G_s, G_r\}$; using G_r being the factor whose impact on the time required for reaching the end point we investigate.

With these assumptions in mind, in what follows, we derive equations for computing the robot trajectories on a plane and the traveling times with and without rollovers. Then we solve the optimization problem and compare the results for these two options.

3   Robot Motion without Rollover

3.1   Robot 2D Kinematics

First we assume the robot that is incapable of rolling over. Even with the limited set of possible gaits the robot has infinite number of options. Indeed, it can make any combination of turns, dashes or pulls and/or sidesteps to make its way to the final pose. Finding a rigorous method for minimizing the time could be subject of a stand-alone study; here we are using heuristic approach.

As it was explained in Section 2, if robot is limited to making turns in points A and B only, this would not affect the time given the assumptions we have made. This substantially simplifies the problem by leaving finite set of just seven options which all can be explicitly evaluated before making a choice (Fig.5). The first three are either *turn-dash-turn*, or *turn-pull-turn*, or *turn-sidestep-turn*. It is also possible for the robot to move with only one turn combined with two translations. There are total four choices of such trajectories; two shown on the right-hand pane in Fig.5. In each, sidestepping is always applied as the second translation.

Because we are interested in minimizing time, of the six parameters of the robot motion optimization problem, $(x_i, y_i, \delta_i, x_f, y_f, \delta_f)$, only three are independent. In particular, for the *turn-dash-turn* combination (Fig.6), these three independent parameters could be:

$$displacement = \sqrt{(x_i - x_f)^2 + (y_i - y_f)^2} \qquad (1)$$

$$angle1 = \beta - \delta_i, \qquad (2)$$

$$angle2 = \delta_f - \beta, \qquad (3)$$

where β is the orientation angle of the displacement vector, which is determined by the two points, (x_i, y_i) and (x_f, y_f) (Fig. 1).

For the *turn-pull-turn* movement, the two angles are:

$$angle1 = (\beta + \pi) - \delta_i, \tag{4}$$

$$angle2 = \delta_f - (\beta + \pi), \tag{5}$$

For *turn-sidestep-turn*, these angles can be calculated as:

$$angle1 = (\beta + \pi/2) - \delta_i, \tag{6}$$

$$angle2 = \delta_f - (\beta + \pi/2), \tag{7}$$

It is assumed that the turning angles returned by (2)-(7) are normalized to the range $[-\pi, -\pi)$. Note that these angles differ for different robot translation modes.

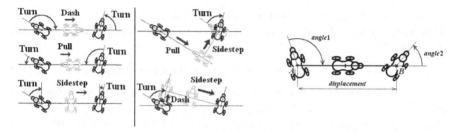

Fig. 5. Five of the total of seven robot movement without rollovers

Fig. 6. Movement parameters

For the two motions shown on the right-hand pane in Fig5, there is just one turning angle:

$$angle1 = \delta_f - \delta_i \tag{8}$$

and two displacements for the first and second translation:

$$displacement1 = \begin{cases} displacement \cdot |\sin(\delta_f - \beta)|, & \text{if turn at } A \\ displacement \cdot |\cos(\delta_i - \beta)|, & \text{if turn at } B \end{cases}, \tag{9}$$

$$displacement2 = \begin{cases} displacement \cdot |\cos(\delta_f - \beta)|, & \text{if turn at } A \\ displacement \cdot |\sin(\delta_i - \beta)|, & \text{if turn at } B \end{cases}, \tag{10}$$

There are also two similar motions with the robot turning in the direction opposite to the orientation required in the final pose.

3.2 Minimal-Time Optimal Robot Motion Planning without the Rollover

Thus we have just a set of seven options to evaluate and to choose one having minimal time for the legged robot to move from A to B. Because of the inborn asymmetry, this time with different gaits may substantially differ. Normally, dash is designed as the fastest of the three translation gaits. Also, depending on translation gait chosen, the total amount of time spent on turns also differs. In the example in Fig.5, the sum of two turns is minimal if the robot chooses to sidestep. If sidestepping is not too slow indeed, in this example it could be best choice in this particular case. So we can expect that, in general, of the seven options that robot has to choose from, some may require noticeably shorter time than others.

With robot dynamics neglected, we assume that the angular speed of the robot is ω and its translation speeds for dashing, pulling, and sidestepping are v_d, v_p, v_s, respectively.

So the time needed to change the robot state from (x_i, y_i, δ_i) to (x_f, y_f, δ_f) using the translation gait G_k is,

$$time_k = (|angle1_k| + |angle2_k|) / \omega + displacement / v_k, \tag{11}$$

where $\mathbf{k}$ is one of the identifiers $\mathbf{d}$, $\mathbf{p}$, $\mathbf{s}$, of the translation gaits.

Calculating time with just one turn and two translation gaits can be done in the similar way:

$$time_{jk} = |angle1| / \omega + displacement1_{jk} / v_k + displacement2_{jk} / v_s, \tag{12}$$

where $\mathbf{j}$ indicates whether the turn was made in A or B and $\mathbf{k}$ is the identifier of the translation gait applied first (either $\mathbf{d}$ or $\mathbf{p}$); the rest parameters are given by (8)-(10).

Thus to find the set of actions for changing its state from $A(x_i, y_i, \delta_i)$ to $B(x_f, y_f, \delta_f)$ in minimal time without rollover, the robot should determine the time for each of the seven options. Then the movement mode delivering the minimal time is selected.

4 Robot Motion with Rollover

4.1 Robot Rollover Kinematics

The rollover kinematics is determined by three parameters, the first being the number of full revolutions N. In what follows, we identify the rest two.

For the purpose of modeling rolls, a legged robot is regarded a truncated cone. For example, because the Sony Aibo robot has a head, its effective diameter at shoulders is greater that that at the hips. For this reason, in the general case, the rolling path is an arc whose radius is R.

Let $chord_1$ be the linear displacement of the robot center, and θ is the angle by which the robot direction changes per full revolution (Fig.7).

Thus we have

$$chord_1 = 2 \cdot R \sin \frac{\theta}{2}. \tag{13}$$

In what follows, we do not call this parameter 'displacement' because the latter designates the distance between points A and B. (In the particular case when the robot could be approximated by a cylinder, R is infinite and θ is zero; $chord_1$ is just the circumference of the cylinder.)

For any number N of revolutions, the chord connecting the initial and final points is:

$$chord_N = 2 \cdot R \sin \frac{N\theta}{2} .$$ (14)

It is important to make sure that the following constraint is satisfied:

$$N \leq \frac{\pi}{\theta} .$$ (15)

Indeed, if $N = \pi/\theta$, robot rolls exactly half circumference whose radius is R. Further increments of N would result in that the robot would be approaching back to the initial point A.

For the purpose of analysis, we have found it more convenient not using R as a parameter of the robot kinematics. Instead, we introduce θ as the second and $chord_1$ as the third independent parameters; both can be easily measured experimentally. By solving (13) and (14), we eliminate R. So the final formula for the straight distance covered by N rollovers is:

$$chord_N = chord_1 \cdot \frac{\sin \dfrac{N\theta}{2}}{\sin \dfrac{\theta}{2}}$$ (16)

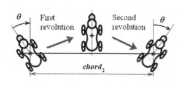
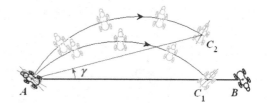

Fig. 7. Robot direction changes by θ after each revolution

Fig. 8. Alternative rollover paths leading close to the final point B

4.2 Minimal-Time Optimal Robot Motion Planning: The General Case

Rollover adds two degrees of freedom in the planning task. They are the number of revolutions N and the direction γ from the initial point to the end point of the rollover (Fig. 8).

These action parameters are not all uniquely determined by the initial and final states of the robot. We simplify the problem by assuming that the rollover trajectory is always selected by setting $\gamma = 0$; i.e. the end point C of the last rollover must be on line AB. However, this is just a near-optimal solution; by carefully choosing γ, in some cases it is indeed possible to reach the final pose faster. However, this gain in speed, if any, would be negligibly small.

This simplification means that of the continuum of potential trajectories we will be selecting one of just two possible pairs with the end points C_1 and C_2 shown in Fig.9. They require N or $(N+1)$ revolutions, respectively. After the robot reaches any of these end points, it faces the task to reach the final state B from C_1 or C_2 in minimal time. To accomplish this, robot applies the optimization method without rollovers described in section 3.2.

In Fig.9 all four paths have different time to reach the final point. The paths differ in the initial and final turning angles; also different is the number of revolutions and the residual paths after rolling.

In the nutshell, to calculate time to move from A to B in the general case, robot first determines the required time for the following options:

- without rollovers (seven options),
- two options with N rollovers and seven options to move from C_1 to B, and
- two options with $N+1$ rollovers and seven options to move from C_2 to B.

Then of the total of $7\cdot3+2+2=25$, the option having the minimal time is selected.

The time to reach the end point C_1 (C_2) with N rollovers along k-th trajectory is:

$$timeR_k = |angle1_k| / \omega + N\cdot revtime, \tag{17}$$

where $angle1_k$ is the first turn angle and $revtime$ is time to execute one full revolution.

The number of revolutions N is determined from the inequality

$$chord_N \leq displacement \leq chord_{N+1}, \tag{18}$$

where $chord_N$ is given by (14).

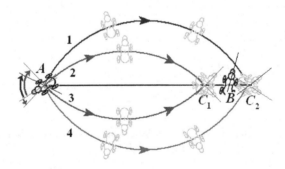

Fig. 9. The two pairs of alternative paths with rollovers. Note different turning angles before and after rolling.

For each trajectory in Fig.9, the turning angle $angle1_k$ can be found as follows:

$$angle1_1 = (\delta_i - \beta) - (\pi + (N+1)\cdot\theta)/2, \tag{19}$$

$$angle1_2 = (\delta_i - \beta) - (\pi + N\cdot\theta)/2, \tag{20}$$

$$angle1_3 = (\delta_i - \beta) - (3\pi - N \cdot \theta)/2, \tag{21}$$

$$angle1_4 = (\delta_i - \beta) - (3\pi - (N + 1) \cdot \theta)/2. \tag{22}$$

Upon arrival in point C_1 (C_2), the robot coordinates and body direction are

$$x_C = x_i + chord_N \cdot \cos\beta, \tag{23}$$

$$y_C = y_i + chord_N \cdot \sin\beta, \tag{24}$$

$$\delta_C = \delta_i + angle1_k \pm N \cdot \theta, \tag{25}$$

where in (25) the minus sign applies to trajectories 1, 2 and the plus sign to trajectories 3, 4.

Then using the new robot initial state $C(x_C, y_C, \delta_C)$ we can find the time needed to cover the remaining path to $B(x_f, y_f, \delta_f)$, by applying the algorithm described in Section 3.2.

This provides the full description of the algorithm for calculating time to change the robot pose $A(x_i, y_i, \delta_i)$ to pose $B(x_f, y_f, \delta_f)$ for any given set of gaits that satisfy our assumptions.

5 Experiments and Performance

The authors came to the idea to use rollovers in late 2005 soon after the first experience with Sony ERS-7 robots equipped with the well-known *Tekkotsu* software package [6]. The walking gaits that come with this software appeared to be too slow. On the other hand, the rounded body of the robot and its ability to place its head in rather low position rendered themselves for rolling possibly faster than walking. So the authors implemented the rollover in ERS-7 and conducted a set of experiments to measure its parameters (Fig.10). After completing this study, the authors discovered that the rollover indeed had been earlier implemented as part of a student project an undergraduate course in robotics at Carnegie-Mellon. However, in that case was used different robot ERS-210. Presumably because of its boxy shape rolling appeared to be clumsy and noticeably slower than our ERS-7.

Fig. 10. A Sony Aibo ERS-7 robot making a rollover

The rollover 'gait' in our experiments was designed using only common sense and thus cannot be regarded fully optimized. Yet we have gained same translation speed (31.0 cm/s in our rollover vs. 30.9 cm/s in *Tekkotsu* dashing). Table 1 provides the experimental values of the main parameters of the ERS-7 robot gaits as they were found in *Tekkotsu*; the rollover parameters are also included.

Table 1. Sony Aibo ERS-7 robot parameter estimates

Gait	Parameter	Mean value	Standard deviation
Straight line dashing	speed, cm/s	30.9	4.5
Side stepping	speed, cm/s	20.4	1.65
Pulling	speed, cm/s	14.0	1.42
Turning	angular speed, degrees/s	205	19.1
Rollover:	Chord, cm	36.6	1.3
	Revolution time, s	1.18	0.06
	Angle increment, degrees	22.5	1.8
	Calculated translation speed, cm/s	31.0	N/A

Using the averages of these parameters, we have implemented a software model to estimate the robot motion performance with and without rollovers. Based on expressions (1)-(25), this model calculates the anticipated time for the robot to change its pose A to pose B. Table 2 presents the summary of the calculation results.

The calculations have been made for 100 values of *alpha*1 in the range (-π, π), combined with 100 values of *displacement* in the range (30, 180) cm, and 100 values *alpha*2 in the range (0, π), thus making the total of 10^6 points uniformly distributed in the 3D parameter space. Distances greater than 180 cm were not considered because from (15) we get the maximal number of revolutions $N=8$; thus from (14) $chord_8=187.6$ cm, which is the maximal reach by rolling over for this robot.

The central column (highlighted in boldface) shows the results for the robot parameters taken from Table 1. It demonstrates that in 42.6 per cent of cases the option to rollover would reduce time by 0.285 seconds on the average. In a soccer game this time gain means that the robot intercepting the ball would be likely outplaying a non-rolling opponent if all the rest conditions are same. Yet this is a somewhat pessimistic estimate; if for the same values of the model the displacement range is limited to (30, 120) cm, the rollover gives noticeably better average time advantage 0.316 seconds in 56.7 per cent of cases. This is because the contribution of the curvature to the rollover path on shorter distances is less significant.

The two left-hand columns give the idea how the advantage gained with optional rollovers decays with the reduced linear speed of rollovers. At 80 per cent of the dashing speed the gain is achieved just 9.0 per cent time; at 60 per cent the contribution of rollover completely diminishes.

Table 2. Rollover vs. non-rollover motion comparison

Performance indicator	(Rollover linear speed)/(dashing speed)				
	0.60	0.80	**1.00**	1.20	1.40
Average time without rollovers, s	4.24	4.24	**4.24**	4.24	4.24
Average time with rollovers, s	4.24	4.22	**4.12**	3.84	3.50
Per cent of times when rollover is faster	0.001	9.02	**42.6**	74.2	93.2
Average gain when rollover is faster, s	0.001	0.18	**0.285**	0.540	0.791
Relative gain when rollover is faster, per cent	0.01	4.27	**6.71**	12.7	18.6

We believe that in Aibos rolling over is more energy efficient than walking and can be further improved. This is supported by the experimental observation of that the robot dashing speed tended to decrease with distance; hence the relatively high standard deviation of the dashing speed in Table 1. We attribute this to chemical processes in the battery under high load. After some rest, robot could be able to dash with the original speed again for a short time. With rollover, however, we did not observe this negative effect of the battery drain.

Because no attempts have been made to optimize the rollover motion sequence, it would be nice to know what might happen if one had managed to improve the coordinated action of the 15 joints of the robot's neck and legs that are making it rolling over. The two right-hand columns in Table 2 show these projections. If we improve the rollover linear speed by 20 per cent, robot would be faster by 0.54 seconds in 74.2 per cent cases. With 40 per cent speed increase, rolling over would be advantageous 93.2 per cent of the time.

6 Conclusion

The rollover 'gait' adds more flexibility in the legged robot movement. Even when the rollover has about same linear speed as dashing, if introduced in a robot, this extra gait allows in some situations saving time on turns that are thus becoming unnecessary due to the possibility to move in the lateral direction. Because the proposed model is taking into account only high-level features of robot motion, we believe that it is applicable to almost any legged robots. By plugging into our model different parameter values, a robot designer may determine if adding the rollover capability would or would not give any advantage in speed.

Evaluating the potential gains for Sony Aibo robot from rolling over was just an illustration of the proposed method. Our study demonstrates that this gain is tangible; this gives the reason for attempting to implement the rollover it in the legged soccer robots.

Yet we are looking forward to the new developments in the legged RoboCup league; thus we hope that this study would be helpful for making high-level design decisions in the development of new robots.

Acknowledgements

The authors thank Prof. Manuela Veloso of CMU and her Ph.D. student Sonya Chernova, who provided the materials on the student projects in the robotics course taught at their school.

References

1. Alexandre, P.: An Autonomous Micro Walking Machine with Rollover Recovery Capability. In: Workshop II: New Approaches on Dynamic Walking and Climbing Machines of the 8th International Conference on Advanced Robotics, Monterey, CA, USA, pp. 48–55 (1997)
2. Son, Y., Kamano, T., Yasuno, T., Suzuki, T., Harada, F.: Generation of Adaptive Gait Patterns for Quadruped Robot with CPG Network Including Motor Dynamic Model. Electrical Engineering in Japan 155(1), 2148–2154 (2006); Translated from Denki Gakkai Ronbunshi 124-C(10) (2004)
3. Stoeter, S.A., Papanikolopoulos, N.: Kinematic Motion Model for Jumping Scout Robots. IEEE Transactions on Robotics 22(2), 398–403 (2006)
4. Kuniyoshia, Y., Ohmuraa, Y., Teradaa, K., Nagakuboc, A., Eitokua, S., Yamamotob, T.: Embodied basis of invariant features in execution and perception of whole-body dynamic actions - knacks and focuses of Roll-and-Rise motion. Robotics and Autonomous Systems 48, 189–201 (2004)
5. Krause, E.F.: Taxicab Geometry. Dover Publications, Mineola (1987)
6. Touretzky, D.S., Ethan, J., Tira-Thompson, E.J.: Tekkotsu: a Sony AIBO application development framework. The Neuromorphic Engineer 1(2), 12 (2004)

Pareto-Optimal Collaborative Defensive Player Positioning in Simulated Soccer

Vadim Kyrylov[1] and Eddie Hou[2]

[1] Rogers State University; Claremore, OK 74017 USA
vkyrylov@sfu.ca
[2] Simon Fraser University – Surrey; Surrey, British Columbia V3T 0A3 Canada

Abstract. The ability by the simulated soccer player to make rational decisions about moving without ball is a critical factor of success. In this study the focus is placed on the defensive situation, when the ball is controlled by the opponent team in 2D simulated soccer. Methods for finding good defensive positions by the robotic soccer players have been investigated by some RoboCup scholars. Although soccer teams using these methods have proved to be reasonably good, the collaboration issue in defense has been overlooked. In this paper, we demonstrate that collaboration in defense yields better results. In doing so, we treat optimal defensive positioning as a multi-criteria assignment problem and propose a systematic approach for solving it. Besides achieving better performance, this makes it possible to gracefully balance the costs and rewards involved in defensive positioning.

1 Introduction

In real-life soccer game and in simulated soccer likewise players must follow some plan. This plan implements the team strategy and requires collaborative effort in order to attain common goals. In doing so, the only thing that soccer players who are not directly controlling the ball can do is moving to some position. With the total of 22 players, an average player is spending less than 10 percent of the total time on intercepting or handling the ball; the rest accounts to moving somewhere without the ball while not trying to intercept it. This implies the crucial importance of addressing rational positioning without the ball. Thus any improvement in the player behavior would presumably have great impact on the whole game.

In our early study on optimized soccer player positioning in offensive situations, i.e. when the ball is controlled by own team, we proposed a method for determining best positions [1]. Because that required taking into account several optimality criteria, we were using the Pareto optimality principle. Now we concentrate on defensive situations, when the ball is possessed by the opponents. As in the defense players are pursuing different objectives, we expect that the approach to choosing best positions by players should be also different. Still we will be using the Pareto optimality principle as a universal tool for balancing risks and rewards in multi-criteria optimization. Player collaboration is one more special issue that left unaddressed in our previous work. Now we want to propose a mathematical model of optimal player collaboration in defense and show the potential gains.

J. Baltes et al. (Eds.): RoboCup 2009, LNAI 5949, pp. 179–191, 2010.

The objective of this paper is to provide a complete solution to the rational positioning problem for simulated soccer (if taken together with our previous study on offensive positioning).

In particular, we want to improve defensive positioning and to measure improvement. Our method boils down to the adjustment of the default position calculated using very general information about the situation on the field such as ball state and the defender's home position. The adjustment of this default position is based on two ideas. First, we propose that this adjustment should be made with some predicted situation on the field in the player's 'mind'. For this purpose, we extend the concept of the prediction time horizon from our earlier work [1]. Second, given the time horizon, we want to optimize the individual player movements with respect to the global criteria that reflect the team success rather than individual performance of the soccer player. In doing so, we propose a set of optimality criteria and develop an algorithm for finding near-optimal solution. We also measure the performance gain from the proposed methods.

2 Defensive Player Positioning in the Simulated Soccer

Real-life soccer provides some clues for the simulated version. The major objective of soccer player positioning in defense is repelling the attack by the opponent team and creating conditions for launching own attack. However, this requires coordinated effort because individual players acting by themselves are unable to accomplish this goal. Normally the fastest to the ball soccer player in the defending team is trying to approach the opponent player who controls the ball, thus forcing him prematurely trying to score the goal or pass the ball to some teammate. Therefore, the objective of the rest players on the defending side is either to block the way of the ball to own goal or to prevent the opponent players from receiving a pass, or create difficulties for further handling the ball if such pass had occurred. To accomplish this, each available defending player moves to a suitable position on the soccer field near each threatening opponent player. In the professional soccer literature, this is referred to as "marking" and "covering" [2, 3]. Coordination is necessary to guarantee that each potentially dangerous opponent is taken care of and none of the opponents is tended by two or more players because of the limited team size.

We want to implement this rational human player behavior in the RoboCup simulated soccer.

The RoboCup scholars have developed a few methods for player positioning. A good overview can be found in [4]; we have provided some extra details in our previous paper [1]. We deliberately excluded machine learning approach to soccer player positioning investigated by some scholars (e.g. [5, 6]). Some of these methods do not treat positioning as a standalone player skill, which makes theoretical comparisons difficult. More difficulties arise while trying to elicit meaningful decision making rules and especially address the convergence issue of learned rules to optimal decision making algorithms based on explicitly formulated principles.

One of major requirements of positioning is that player behavior must be persistent over several simulation cycles; player is not supposed changing its mind in each cycle, anyway. This implies that the true intelligent soccer player should keep some aspired position in mind that it persistently should be moving to. This position changes substantially if only the situation in the game also changes substantially.

We believe that using a two-layer control structure makes it easier to explain how this desired position should be calculated.

On the higher level, some *default* position is determined based on rather general considerations such as the player role in the team formation, ball state vector, and game situation (attack, scoring the goal, defense). Methods for determining this default position by different authors somewhat differ, but their common feature is that they do not take into account the detail situation on the field such as the state vectors of players. These are some indications that in some RoboCup teams these details are taken into account somehow by adjusting the default position with respect to the surrounding players; yet no systematic approach has been published. One exception is [4], where a mathematical model for player positioning based on the Voronoi diagrams was proposed. The shortcoming of this approach, however, is that the relationship between soccer tactics and the proposed method is not apparent and therefore it is difficult to implement. Moreover, the authors in [4] concentrated on offensive positioning of players.

Here is a typical example how the high-level positioning problem both in attach and defense could be approached. This is a generalization of the ideas originally proposed in two RoboCup team descriptions [7, 8]; both teams had won top places in the world competitions.

Let the team formation prescribe a specific fixed *home* position on the field for each player. This position is reflecting the players' role (e.g. right-wing attacker, central defender, and so on). At any time, the default player position is determined by the three factors: (1) home position, (2) the current location of the ball, and (3) which side is currently controlling the ball.

Assuming that both goals are lying on x-coordinate axis, the method for calculating the default position (x_i, y_i) of i-th player is given by the formulas:

$$x_i = w*xhome_i + (1-w)*xball + x_i,$$

$$y_i = w*xhome_i + (1-w)*yball,$$

(1)

where w is some weight ($0<w<1$), $(xhome_i, yhome_i)$ the fixed home position of the player; $(xball, yball)$ is the current ball position; and x_i is the fixed individual adjustment of x-coordinate whose sign and value is different for the offensive and defensive situations.

So the default position is changing over time with the ball and maintains relative locations of the players in the formation thus implementing the team strategy. This resembles what the human soccer players are doing, especially if the ball is rather far away; they just move towards the default position. Persistence of player positioning is accomplished by the weight w that translates the ball movement with a reduced impact on the default positions; these changes are continuous over time, anyway. Abrupt changes of the default position occur only when the situation changes from attack to defense.

The decision about whether the current situation is attacking or defensive is made when the ball is rolling freely. Each player determines when and where the ball will be likely intercepted by some player. If this player is a teammate, we have an offensive situation; we addressed this case in [1] earlier. If the ball is going to be intercepted by an opponent player, the situation is defensive; this is exactly the case we are discussing here.

Thus on the higher level of control, only the location of the ball and the home position are taken into account; teammates and opponents are ignored.

On the lower level of control, however, other players are the main factor to consider. In the defensive situation, the individual soccer player is constantly fine tuning his position in the vicinity of the default position. In doing so, the player is taking into account the local situation trying to determine the best position that would most likely lead to preventing the nearby opponent player from receiving the ball passed to him by the teammate or shoot at the goal. Because reaching this optimal position takes some time, persistent actions by the defender are required over several simulation cycles. Obviously, we are interested in making sure that the aspired position is optimal not now, but at the future time when it would have been reached and the defender reaches the ball.

From the very brief RoboCup team descriptions it follows that different scholars have approached this issue in different way; yet a systematic approach has not been developed or at least published yet.

Two critical issues are related to player positioning in defense: (1) choosing the time horizon for predicting the situation on the field and (2) player collaboration.

Choosing the time horizon T for predicting the situation. Obviously the set of all possible future positions for each player is infinite; we want to find the way to limiting the size of this set to some tractable finite subset. In doing so, we refer to the set of all positions currently reachable by player in time T as the *feasible set*. This set is approximately a circle whose radius is the distance that given player can cover in time T. As the player changes his own position, this feasible set is moving accordingly. Making decision by the player about where to go is in fact choosing the best position in this set.

Choosing the right time horizon T is worthy of closer consideration. This time certainly cannot be too large, as we do not know how exactly the opponent team is going to act even in the near future, to say nothing about true unpredictable random factors present in the game that are making long-term predictions useless. On the other hand, too short T makes little sense, as only positions that are very close to current location of the defender would be deemed to be feasible; this may result in the lack of persistence of player behavior. Thus we need to use the greatest possible value of T that still maintains reasonable accuracy of prediction.

In our previous paper [1], we have proposed to set the time horizon T equal to the time T_b remaining until the freely moving ball will be intercepted by some player; in defense, this is a member of the opponent team. It is based on the assumption that while the ball is rolling, the situation on the soccer field could be predicted with reasonably high precision based on the logic of the soccer game. Indeed it is reasonable to assume that the two fastest players to the ball from both teams would be trying to intercept it. The rest players would tend to move to their default positions determined by their role in the team formation and the location of the ball interception point. The experience with real-life soccer proves that unless players do so, their team would be at disadvantage. Thus we have rather solid grounds for predicting the situation while the ball is rolling freely. Note that in an average soccer game the ball is rolling about 90-95 per cent of all the time; so we indeed can make such predictions most of the time.

Further analysis has shown that in many cases this time horizon can be extended by some time T whose precise meaning will be explained later. This T is slightly less that the time needed for the opponent player to further pass the ball to his teammate after the interception. (Indeed, the defender wants to prevent this pass from happening or if it happens to intercept the ball.) So in general, the actual time horizon is $T=T_b+T$, i.e. greater than we proposed in [1]. Thus on the second layer of control in a defensive situation the default positions should be adjusted to better fend off the attack by the future time $t=Now+T_b+T$.

Without such adjustment, default positions create conditions for successfully disrupting the opponent's attack only incidentally. Individual adjusting can make this happen more frequently, thus contributing to the success of the team. We want to implement these behaviors in the simulated soccer game.

Player collaboration. Player positioning in defense substantially differs from positioning in attack by the critical importance of player collaboration. To further explain this feature, consider an example. Figure 1 shows the situation when the red team is about to score the yellow team's goal.

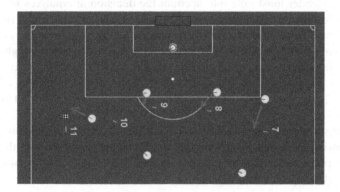

Fig. 1. The red team is attacking. The ball is rolling up the field and could be reached by red #11. Yellow defenders are individually marking closest opponents; thus red #10 is left unattended.

Arrows in the magenta color show the intentions by the yellow team defenders to place themselves to fend off the attack. The ball is rolling up the field and is about to be reached by red #11. The fastest to the ball yellow #4 is also going to intercept it. The yellow defenders #2, #3, and #5 are trying to mark the attackers. They are making these decisions individually without taking into account the decisions made by their teammates, i.e. without collaboration. Thus each defender chooses to mark the nearest opponent. Yellow midfielders #6 and #7 are moving towards their default positions waiting for the outcome of the opponent's attack.

What happens next is shown in Figure 2. Red player #10 is left unattended and is going to receive a pass from red #11 before yellow #4 interferes. Once red #10 receives the pass, it would be able to score the goal or pass the ball to red #9 whose scoring chances would be even better. Thus the red team accomplishes its goal almost for certain.

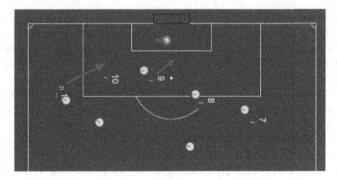

Fig. 2. After about 1 second, red #11 passes the ball to the unattended red #10 before yellow #4 interferes. The red team is likely to score the goal.

The attack by the red team could be more likely fended off if the defenders had collaborated instead of making individual decisions about marking the opponents. Thus each defender must take into account the decision alternatives for the whole team (or at least for a group of closely located teammates) and find a solution that would balance some global optimality criteria. This collaboration is shown in Figure 3. Note that yellow player is going to mark red #10 even though this is not the closest opponent. Yellow midfielder #6 joins the defenders by going to mark red #9 thus contributing to the team effort. The critical condition is that each defending player must know what its teammates are going to do. This requires collaboration.

We want simulated soccer players to exhibit this intelligent collaborative behavior in the defensive situations. So far we have not found any suggestions in the RoboCup community publications that propose a systematic solution to this problem. In what follows, we develop such solution.

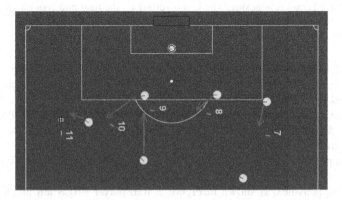

Fig. 3. Defending players have made a collaborative decision that prevents from leaving red #10 unattended. The attack by the red team would likely be fended off.

3 The Proposed Positioning Method in a Nutshell

To contain the complexity of the collaborative defensive positioning problem, we split it into two sub problems: (1) making a collaborative decision for a group of defenders and (2) making the decision about the point to move to by an individual soccer player.

The collaborative decision making concerns optimization of the assignment of n defenders to cover m attackers. This problem could be formulated, as follows:

Let α_{ij} be a decision variable whose value is 1 if i-th defender is assigned to mark j-th opponent ($1 \leq i \leq n$; $1 \leq j \leq m$) and 0 if otherwise:

$$\alpha_{ij} \in \{0,1\}. \tag{2}$$

Thus there are total of $n*m$ such unknown variables. Also let u_{ij} be the 'utility' resulting from the assignment $i \rightarrow j$.

By varying the set $\{\alpha_{ij}\}$, we want to gain the maximal total utility of the collaborative action by n defenders:

$$U(\{\alpha_{ij}\}) = \sum_{i=1}^{n} \sum_{j=1}^{m} \alpha_{ij} u_{ij}. \tag{3}$$

In doing so, besides (2), the following constraints must be observed:

$$\sum_{i=1}^{n} \alpha_{ij} \leq 1, \tag{4}$$

and

$$\sum_{j=1}^{m} \alpha_{ij} \leq 1. \tag{5}$$

Constraints (4) and (5) mean that each defender must be assigned to mark no more than one opponent and each opponent must be marked by no more than one defender.

This problem is referred to as the *Linear Assignment Problem*; its precise solution is delivered by so-called *Hungarian algorithm*, whose complexity is $O((\max(m,n))^4)$. [9].

In the context of our study, however, it is difficult to measure the defender assignment utility with just one criterion. Actually, we have to be balancing rewards and risks; this implies several criteria functions that yet to be specified.

In what follows, we resolve by deriving the criteria functions from soccer tactics and offering an algorithm that would provide an optimal solution of the *multi-criteria assignment problem* that is based on the Pareto optimality principle.

Decision making by the individual soccer player is based on the assignment to take care of the specific attacking opponent. The defending player must find the optimal point to move to by balancing the risks and rewards incurred with such movement. The end point must be reachable within the time horizon T, i.e. before the situation becomes hardly predictable or it is just too late. In what follows, we propose a simple method for finding the optimal point with respect to the limited time balance.

4 Identifying the Feasible Options

While the ball is rolling freely, the defending player can determine the time T_b until the ball will be intercepted and predict the situation rather precisely. In Figure 4 this is the time left until the ball reaches point A. This example shows the two main constraints for the yellow defender #3. The yellow circle is the reachable area in time $T=T_b + T$, where the meaning of T is explained below. The magenta circle is the responsibility area where this player must be staying to maintain the team formation. The center of this area is given by (1). The intersection of these two circles makes the set of feasible positions.

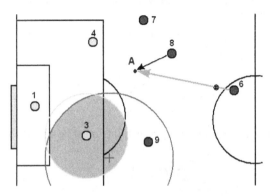

Fig. 4. Feasible alternative positions for the yellow player #3 (shaded area). The ball is rolling freely towards red #8 who is going to intercept it in **A**.

Early studies [3, 7, 10] have shown that, for any given predicted locations of the ball interception point and the opponent player without the ball, it is possible to determine the best location for the defender to mark or cover this opponent. However, the limitation factor of the available time $T_b+ T$ in these studies has been neglected. It turns out that, depending on time balance, we have indeed two different cases shown in Figures 5 and 6, respectively.

One of the new ideas that we claim in this paper is the precise definition of the soccer terms "marking' and "covering" as presented below. The existing professional soccer literature [2, 3] does not even attempt to precisely define these concepts leaving its interpretation up to the reader. As usual, the need in such definitions arises when it comes to mathematical modeling.

Case 1: Marking. If the defender #3 can reach the best marking position before the ball could be sent to the opponent #9 being tended by this defender, the recommended marking point **C** is lying on the line between the anticipated ball interception point **A** and the predicted location **B** of the attacker at time T_b (Figure 5). In this case, T is the time necessary for the ball to reach **C** after it had been passed by red #8 to red #9. Thus the defender himself must reach **C** in time $T_b+ T$.; the shaded feasible area also must be also determined for this extended time horizon. Distance **BC** must be large enough for the yellow player #3 avoiding the interference by red #9 after intercepting the ball.

Case 2: Covering. If the best marking point is beyond the reach of the player in time $T_b+ T$, the recommended defensive position must be lying on the line segment **BG** connecting the predicted position of the opponent and the center of own goal (Figure 6). Instead of marking the opponent we get what is called 'covering' [3]. In this case, T is the time necessary for the ball to reach **B** after it was passed by red #8 to red #9.

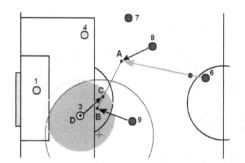

Fig. 5. Yellow defender #3 decides to move from **D** to **C** to block the red opponent #9 from receiving pass made by red #8

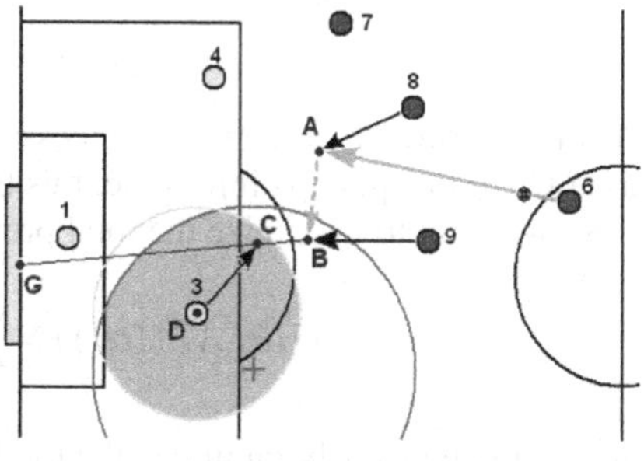

Fig. 6. The time balance would not permit yellow #3 to reach the position from that it could block red #9. Point **C** allows covering the direction to own goal and likely repossessing the ball from red #9 if it receives a pass.

While planning collaborative defensive positioning, the recommended point C_{ij} is calculated for each pair of i-th defender and j-th opponent. Distance **BC** must be small enough for the yellow defender #3 to prevent the red attacker #9 from freely handling the ball. The latter requirement cannot be satisfied if the opponent time advantage is too large. If this is the case for i_k-th defender and j_s-th opponent, pair (i_k, j_s) is eliminated from the set of feasible assignments $\{i \rightarrow j\}$.

5 Criteria for Collaborative Decision Making and the Optimization Algorithm

The utility of assigning n defenders on m attackers is difficult to express it terms of single criterion because there are several conflicting factors. We consider just two: gain and cost. Gain could be measured in terms of the threat prevented by taking care of an opponent player. Cost is the required time to implement this action; on the intuitive level, this time is directly related to risk. We want to maximize total gain and minimize total cost (risk) simultaneously.

We measure the threat imposed by j-th attacking player by using a heuristic criterion that takes into account three factors:

(1) the angular size $\beta(x_j, y_j)$ of own goal from the predicted opponent's location (x_j, y_j);

(2) distance from the opponent's location to own goal $d_{goal}(x_j, y_j)$; and

(3) distance between the ball and the opponent's location $d_{ball}(x_j, y_j)$.

We mean that the threat increases with the first factor and decreases with the other two. Distance to the goal contributes to threat more than distance to the ball. Thus we get:

$$Threat_j = \frac{\sqrt{\beta(x_j, y_j)}}{d_{goal}(x_j, y_j) * \sqrt{d_{ball}(x_j, y_j)}} \tag{6}$$

Cost is measured by the time T_{ij} necessary for i-th defender to reach the recommended point $\mathbf{C}_{ij}$; if this point is infeasible, this time is set to infinity.

So we want to maximize the total prevented threat:

$$THREAT(\{\alpha_{ij}\}) = \sum_{j=1}^{m} Threat_j \sum_{i=1}^{n} \alpha_{ij}, \tag{7}$$

and to simultaneously minimize the total time expenditure:

$$TIME(\{\alpha_{ij}\}) = \sum_{i=1}^{n} \sum_{j=1}^{m} \alpha_{ij} T_{ij}. \tag{8}$$

Balancing these two optimality criteria requires following the Pareto optimality principle. Unfortunately, it looks like an algorithm to precisely solve this problem has not been developed as yet; all methods that we have found in the literature so far are dealing with special cases only [11, 12, 13]. Thus, assuming so-called preemptive priorities of our criteria, we can use one of these methods [12]. By following this approach, we assume that threat is preemptive over time. So we select the opponents one at a time in the descending order of the anticipated prevented threat. For each unassigned opponent, we find the defender having the minimal time T_{ij} to mark or cover it. The process ends when all available opponents or all defenders are allocated. Thus it is unlikely that the defeat shown in Figure 2 would have ever happened.

The critical assumption in the method described above is that each defender has same world model. If this is indeed the case, each collaborating player would be getting same solution to the multi-criteria assignment problem. However, if the knowledge of the situation is imperfect, the decisions made by different defenders would not necessarily match, thus disrupting the collaboration. This is exactly the case with 2D RoboCup simulated soccer. In our experiments we have found that the differences in the decisions made by different players are frequently incompatible; thus no collaboration is possible. This difficulty can be overcome by leaving decision making up to one player whose world model is the best. Implementing this decision is all what is left up to the rest players. In the defensive situations the player with the best world model is normally the goalie. This player assumes the role of the coordinator and broadcasts the decisions to teammates via the simulated aural communication channel. So once i-th defender received from the coordinator the assignment $i \rightarrow j$, this defender starts moving straight towards the recommended position $\mathbf{C}_{ij}$. The obvious cost that we have to pay for this collaboration is the one-cycle delay in decision making.

6 Experimental Results and Conclusion

We have conducted experiments with the purpose to estimate the sole contribution of the proposed method for the lower-level optimized player positioning compared with only strategic, higher-level positioning.

Measuring the player performance using existing RoboCup teams is difficult because new features always require careful fine tuning with the existing ones. For this reason, we decided to compare two very basic simulated soccer teams. The only difference was that the experimental team had player positioning on two levels and the control team just on one level. Players in both teams had rather good direct ball passing and goal scoring skills and no dribbling or holding the ball at all. Thus any player, once gaining the control of the ball, was forced to immediately pass it to some teammate. In this setting, the ball was rolling freely more than 95 per cent of the time, thus providing ideal conditions for evaluating the proposed method.

To further isolate the effects of imperfect sensors, we decided to use *Tao of Soccer*, the simplified soccer simulator with complete information about the world; it is available as the open source project [14]. Using the RoboCup simulator would require prohibitively long running time to sort out the effects of improved player positioning among many ambiguous factors.

The higher-level player positioning was implemented as presented in expression (1), which is similar to used in *UvA Trilearn* [7]; this method proved to be reasonably good indeed.

Because players in the control team were moving to the reference positions without any fine tuning, the interception of opponent ball passing by defenders was occurring as a matter of chance. In the experimental team, rather, players were adjusting their location about the default positions as described in this paper. As the result, they were purposefully attempting to mark the opponents or cover own goal.

Like in our previous paper [1], the sole purpose of this experiment was to demonstrate that using smart defensive positioning is in principle better that using just default positions calculated as (1). Precise measurements of performance, however, are only possible with full set of advanced features implemented in the artificial players.

The team performance was measured by the score difference. Figure 7 shows the histogram based on 100 games each 10 minutes long.

The experimental team has the average score greater by 1.63; however, this difference has too low statistical significance, as with only 100 games, the distribution appears to have too long tails. A little more cautious claim about the score difference, however, is significant. Indeed, at 95% confidence level, the experimental team with advanced defensive positioning scores on average at least 1.3 extra goals per game. This difference is indeed worthy of trying to achieve!

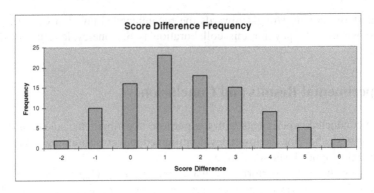

Fig. 7. A histogram of the score difference in 100 games

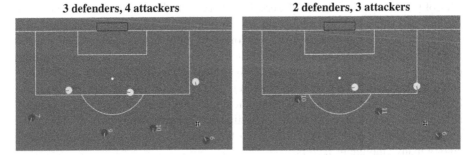

Fig. 8. Two custom-designed scenarios with defenders outnumbered by attackers

Because the defensive situations in which our improvements are showing up happen rather infrequently, we also created two scenarios in that proper marking and covering opponents is very critical for the defending team. Figure 8 shows two situations where defenders are outnumbered by attackers who are about to score the goal. In these experiments, no goalie was at the goal line for the defending team; the ball was randomly placed in front of an attacker, with randomly positioned attackers. The situation, whose duration was 200 cycles (10 seconds), was repeated 500 times. In the end of each repetition the statistics on the goal scored was gathered.

The results for the control and experimental team are shown in Table 1.

Table 1. Experimental results with special scenarios

Defending team	Average goals scored	Standard deviation
3 defenders, 4 attackers		
Control	0.400	0.015
Experimental	0.304	0.015
2 defenders, 3 attackers		
Control	0.568	0.016
Experimental	0.280	0.014

These results indicate that there is a statistically significant (at more than 99% confidence) improvement in the team performance. Indeed, the modifications made to the experimental team reduced the probability of scoring the goal by the attackers 1.3-2.0 times. Best results are achieved when the numerical superiority of attackers is greater.

This proves the viability of the proposed method for defensive positioning in simulated soccer.

References

1. Kyrylov, V., Razykov, S.: Pareto-Optimal Offensive Player Positioning in Simulated Soccer. In: Visser, U., Ribeiro, F., Ohashi, T., Dellaert, F. (eds.) RoboCup 2007. LNCS (LNAI), vol. 5001, pp. 228–237. Springer, Heidelberg (2008)
2. Beim, G.: Principles of Modern Soccer. Houghton Mifflin Company, Boston (1977)
3. Bangsbo, J., Peitersen, B.: Defensive Soccer Tactics. Human Kinetics, Champaign (2002)
4. Dashti, H.T., Kamali, S., Aghaeepour, N.: Positioning in Robots Soccer. In: Lima, P. (ed.) Robotic Soccer, Vienna, Austria, pp. 29–44. I-Tech Education and Publishing (2007)
5. Andou, T.: Refinement of Soccer Agents' Positions Using Reinforcement Learning. In: Kitano, H. (ed.) RoboCup 1997. LNCS, vol. 1395, pp. 373–388. Springer, Heidelberg (1998)
6. Riedmiller, M., Gabel, T., Knabe, J., Strasdat, H.: Brainstormers 2D – Team Description 2005. In: Bredenfeld, A., Jacoff, A., Noda, I., Takahashi, Y. (eds.) RoboCup 2005: Robot Soccer World Cup IX. Springer, Heidelberg (2006) (CD Supplement)
7. Kok, J., de Boer, R., Vlassis, N., Groen, F.: UvA Trilearn 2002 Team Description. In: Kaminka, G.A., Lima, P.U., Rojas, R. (eds.) RoboCup 2002. LNCS (LNAI), vol. 2752, p. 549. Springer, Heidelberg (2003)
8. Reis, L.P., Lau, N.: FC Portugal Team Description: RoboCup 2000 Simulation League Champion. In: Stone, P., Balch, T., Kraetzschmar, G.K. (eds.) RoboCup 2000. LNCS (LNAI), vol. 2019, pp. 29–40. Springer, Heidelberg (2001)
9. Kocay, W., Kremer, D.: Graphs, Algorithms, and Optimization. Chapman & Hall/CRC Press, Boca Raton (2006)
10. Hunter, M., Kostiadis, K., Hu, H.: A Behaviour-based Approach to Position Selection for Simulated Soccer Agents. In: 1st European Workshop on RoboCup, Amsterdam, 28 May - 2 June (2000)
11. Larichev, O.I., Kozhukharov, A.N.: Multiple Criteria Assignment problem: Combining the Collective Criterion with Individual Preferences. Matematiques et Sciences Humaines 68, 63–77 (1979)
12. Wilamowsky, Y., Epstein, S., Dickman, B.: Multicriteria assignment problems with preemptive priorities. Mid-Atlantic Journal of Business 30(1), 113–120 (1994)
13. Scarelli, A., Narula, S.C.: A Multicriteria Assignment Problem. Journal of Multi-Criteria Decision Analysis 11(2), 65–74 (2003)
14. Zhan, Y.: Tao of Soccer: An Open Source project (2006), https://sourceforge.net/projects/soccer/

A Novel Camera Parameters Auto-adjusting Method Based on Image Entropy

Huimin Lu, Hui Zhang, Shaowu Yang, and Zhiqiang Zheng

College of Mechatronics and Automation,
National University of Defense Technology, Changsha, China
{lhmnew,huizhang_nudt,ysw_nudt,zqzheng}@nudt.edu.cn

Abstract. How to make vision system work robustly under dynamic light conditions is still a challenging research focus in robot vision community. In this paper, a novel camera parameters auto-adjusting method based on image entropy is proposed. Firstly image entropy is defined and its relationship with camera parameters is verified by experiments. Then how to optimize the camera parameters based on image entropy is proposed to make robot vision adaptive to the different light conditions. The algorithm is tested using the omnidirectional vision system in indoor RoboCup Middle Size League environment and outdoor RoboCup-like environment, and the results show that our method is effective and color constancy to some extent can be achieved.

1 Introduction

How to make vision system work robustly under dynamic light conditions is still a challenging research focus in computer vision/robot vision community [1]. There are mainly three approaches to achieve this goal. The first one is to process and transform the images to achieve some kind of constancy, such as color constancy by Retinex algorithm [2]. The second one is to analyze and understand the images robustly, such as designing adaptive or robust object recognition algorithms [3, 4]. These two approaches have attracted lots of researchers' interest, and lots of progresses have been achieved. The third one is always ignored by researchers, which is to output the images to describe the real scene as consistently as possible in different light conditions by auto-adjusting the camera parameters (in this paper, camera parameters are the image acquisition parameters, not the intrinsic or extrinsic parameters in camera calibration). In the digital still cameras and consumer video cameras, many parameters adjusting mechanisms have been developed to achieve good imaging results, such as auto exposure by changing the iris or the shutter time [5], auto white balance [6], and auto focus [7]. In some special multiple slope response cameras, the response curve can be adjusted to adapt the dynamic response range to different light conditions by automatic exposure control [8]. But these methods are always on the camera hardware level, and we can not do these things or make modification on most cameras used in robot vision system except some special hardware-support cameras.

The RoboCup Middle Size League (MSL) competition is a standard real-world test bed for robot vision and other related research subjects. It is still a color-coded environment, though some great changes have taken place in the latest competition rules,

J. Baltes et al. (Eds.): RoboCup 2009, LNAI 5949, pp. 192–203, 2010.

such as replacing the blue/yellow goals with white goal nets, no color flag post any more. The final goal of RoboCup is that robot soccer team defeats human champion, so robots will have to be able to play competition in highly dynamic light conditions even in outdoor environment. So designing robust vision system to recognize color-coded objects is a research focus in RoboCup community. Besides adaptive color segmentation methods [3], color online learning algorithms [9, 10], and object recognition methods independent on color information [11, 12], several researchers also have tried to apply the third approach to help achieving the robustness of vision sensors. Paper [13] defined the camera parameters adjustment as an optimization problem, and used the genetic meta-heuristic algorithm to solve it by minimizing the distance between the color values of some image areas and the theoretic values in color space. The theoretic color values were used as referenced values, so the effect from illumination could be eliminated, but the special image areas needed to be selected manually by users in the method. Paper [14] used a set of PID controllers to modify the camera parameters like gain, iris, and two white balance channels according to the changes of a white reference color always visible in the omnidirectional vision system. Paper [15] adjusted the shutter time by designing a PI controller to modify the reference green field color to be the desired color values.

In this paper, we try to use the third approach to achieve the robustness and adaptability of camera's output under different light conditions. We also want to provide an objective method for vision/camera setup by this research, for the cameras are usually set manually according to user's subjective experiences when coming to a totally new working environment. We define the image entropy as the optimizing goal of camera parameters adjustment, and propose a novel camera parameters auto-adjusting technique based on image entropy. We use our omnidirectional vision systems [16] and the RoboCup MSL environment as the test bed for our algorithm.

In the following part, we will firstly present the definition of image entropy and verify that the image entropy is valid to represent the image quality for image processing and to indicate that whether the camera parameters are properly set by experiments in section 2, and then propose how to auto-adjust the camera parameters based on image entropy to adapt to the different illumination in section 3. The experiment results in indoor and outdoor environment and the discussions will be presented in section 4 and section 5 respectively. The conclusion will be given in section 6 finally.

2   Image Entropy and Its Relationship with Camera Parameters

The setting of camera parameters affects the quality of outputting images greatly. Taking the cameras of our omnidirectional vision system as the example, only exposure time and gain can be adjusted (auto white balance has been realized in the camera, so we don't consider white balance). Several images captured under different parameters are shown in figure 1. The quality of images in figure 1(a) and (c) are much worse than that in figure 1(b), because they are less-exposed and over-exposed respectively, and the image in figure 1(b) is well exposed. The two images in figure 1(a) and (c) can't represent the environments well, and we can say that the information content in these two images is less than that in figure 1(b). So both less-exposure and over-exposure will cause the loss of image information [17].

According to Shannon's information theory, the information content can be measured by entropy, and entropy increases with the information content. So we use image entropy to measure the image quality, and we also assume that the entropy of outputting images can indicate that whether the camera parameters are properly set. In the following part of this section, we will firstly present the definition of the image entropy, and then verify this assumption by analyzing the distribution of image entropy with different camera parameters.

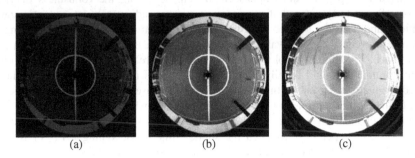

(a)   (b)   (c)

Fig. 1. The images captured by our omnidirectional vision system with different exposure time. The gain is always 18. (a) The exposure time is 5ms. (b) The exposure time is 18ms. (c) The exposure time is 40ms.

2.1 The Definition of Image Entropy

We use Shannon's entropy to define the image entropy. So the image entropy can be expressed as follows:

$$Entropy = -\sum_{i=0}^{L-1} p_{Ri} \log p_{Ri} - \sum_{i=0}^{L-1} p_{Gi} \log p_{Gi} - \sum_{i=0}^{L-1} p_{Bi} \log p_{Bi} \tag{1}$$

Where $L = 256$ is the discrete level of RGB color channels, and p_{Ri}, p_{Gi}, p_{Bi} are the probability of color Ri, Gi, Bi existing in the image, and they can be replaced with frequency approximately and then calculated according to the histogram distribution of RGB color channels.

According to the definition in (1), $0 = Min(Entropy) \leq Entropy \leq Max(Entropy) =$

$-3 * \sum_{i=0}^{256-1} \frac{1}{256} \log \frac{1}{256} = 16.6355$, and the entropy will increase monotonously with the

degree of average distribution of color values.

2.2 Image Entropy's Relationship with Camera Parameters

We capture a series of panoramic images by using our omnidirectional vision system with different exposure time and gain in indoor environment and outdoor environment, and then calculate the image entropy according to equation (1) to see how image entropy varies with camera parameters. The indoor environment is a standard RoboCup MSL field with dimension of 18m*12m, but the illumination is not only determined by the artificial lights, but also influenced greatly by natural light through lots of windows. The outdoor environment includes two blue patches and several components of the indoor environment such as a piece of green carpet, two orange balls and black obstacles. All the experiments of this paper are performed in these two environments. Furthermore, because the illumination in two environments is totally different and the dynamic response range of our cameras is limited, so we use two omnidirectional vision systems (two robots) with different iris setting (the iris can be adjusted only manually) in the two environments.

In the experiment of indoor environment, the range of exposure time is from 5ms to 40ms and the range of gain is from 5 to 22. The experiment time of this section is evening, so the illumination is not affected by natural light. In the experiment of outdoor environment, the range of exposure time is from 1ms to 22ms and the range of gain is from 1 to 22. The weather is cloudy, and the experiment time is midday. The minimal adjusting step of the two parameters is 1ms and 1 respectively. We captured one image with each group of parameters. The image entropies changing with different camera parameters are shown in figure 2 and figure 3 in the two experiments.

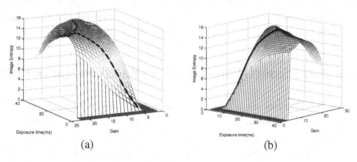

(a)   (b)

Fig. 2. The image entropies changing with different exposure time and gain in indoor environment. (a) and (b) are the same result viewed from two different view angles.

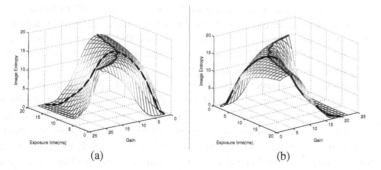

(a)   (b)

Fig. 3. The image entropies changing with different exposure time and gain in outdoor environment. (a) and (b) are the same result viewed from two different view angles.

From figure 2 and 3, we can find that the manner in which the image entropy varies with camera parameters is the same in the two experiments, and there is ridge curve (the blue curve in figure 2 and 3). Along the ridge curve, the image entropies are almost the same in each experiment, and there is not obvious maximal value. So which image entropy along the ridge curve indicates the best image, or whether all the images related to the image entropy along the ridge curve are good?

For the images are used to processed and analyzed to realize object recognition, self-localization or other robot vision task, we test the quality of images by using the same color calibration result learned from one image [18] corresponding to a certain entropy on the ridge curve to segment the images corresponding to all the entropies along the ridge curve and detect the white line points using the algorithm proposed in paper [4]. The typical images along the ridge curve and the processing results in the two experiments are demonstrated in figure 4 and figure 5.

As shown in the two figures, the images can be well segmented by the same color calibration result in each experiment, and object recognition can be realized successfully for soccer robots. The same processing results are achieved in all the other images related to the image entropy along the ridge curve. So all these images are good for robot vision, and there is some kind of color constancy in these images, though they are captured under different camera parameters. It also means that all the setting of exposure times and gains corresponding to the image entropy along the ridge curve are acceptable for robot vision. So the assumption is verified that the image entropy can indicate that whether the camera parameters are properly set.

3   Camera Parameters Auto-adjusting Based on Image Entropy

According to the experiments and analysis in last section, image entropy can indicate the image quality for robot vision and that whether the camera parameters are properly set, so camera parameters adjustment can be defined as an optimization problem, and image entropy can be used as optimizing goal. But as is shown in figure 2 and 3, the image entropies along the blue ridge curve are almost the same, and it is not easy to search the global optimal solution. Furthermore, camera parameters themselves will affect the performance of vision systems. For example, the real-time ability will decrease as exposure time increases, and the image noise will increase as gain increases. So exposure time and gain themselves have to be taken into account in this optimization problem. But it is difficult to measure the degree of these parameters' effect, so it is almost impossible to add some indicative or constraint function to image entropy directly for the optimization problem.

Considering that the images related to the image entropies along the ridge curve are all good for robot vision, we turn the two-dimension optimization problem to be one-dimension one by defining some searching path. For RoboCup MSL competition is a highly dynamic and color-coded environment, the exposure time and gain should not be too high for soccer robots. So we define the searching path as exposure time=gain (just equal in number value, fo`r the unit of exposure time is ms, and there is no unit for gain) to search the maximal image entropy in this path, and the camera parameters corresponding to the maximal image entropy are best for robot vision in current environment and current light condition. The searching path is shown as the black curve in figure 2 and 3 respectively in indoor environment and outdoor environment. The distributions of image entropy along the path in the two environments are demonstrated in figure 6.

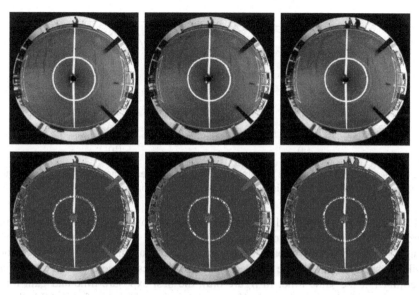

Fig. 4. The typical images along the ridge curve and the processing results in indoor experiment. (top) are the typical images. (bottom) are the processing results, and the red points are the detected white line points. The camera parameters are as follows: (left) exposure time: 34ms, gain: 13. (middle) exposure time: 18ms, gain: 18. (right) exposure time: 14ms, gain: 21.

Fig. 5. The typical images along the ridge curve and the processing results in outdoor experiment. In this experiment, there are not white lines to detect. (top) are the typical images. (bottom) are the processing results. The camera parameters are as follows: (left) exposure time: 17ms, gain: 5. (middle) exposure time: 9ms, gain: 9. (right) exposure time: 2ms, gain: 18.

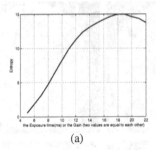
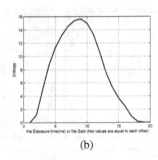

(a)   (b)

Fig. 6. The distribution of image entropy along the defined searching path. (a) The distribution in indoor environment. (b) The distribution in outdoor environment.

From figure 6, a very good property of image entropy can be found that the image entropy will increase monotonously to the peak and then decrease monotonously along the defined searching path. So the global maximal image entropy can be found easily by searching along the defined path, and the best camera parameters are also determined at the same time. In figure 6(a), the best exposure time and gain for the omnidirectional vision system are 18ms and 18 respectively; in figure 6(b), the best exposure time and gain are 9ms and 9 respectively.

According to the special character of omnidirectional vision, the robot itself will be imaged in the central area of the panoramic images. So in the real application, robot can judge that whether it comes into a totally new environment or the illumination changes in the current environment by calculating the mean brightness value on the central part of panoramic image. If the increase of the mean value is higher than a threshold, the robot will consider that the illumination becomes stronger, and the optimization of camera parameters will be run towards the direction that exposure time and gain reduce and along the searching path. Similarly, if the decrease of the mean value is higher than the threshold, the optimization will be run towards the direction that exposure time and gain raise and along the searching path. In our experiment, we set the threshold as 20. In the optimizing process, a new group of parameters will be set into the camera, and then a new image will be captured and the image entropy can be calculated according to equation (1). The new entropy will be compared with the last one to check whether the maximal entropy has reached. This iteration will go on and on until the maximal entropy is reached. About how to choose new parameters, the technique of varying optimizing step could be used to accelerate the optimization process. When the current entropy is not far from $Max(Entropy)$, the optimizing step could be 1, which means that the change of exposure time is 1ms and the change of gain is 1. When the current entropy is far from $Max(Entropy)$, the optimizing step could be 2 or 3.

The searching path can be changed according to different requirement about the vision system in different application. In some cases, the signal noise ratio of image is required to be high and the real-time performance is not necessary, so the searching path could be exposure time= α *gain (also just equal in number value), and $\alpha > 1$. In some other application, the camera is required to output image as soon as possible and the image noise is not restricted too much, so the searching path could be exposure time= α *gain (also equal in number value), and $\alpha < 1$.

4   The Experimental Results under Different Light Conditions

In this section, we test our novel camera parameters auto-adjusting algorithm proposed in last section under different light conditions in indoor environment and outdoor environment respectively. We verify that whether the camera parameters are properly set successfully by processing the images using the same color calibration result learned in the experiments of section 2. We also evaluate the robot's self-localization based on omnidirectional vision after the camera parameters are optimized in different illumination.

4.1   The Experiment in Indoor Environment

In this experiment, the weather is cloudy, and the experiment time is midday, so the illumination is influenced by artificial and natural light. We also turn off some lamps gradually to change the illumination. We use the color calibration result in the indoor experiment of section 2 to process the images for soccer robots. The outputting image and the processing result are shown in figure 7 when camera is set with the best parameters in section 2. The image is over-exposed, and processing result is terrible. After the parameters have been optimized by our method, the outputting image and the processing result are demonstrated in figure 8(a) and (b). The distribution of image entropy along the searching path is shown in figure 8(c). The optimal exposure time is 14ms and gain is 14, so the image is well-exposed, and the processing result is also good. When the illumination changes gradually, the similar results are achieved.

4.2   The Experiment in Outdoor Environment

In this experiment, the weather is sunny, and the experiment time is from midday to dusk, so the illumination is from bright to dark decided by natural light. We also use the same color calibration result in the outdoor experiment of section 2 to process the images for soccer robots. The outputting image and the processing result are shown in figure 9 when camera is set with the best parameters in section 2. The image is also over-exposed, and processing result is unacceptable for robot vision. After the parameters have been optimized, the outputting image and the processing result are demonstrated in figure 10(a) and (b). The distribution of image entropy along the searching path is shown in figure 10(c). The optimal exposure time is 3ms and gain is 3, so the image is well-exposed, and the processing result is also good. When the experiment is run in different time from midday to dusk, all images can be well-exposed and well processed after the camera parameters have been optimized.

4.3   Comparison of Robot's Localization under Different Illumination

In this experiment, we compare the robot's self-localization results based on omnidirectional vision with optimized camera parameters in indoor RoboCup MSL standard

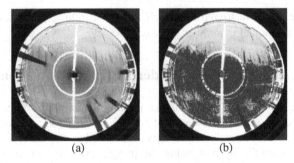

Fig. 7. (a) The outputting image when the camera parameters have not been optimized in indoor environment. The best parameters in section 2 are used. (b) The processing result.

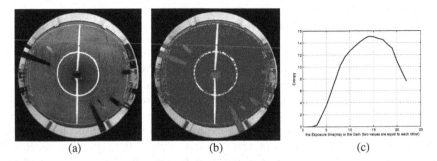

Fig. 8. (a) The outputting image after camera parameters have been optimized. (b) The processing result. (c) The distribution of image entropy along the searching path.

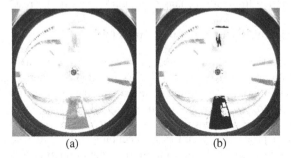

Fig. 9. (a) The outputting image when the camera parameters have not been optimized in outdoor environment. The best parameters in section 2 are used. (b) The processing result.

environment under very different illumination. The first light condition is the same as that in the indoor experiment of section 2. The second one is that the illumination is affected by strong sun's rays in a sunny day, and the optimal exposure time and gain are 12ms and 12 respectively. The robot's self-localization results by the method proposed in [19] under these two light conditions are demonstrated in figure 11. In this experiment, the robot is pushed by human to follow some straight traces on the

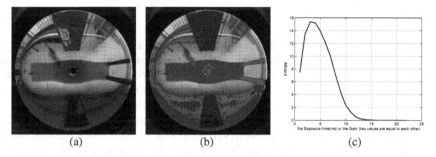

Fig. 10. (a) The outputting image after camera parameters have been optimized. (b) The processing result. (c) The distribution of image entropy along the searching path.

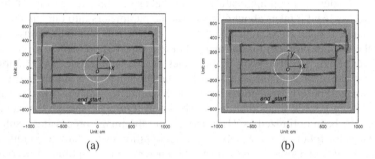

Fig. 11. (a) The robot's localization result when the illumination is not affected by natural light. (b) The robot's localization result when the illumination is affected greatly by strong sun's rays.

Table 1. The statistic of robot's self-localization error under different illumination. In this table, x , y , θ are the self-localization coordinate related to the location x, y and orientation.

	Under the first light condition			Under the second light condition		
	mean error	standard dev	maximal error	mean error	standard dev	maximal error
x (cm)	5.907	7.334	30.724	6.416	12.431	95.396
y (cm)	5.967	7.117	35.595	5.544	7.381	33.063
θ (rad)	0.044	0.052	0.286	0.067	0.093	0.580

field shown as black lines in figure 11. The statistic of localization errors is shown in Table 1. The robot can achieve good localization results with the same color calibration result even under very different light conditions, though sometimes the effect from sun's rays is so strong that the maximal localization error under the second light condition is much larger. This experiment also verifies that our camera parameters adjusting method is effective.

5 Discussion

According to the analysis and the experimental results in above sections, our method can make the camera's output adaptive to different light conditions, so the images can describe the real world as consistently as possible. Our method also provides an objective camera setup technique when robots come into a totally new environment, so users don't need to adjust camera parameters manually according to experience.

Although only exposure time and gain are adjusted in our experiments, our method can be extended to adjust more parameters (if supported by hardware). Besides omnidirectional vision, our method can also be applied in other vision systems, but maybe some special object should be recognized and then used as reference image area to judge whether the illumination changes for camera parameters auto-adjustment.

About the real-time performance of our method, for the light condition will not change too suddenly in real application, it only takes several cycles to finish the optimizing process. And it takes about 40ms to set the parameters into our camera for one time. So camera parameters adjustment can be finished in maximal several hundred ms, and there is not problem for our method in real-time requirement.

However, there are still some deficiencies in our algorithm. For example, our method can not deal with the situation that the illumination is highly not uniform. Because image entropy is a global appearance feature for image, it may be not the best optimizing goal in this situation. As shown in figure 12, though the camera parameters have been optimized as 21ms and 21, but the image processing result is still unacceptable for robot vision. Maybe object recognition or tracking technique should be integrated in our method, and camera parameters can be optimized according to local image features near the object area on the images.

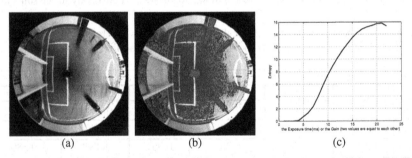

(a)   (b)   (c)

Fig. 12. (a) The outputting image with optimal parameters when illumination is highly not uniform. (b) The processing result. (c) The distribution of image entropy along searching path.

6 Conclusion

In this paper, a novel camera parameters auto-adjusting method is proposed to make the output of robot vision adaptive to different light conditions. Firstly we present the definition of image entropy, and use image entropy as optimizing goal for the optimization problem of camera parameters after verifying that image entropy can indicate whether the camera parameters are properly set by experiments. Then how to optimize the camera parameters based on image entropy is proposed to adapt to different illumination. The experiments in indoor RoboCup MSL standard field and outdoor RoboCup-like environment show that our algorithm is effective and the color constancy to some extent in the output of vision systems can be achieved.

References

1. Mayer, G., Utz, H., Kraetzschmar, G.K.: Playing Robot Soccer under Natural Light: A Case Study. In: Polani, D., Browning, B., Bonarini, A., Yoshida, K. (eds.) RoboCup 2003. LNCS (LNAI), vol. 3020, pp. 238–249. Springer, Heidelberg (2004)
2. Mayer, G., Utz, H., Kraetzschmar, G.K.: Towards Autonomous Vision Self-calibration for Soccer Robots. In: Proceedings of the 2002 IEEE/RSJ IROS, pp. 214–219 (2002)
3. Gönner, C., Rous, M., Kraiss, K.: Real-Time Adaptive Colour Segmentation for the RoboCup Middle Size League. In: Nardi, D., Riedmiller, M., Sammut, C., Santos-Victor, J. (eds.) RoboCup 2004. LNCS (LNAI), vol. 3276, pp. 402–409. Springer, Heidelberg (2005)
4. Lu, H., Zheng, Z., et al.: A Robust Object Recognition Method for Soccer Robots. In: Proceedings of the 7th World Congress on Intelligent Control and Automation, pp. 1645–1650 (2008)
5. Kuno, T., Sugiura, H., Matoba, N.: A New Automatic Exposure System for Digital Still Cameras. IEEE Transactions on Consumer Electronics 44, 192–199 (1998)
6. Chikane, V., Fuh, C.: Automatic White Balance for Digital Still Cameras. Journal of Information Science and Engineering 22, 497–509 (2006)
7. Chern, N.K., Neow, P.A., Ang Jr., M.H.: Practical Issues in Pixel-Based Autofocusing for Machine Vision. In: Proceedings of the 2001 IEEE ICRA, pp. 2791–2796 (2001)
8. Gooßen, A., Rosenstiel, M., Schulz, S., Grigat, R.-R.: Auto Exposure Control for Multiple-Slope Cameras. In: Campilho, A., Kamel, M.S. (eds.) ICIAR 2008. LNCS, vol. 5112, pp. 305–314. Springer, Heidelberg (2008)
9. Anzani, F., Bosisio, D., Matteucci, M., Sorrenti, D.G.: On-Line Color Calibration in Non-Stationary Environments. In: Bredenfeld, A., Jacoff, A., Noda, I., Takahashi, Y. (eds.) RoboCup 2005. LNCS (LNAI), vol. 4020, pp. 396–407. Springer, Heidelberg (2006)
10. Heinemann, P., Sehnke, F., Streichert, F., Zell, A.: Towards a Calibration-Free Robot: The ACT Algorithm for Automatic Online Color Training. In: Lakemeyer, G., Sklar, E., Sorrenti, D.G., Takahashi, T. (eds.) RoboCup 2006. LNCS (LNAI), vol. 4434, pp. 363–370. Springer, Heidelberg (2007)
11. Hanek, R., Schmitt, T., Buck, S., Beetz, M.: Towards RoboCup without Color Labeling. In: Kaminka, G.A., Lima, P.U., Rojas, R. (eds.) RoboCup 2002. LNCS (LNAI), vol. 2752, pp. 179–194. Springer, Heidelberg (2003)
12. Treptow, A., Zell, A.: Real-time object tracking for soccer-robots without color information. Robotics and Autonomous Systems 48, 41–48 (2004)
13. Grillo, E., Matteucci, M., Sorrenti, D.G.: Getting the most from your color camera in a color-coded world. In: Nardi, D., Riedmiller, M., Sammut, C., Santos-Victor, J. (eds.) RoboCup 2004. LNCS (LNAI), vol. 3276, pp. 221–235. Springer, Heidelberg (2005)
14. Takahashi, Y., Nowak, W., Wisspeintner, T.: Adaptive Recognition of Color-Coded Objects in Indoor and Outdoor Environments. In: Visser, U., Ribeiro, F., Ohashi, T., Dellaert, F. (eds.) RoboCup 2007. LNCS (LNAI), vol. 5001, pp. 65–76. Springer, Heidelberg (2008)
15. Lunenburg, J.J.M., Ven, G.V.D.: Tech United Team Description. In: RoboCup 2008, Suzhou, CD-ROM (2008)
16. Lu, H., Zhang, H., Xiao, J., Liu, F., Zheng, Z.: Arbitrary ball recognition based on omni-directional vision for soccer robots. In: Iocchi, L., Matsubara, H., Weitzenfeld, A., Zhou, C. (eds.) RoboCup 2008. LNCS (LNAI), vol. 5399, pp. 133–144. Springer, Heidelberg (2009)
17. Goshtasby, A.A.: Fusion of Multi-exposure images. Image and Vision Computing 23, 611–618 (2005)
18. Liu, F., Lu, H., Zheng, Z.: A Modified Color Look-Up Table Segmentation Method for Robot Soccer. In: Proceedings of the 4th IEEE LARS/COMRob 2007 (2007)
19. Zhang, H., Lu, H., et al.: NuBot Team Description Paper 2008. In: RoboCup 2008 Suzhou, CD-ROM (2008)

Object Recognition with Statistically Independent Features: A Model Inspired by the Primate Visual Cortex

Mohsen Malmir and Saeed Shiry

Computer Engineering Department, Amirkabir University of Technology, 424 Hafez Ave.,
Tehran, Iran
{mmalmir,Shiry}@aut.ac.ir

Abstract. Human can perform object recognition with high accuracy under a
variety of object rotations and translations. The structure and function of the
visual cortex has inspired many models for invariant object recognition. In this
paper, we propose a hierarchical model for object recognition based on the two
well-known properties of the visual cortex neurons: invariant responses to
stimulus transformations and redundancy reduction. We used the trace learning
rule to provide the neurons in the model with invariant responses to object
transformations. In hierarchical neural networks, neighboring neurons are tuned
to similar features because their receptive fields in the image overlap. This similarity results in a form of redundancy in neuronal responses. We used a variant
of divisive normalization mechanism to increase the efficiency of responses of
neurons in the model. Results of experiments demonstrate the high recognition
rates of the proposed model.

Keywords: Invariant Object Recognition, Visual Cortex, Redundancy Reduction, Trace Learning Rule.

1 Introduction

One of the major challenges for computer vision is to recognize objects from different
view-points and distances. The need for automatic object recognition in a wide range
of applications including robotics has lead to an increase in the amount of research on
the development of computer based object recognition systems. Several methods have
been proposed that can recognize objects with high accuracy. However these methods
have gained limited popularity because their demands do not match the real world
applications. One line of research that has attracted much attention in the computer
vision is to implement a model based on the studies of object recognition in the primate vision system. Human can recognize objects under different transformations
with accuracies that exceed almost all of the object recognition systems currently in
use.

In the primate brain, optic nerves enter the primary visual cortex in the occipital
lobe. Data about the image we see goes through several areas in the visual cortex
and its neuronal representation changes from neuronal responses corresponding to
small edges to signals that inform the presence of specific objects in the image. Different areas in the primate visual cortex and their function have inspired many object

J. Baltes et al. (Eds.): RoboCup 2009, LNAI 5949, pp. 204–214, 2010.
© Springer-Verlag Berlin Heidelberg 2010

recognition systems. Perhaps the most influential work in the study of visual cortex is the classic paper of Hubel and Wiesel in which they explained the function of visual cortex as a hierarchy of feature detector neurons [1]. So far, a large amount of information has been collected about the selectivity and connectivity of these neurons. A prominent hypothesis is that the object recognition pathway has a hierarchical structure along which the neurons extract features with increasing complexities [2]. In this hierarchy, neurons in a cascade of areas extract features with different complexities ranging from simple contrast changes to faces and hands [1], [3]. Several authors have explained the work of these neurons in the statistical framework of efficient coding [4], [5].

In this paper, we propose a hierarchical model for object recognition which is inspired directly from the structure of primate visual cortex. We aim to develop a network in which neuronal responses have high information content and are invariant to object transformations. Neurons in the visual cortex exhibit responses that are invariant to object transformation [6], [7]. It is proposed that visual neurons learn their invariant selectivity from sequences of images that contain the transformation of an object. We implement the trace learning rule in the model to provide invariant selectivity to 3D object rotations. Moreover, we used horizontal connections which weights are learned to increase the efficiency of neuronal responses [4], [5], [8]. Results of object recognition on custom images and a standard dataset demonstrate the ability of this model to perform robust object recognition under a variety of object transformations.

2 Literature Review

One of the simplest methods for object recognition is to use correlation based template matching [9]. However, models that use this technique are very sensitive to object transformations and therefore are ineffective for most of the applications. To overcome this limit, component based methods have been proposed that extract object components for recognition [10], [11], [12]. In these models, a trade-off between the selectivity and invariance is unavoidable. For example, histogram based models are very robust to object transformations but cannot differentiate between similar objects from the same category [13]. Methods that use grayscale patches of objects are very selective but cannot recognize transformed views of the same object [10], [12].

Starting with Neocognitron [14], several models have been proposed for invariant object recognition based on the features of visual cortex and were successfully applied to specific objects like faces and cars [10], [12]. These models are mainly based on the idea of Perrett and Oram, who proposed that transformation invariance can be achieved by pooling over units that are tuned to different views of the same feature [15]. Convolutional neural networks are a subclass of hierarchical models that perform face and generic object recognition with high accuracy [16], [17].

The models mentioned above cannot be mapped into areas of the visual cortex and therefore are not biologically plausible. There are other models which have been directly inspired from different areas of the visual cortex and have predicted different properties of neurons in these areas. A series of models of VisNet have been proposed that perform object recognition under translation, transformation and lighting variations [18], [19], [20]. Lissom is a set of hierarchically connected areas in which long range inhibitory and excitatory connections provide selectivity similar to the neurons of different areas of the visual cortex [21]. Serre and Poggio introduced a hierarchical architecture based on the HMax model that performs object recognition in cluttered environments with high accuracy [22]. The main characteristic of their model was the alternating layers of simple and complex cells that provide recognition and invariance respectively.

In this paper, we propose a model for object recognition by combining two different approaches to the study of the visual cortex. We use the trace learning rule that is based on the behavior of the visual cortex neurons to provide invariant selectivity. Moreover, to optimize the responses of neurons in each layer, a redundancy reduction mechanism based on the statistical properties of natural images is used. We show that a model based on these mechanisms can achieve high recognition rates for different objects. In the next section, we describe the properties of the proposed model.

3 Model Description

The proposed model includes a set of areas analogous to the visual areas V1, V2 and V4 which are connected in a hierarchical organization (Fig. 1). Neurons in each layer receive bottom-up input from the previous layer and horizontal input from the same layer. Units in the first layer receive bottom-up input from small regions in the input image. Bottom-up input to each model neuron is excitatory and determines its primary form of selectivity, while horizontal input is inhibitory and facilitates the extraction of optimal features in images.

It was shown that Gabor filters provide an appropriate model of V1 neuronal selectivities [23]. Therefore, we used Gabor filters with different orientations to model the V1 neurons:

$$
\begin{aligned}
F(x, y) &= \exp\!\left(-\left(x_0^2 + \gamma^2 y_0^2\right)\big/\left(2\sigma^2\right)\right)\cos\!\left((2\pi x_0)/\lambda\right) \\
x_0 &= x\cos(\theta) + y\sin(\theta) \\
y_0 &= -x\sin(\theta) + y\cos(\theta)
\end{aligned}
\tag{1}
$$

Where x and y are the position of filter in the image, γ is the aspect ratio, θ is the orientation, σ is the effective width and λ is the wavelength of the filter. We used a set of filters with 6 different orientations and a single spatial frequency (Fig. 2).

Bottom-up activity for V2 and V4 neurons is calculated as a weighted sum of responses of neurons in their receptive field:

$$
y_i^{bup} = \sum_{j \in RF_i} w_{ij}^{bup} x_j
\tag{2}
$$

here, y_i^{bup} is the bottom-up activity of neuron i and RF_i is its receptive field in the previous layer, x_j is a neuron in RF_i and w_{ij}^{bup} is the weight of bottom-up connection between neuron i and neuron j. Table 1 displays the parameters of the proposed model.

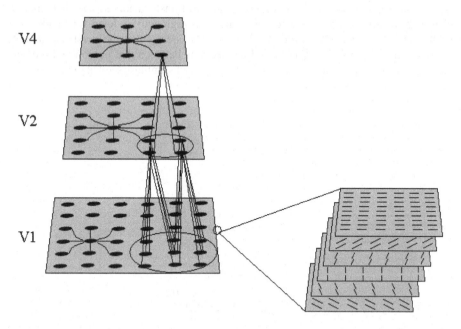

Fig. 1. The proposed model. Neurons in the first layer are modeled by Gabor filters with different orientations from 0° to 150° with 30° steps. Units in each layer receive two sets of input through bottom-up and horizontal connections. Horizontal and bottom-up connections are displayed for different neurons for better visualization.

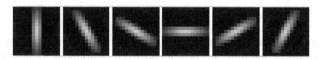

Fig. 2. The set of Gabor filters with 6 different orientations used to model V1 neurons

Table 1. The Model Parameters

Layer	Dimensions	Bottom-up RF	Horizontal Connections
V1	128×128	7×7	21×21
V2	104×104	21×21	21×21
V4	80×80	41×41	21×21

3.1 Bottom-Up Connections and the Trace Learning Rule

The receptive field size of neurons along the ventral visual pathway gradually increases and their preferred stimuli become more complex [2]. According to electrophysiological experiments, neurons in V2 and V4 exhibit invariant selectivity to transformations of their preferred stimuli [6], [7]. It has been suggested that complete invariant selectivity in IT neurons can be developed based on partial invariant responses of V2 and V4 neurons [24]. A neural mechanism has been proposed to achieve invariant response to the preferred stimuli transformations [25]. This mechanism is based on the fact that neurons retain their high activity level for hundreds of milliseconds. Fig. 3 explains this mechanism. The idea is that the persistent activity level of a neuron results in established connections between that neuron and the neurons representing transformed version of its preferred stimulus (see legend of Fig. 3).

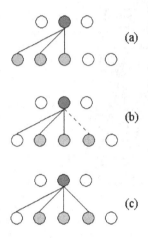

Fig. 3. The idea of continuous transformation. (a) Neuron shown in dark gray is activated by a set of inputs from gray neurons and the connections between them are established. (b) As the stimulus transforms, the set of active neurons in the lower layer changes. The dark gray neuron still preserves its high activity level. (c) The connection between the dark gray neuron and the neuron that represents the transformed version of its preferred stimulus is established because both neurons are active.

Földiák proposed the Trace learning rule to describe the mechanism in which invariant selectivity of neurons are developed from continuous spatial and temporal transformations of the objects [26]. The rule extends the simple Hebbian learning to contain a trace of the previous responses of neuron in its weight update:

$$\nabla w_{ij}^t = \lambda \bar{y}_j^t \left(x_i^t - w_{ij}^{t-1} \right) \tag{3}$$

where $\bar{y}_j^t$ is the trace value for neuron j in the iteration t, w_{ij} is the weight of connection between neuron i and neuron j, x_i^t is the input i in iteration t, and λ is the learning rate. The term $-\lambda \bar{y}_j^t w_{ij}^{t-1}$ is added to avoid unlimited increases in the weights of connections. Trace value for a neuron is calculated using (4):

$$\bar{y}_j^t = \eta \bar{y}_j^{t-1} + (1-\eta) y_j^t \tag{4}$$

here $0 < \eta < 1$ is the trace constant and y_j^t is the bottom-up activity of neuron j in the iteration t.

The trace learning rule is applied to develop invariant selectivity of neurons to their bottom-up input. The bottom-up activity of each neuron is calculated as a weighted

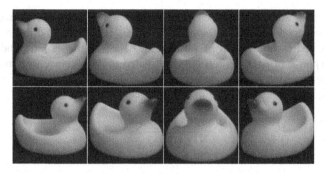

Fig. 4. An example object from coil100 dataset. Images in this dataset are taken from objects rotated with 5° steps.

sum of activities of neurons in its receptive field. We used a subset of images from the coil100 dataset [35] to train the connectivity between different layers of the model. Coil100 dataset contains images of objects from different view-points and therefore is an appropriate training set for model neurons (Fig. 4).

3.2 Extracting Statistically Independent Features with Redundancy Reduction

For a feature-based object recognition system, efficiency of the features extracted from images is very important. A model that simply tries to learn invariant responses cannot achieve high recognition rates for different objects. Features with high information content should be used that can discriminate between objects. The redundancy reduction becomes an important issue when dealing with different objects from different categories. In the proposed model, we used a redundancy reduction mechanism that is based on the statistical properties of natural images. This mechanism is used in different layers to provide a set of globally optimal features for object recognition.

Primary studies on redundancy in natural images revealed that linear filters like Gabor or Wavelet filters are optimal features considering first order statistics [27], [28]. However dependencies in natural images are nonlinear and hidden in their first order statistics. For example, it was observed that there exists a special form of dependency between responses of filters, i.e. their variances are dependent on the responses of other filters [29], [30]. Wainwright and Simoncelli proposed that Gaussian scale mixtures model can provide an explanation for variance dependency [31]. Schwartz and Simoncelli used divisive normalization to produce the independent responses over a set of patches selected from natural images [32]. We extended this model to a hierarchical architecture and developed neurons with similar properties to neurons in the higher order visual area V2 [33], [34]. This model is used in the proposed hierarchy to extract the most efficient features from natural images.

Neurons in each layer of the proposed model are connected to a set of neighboring neurons with horizontal connections. The weights of these connections are learned to predict the variance of responses of their corresponding neuron:

$$\mathrm{var}\left(L_x \mid L_y, y \in C_x\right) = \sum_{y \in C_x} w_{xy} L_y^2 + \sigma_x^2 \tag{5}$$

where L_x and L_y are the responses of neurons x and y respectively and w_{xy} is the weight of horizontal connection between them, C_x is the neighboring region around the neuron x and σ_x^2 is the part of variance of neuron x that is independent of other neurons. Response of each neuron is then divided to its variance to remove the variance dependency. This mechanism is repeated in different layers of the model to produce neuronal responses that are globally independent:

$$R_x = \frac{L_x^2}{\sum_{y \in C_x} w_{xy}^i L_y^2 + \sigma_x^2} \tag{6}$$

the result of (6) is the independent response R_x of filter x.

The weights of horizontal connections for each neuron are learned in a training procedure after the bottom-up connections to that neuron has been established. The goal is to predict the variance of responses of each neuron using responses of its neighboring neurons. An unbiased estimate for variance of responses of a neuron is:

$$\bar{\mu}_x = \frac{1}{N} \sum_{t=1}^{N} L_x^t$$

$$\bar{\sigma}_x^2 = \frac{1}{N-1} \sum_{t=1}^{N} \left(L_x^t - \bar{\mu}_x \right)^2 \tag{7}$$

An iterative form for the above equations is:

$$\bar{\mu}_x^t = \left(\bar{\mu}_x^{t-1} \times (t-1) + L_x^t \right) / t$$

$$\left(\bar{\sigma}_x^t \right)^2 = \left(\left(\bar{\sigma}_x^{t-1} \right)^2 \times (t-2) + \left(L_x^t - \bar{\mu}_x^t \right)^2 \right) / (t-1) \tag{8}$$

We used gradient descent to minimize the MSE between variance estimates of (5) and (8). The update rules for horizontal connections are as follows:

$$E^2 = \frac{1}{N} \sum_{i=1}^{N} \left(Var\left(L_x^i \right) - \left(\bar{\sigma}_x^i \right)^2 \right)^2$$

$$\frac{\partial E^2}{\partial w_{xy}} = \frac{2}{N} \sum_{i=1}^{N} \left(Var\left(L_x^i \right) - \left(\bar{\sigma}_x^i \right)^2 \right) \times \frac{\partial Var\left(L_x^i \right)}{\partial w_{xy}} = \frac{2}{N} \sum_{i=1}^{N} \left(Var\left(L_x^i \right) - \left(\bar{\sigma}_x^i \right)^2 \right) \times \left(L_y^i \right)^2 \tag{9}$$

$$w_{xy}^t = w_{xy}^{t-1} - \eta \frac{\partial E^2}{\partial w_{xy}^{t-1}} = w_{xy}^{t-1} - \eta \times \left(Var\left(L_x^t \right) - \left(\bar{\sigma}_x^t \right)^2 \right) \times \left(L_y^t \right)$$

$$\frac{\partial E^2}{\partial \sigma_x^2} = 2 \sum_{i=1}^{N} \left(Var\left(L_x^i \right) - \left(\bar{\sigma}_x^i \right)^2 \right) \times \frac{\partial Var\left(L_x^i \right)}{\partial \sigma_x^2} = 2 \sum_{i=1}^{N} \left(Var\left(L_x^i \right) - \left(\bar{\sigma}_x^i \right)^2 \right) \times 2\sigma_x \tag{10}$$

$$\sigma_x^t = \sigma_x^{t-1} - \eta \frac{\partial E^2}{\partial \sigma_x^{t-1}} = \sigma_x^{t-1} - \eta \times \left(Var\left(L_x^t \right) - \left(\bar{\sigma}_x^t \right)^2 \right) \times \sigma_x^{t-1}$$

4 Experimental Results

We performed a set of experiments to evaluate the performance of the proposed model for object recognition under transformations and with different backgrounds. We used the coil100 dataset which contains images of different objects under different viewpoints to measure the generalization of the proposed model. The results are shown in Table. 2. It can be seen that the proposed model is superior to other models reported in [21].

Table 2. Recognition rates of the proposed model on coil100 dataset compared to Lissom2 and SNoW methods [21]

Number of views used for training	SNoW one against all	LISSOM2	Our Model
4	0.760	0.798	**0.889**
8	0.819	0.816	**0.917**
18	0.913	0.845	**0.934**

Images in the coil100 dataset do not challenge the ability of the proposed model to recognize objects in different lighting conditions and backgrounds. Therefore, we generated a set of images from 6 objects in different distances, viewpoints and backgrounds to examine the model (Fig. 5). Samples images for an object are shown in Fig. 6. For each object, we used two third of images for training and one third for test. In the first experiment, images for training were selected from all backgrounds. Results are shown in the second column of table. 3. In the second experiment, we used images of two backgrounds for training and the third background was only used for test. Results are shown in the third column of table. 3. It can be seen that the performance of the model is on average above 95%.

Fig. 5. Six objects used to generate a set of images with different view-points, backgrounds and distances

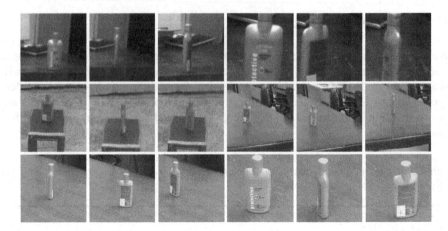

Fig. 6. Samples of object 2 images from different view-points, distances and backgrounds

Table 3. Recognition rates of the model for the six objects of Fig. 5

Object	Train and test with different backgrounds	Train with two backgrounds and test with the third one
Object1(Pig)	0.957	0.956
Object2(Shampoo)	0.916	0.975
Object3(Coffee)	0.975	0.963
Object4(Ball)	0.971	0.970
Object5 (Blue)	0.971	0.943
Object6(Apple)	0.923	0.920

5   Conclusion

In this paper, we proposed a hierarchical model for object recognition based on the well known properties of neurons in the primate visual cortex. Two key characteristics of the proposed model are the invariance of neuronal responses to object transformations and the extraction of efficient features from images that enhance the object recognition accuracy. We used the trace rule to develop neurons with invariant responses to object transformation. In order to increase the accuracy of recognition, we used a model of redundancy reduction previously used to develop neurons with selectivities similar to neurons in the visual cortex. We examined the performance of the model with a set of images of objects from different view-points and backgrounds. On average, the recognition rate was higher than 95% on custom images.

Previous studies on redundancy reduction tried to develop neuronal selectivities similar to that of the primary visual cortex. In a recent study, redundancy reduction was extended to simulate the selectivity of extrastriate neuronal selectivities [34]. In this paper, we proposed a hierarchical architecture for object recognition with features similar to the selectivities of neurons in V2 and V4. Neurons in these areas exhibit some degrees of response invariance to stimuli transformations. Previous studies reported a high recognition rate with features inspired by the visual cortex. In this paper, we provided a more biologically plausible model for object recognition with neurons similar to the visual cortex neurons.

References

1. Hubel, D.H., Wiesel, T.N.: Receptive fields, binocular interaction and functional architecture in the cat's visual cortex. J. Physiology 160, 106–154 (1962)
2. Van Essen, D.C., Maunsell, J.H.: Hierarchical organization and functional streams in the visual cortex. Trends in Neuroscience 6, 370–375 (1983)
3. Tanaka, K.: Inferotemporal cortex and object vision. Ann. Rev. Neuroscience 19, 109–139 (1994)
4. Atick, J.: Could information theory provide an ecological theory of sensory processing? Network: Computation in Neural Systems 3, 213–251 (1992)
5. Barlow, H.B.: Possible principles underlying the transformation of sensory messages. In: Rosenblith, W.A. (ed.) Sensory Communication, pp. 217–234 (1961)
6. Desimone, R., Schein, S.J.: Visual properties of neurons in area V4 of the macaque: sensitivity to stimulus form. Journal of Neurophysiology 57(3), 835–868 (1987)
7. Kobatake, E., Tanaka, K.: Neuronal selectivities to complex object features in the ventral visual pathway of the macaque cerebral cortex. J. Neurophysiology 71, 856–867 (1994)
8. Laughlin, S.: A simple coding procedure enhances a neuron's information capacity. Z. Naturforsch 36, 910–912 (1981)
9. Vaillant, R., Monrocq, C., Le Cun, Y.: An original approach for the localization of objects in images. In: International Conference on Artificial Neural Networks, pp. 26–30 (1993)
10. Heisele, B., Serre, T., Pontil, M., Vetter, T., Poggio, T.: Categorization by Learning and Combining Object Parts. In: Advances in Neural Information Processing Systems, vol. 14 (2002)
11. Ullman, S., Vidal-Naquet, M., Sali, E.: Visual Features of Intermediate Complexity and Their Use in Classification. Nature Neuroscience 5(7), 682–687 (2002)
12. Leung, T.K., Burl, M.C., Perona, P.: Finding faces in cluttered scenes using random labeled graph matching. In: Proc. Int'l Conf. Computer Vision, Cambridge, MA, pp. 637–644 (1995)
13. Lowe, D.G.: Object Recognition from Local Scale-Invariant Features. In: Proc. Int'l Conf. Computer Vision, pp. 1150–1157 (1999)
14. Fukushima, K.: Neocognitron: A self-organizing neural network model for a mechanism of pattern recognition unaffected by shift in position. Biological Cybernetics 36(4), 93–202 (1980)
15. Perrett, D., Oram, M.: Neurophysiology of shape processing. Imaging Vis. Comput. 11, 317–333 (1993)
16. LeCun, Y., Huang, F.J., Bottou, L.: Learning Methods for Generic Object Recognition with Invariance to Pose and Lighting. In: Proc. IEEE Conf. Computer Vision and Pattern Recognition (2004)
17. Chopra, S., Hadsell, R., LeCun, Y.: Learning a Similarity Metric Discriminatively, with Application to Face Verification. In: Proc. IEEE Conf. Computer Vision and Pattern Recognition (2005)
18. Rolls, E.T., Milward, T.: A model of invariant object recognition in the visual system: learning rules, activation functions, lateral inhibition, and information-based performance measures. Neural Computation 12, 2547–2572 (2000)

19. Rolls, E.T., Stringer, S.M.: Invariant visual object recognition: A model, with lighting invariance. J. Physiology – Paris 100, 43–62 (2006)
20. Perry, G., Rolls, E.T., Stringer, S.M.: Spatial vs. temporal continuity in view invariant visual object recognition learning. Vision Research 46, 3994–4006 (2006)
21. Plebe, A., Domenella, R.G.: Object recognition by artificial cortical maps. Neural Networks 20, 763–780 (2007)
22. Serre, T., Wolf, L., Bileschi, S., Riesenhuber, M., Poggio, T.: Robust Object Recognition with Cortex-Like Mechanisms. IEEE Transactions on Pattern Analysis and Machine Intelligence 29(3) (2007)
23. Jones, J.P., Palmer, L.A.: An evaluation of the two-dimensional Gabor filter model of simple receptive fields in cat striate cortex. Journal of Neurophysiology 58(6), 1233–1258 (1987)
24. Riesenhuber, M., Poggio, T.: Hierarchical models of object recognition in cortex. Nature Neuroscience 2, 1019–1025 (1999)
25. Stringer, S.M., Perry, G., Rolls, E.T., Proske, J.H.: Learning invariant object recognition in the visual system with continuous transformations. Biological Cybernetics 94, 128–142 (2006)
26. Földiák, P.: Learning Invariance from Transformation Sequences. Neural Computation 3(2), 194–200 (1991)
27. Olshausen, B.A., Field, D.J.: Emergence of simple cell receptive field properties by learning a sparse code for natural images. Nature 381, 607–609 (1996)
28. Bell, A.J., Sejnowski, T.J.: The 'independent components' of natural scenes are edge filters. Vision Research 37(23), 3327–3338 (1997)
29. Zetzsche, C., Wegmann, B., Barth, E.: Nonlinear Aspects of Primary Vision: Entropy Reduction Beyond Decorrelation. In: Morreale, J. (ed.) Proceedings of the SID, Society for Information Display, Playa del Ray, CA, pp. 933–936 (1993)
30. Simoncelli, E.P.: Statistical Models for Images: Compression, Restoration and Synthesis. In: 31st Asilomar Conf. on Signals, Systems and Computers (1997)
31. Wainwright, M.J., Simoncelli, E.P.: Scale Mixtures of Gaussians and the Statistics of Natural Images. In: Advances in Neural Information Processing Systems, vol. 12, pp. 855–861 (2000)
32. Schwartz, O., Simoncelli, E.P.: Natural signal statistics and sensory gain control. Nature Neuroscience 4, 819–825 (2001)
33. Malmir, M., Shiry, S.: A Model of Angle Selectivity in Area V2 with Local Divisive Normalization. In: IEEE Symposium Series on Computational Intelligence, Nashville, TN, USA (2009)
34. Malmir, M., Shiry, S.: Class Specific Redundancies in Natural Images: a Theory of Extrastriate Visual Processing. In: International Joint Conference on Neural Networks (IJCNN), Atlanta, GA, USA (2009)
35. COIL100 image dataset,
    http://vision.ai.uiuc.edu/mhyang/object-recognition.html

Using Genetic Algorithms for Real-Time Object Detection

J. Martínez-Gómez[1], J.A. Gámez[1], I. García-Varea[1], and V. Matellán[2]

[1] Computing Systems Department, University of Castilla-La Mancha, Spain
[2] Dept. of Mechanical and Computer Engineering, University of León, Spain
{jesus_martinez,jgamez,ivarea}@dsi.uclm.es, vicente.matellan@unileon.es

Abstract. This article presents a new approach to mobile robot vision based on genetic algorithms. The major contribution of this proposal is the real-time adaptation of genetic algorithms, which are generally used offline. In order to achieve this goal, the execution time must be as short as possible. The scope of this system is the Standard Platform category of the RoboCup[1] soccer competition. The system developed detects and estimates distance and orientation to key elements on a football field, such as the ball and goals. Different experiments have been carried out within an official RoboCup environment.

1 Introduction

For mobile robotics, image processing has become one of the most important elements. Intelligent robots need to retrieve information from the environment in order to interact with it. Vision cameras are one of a robot's key devices. The images taken by the robot's camera need to be processed in real time with limited processing resources. The systems developed need to cope with noisy and low quality images, and in order to process the maximum number of images by second, the algorithms must be as efficient as possible.

In the RoboCup[1] environment different solutions have been proposed over the last years. These proposals use the information obtained with colour filtering processes[2]. Scan-lines[3] and edge-based[4] solutions have been one of the most widely-used for the RoboCup competition.

The approach presented here carries out object recognition by using real-time genetic algorithms[5](GAs). The number of iterations and individuals for the GA must be reduced as much as possible in order to improve efficiency (some authors propose the use of cellular GAs[6] instead of reducing the number of individuals and iterations). This is necessary because the system has to be applied in real time. In order to prevent system performance from being affected by this reduction, the individuals will be initialized using all the available information. This initial information can be obtained from previous populations and from the colour filtering process applied to the last image taken by the robot's camera. After an image showing an object o, the next image has a high probability of

---

[1] http://www.robocup.org/

J. Baltes et al. (Eds.): RoboCup 2009, LNAI 5949, pp. 215–227, 2010.

showing the same object. The information obtained from previous populations allows us to take advantage of the high similarity between consecutives images taken by the camera.

Our hypothesis is that the similarity between captured images, and the information obtained with the filtering process, can be used to develop a real-time vision system based on genetic algorithms. Different tests in real scenarios using the biped robot Nao have been carried out to evaluate our proposal. These tests show the object (ball and goals) recognition process on the official RoboCup football field.

The article is organized as follows: problem restrictions are outlined in Section 2. We describe the full vision system in Section 3, and in Section 4 we explain the experiments performed and the results obtained. Finally, the conclusions and areas for future work are given in Section 5.

2   Problem Restrictions

The vision system has to be valid for use in the Standard Platform category. Robot Nao[2] is the official platform for this category, and its camera takes 30 (320 x 240 pixels) frames per second. The camera's native colour space is YUV[7].

In order to reduce the amount of information to work with, the captured images are filtered. This processing removes the pixels that do not pass a colour filter. The key colours in the RoboCup environment are yellow and blue for the goals, green for the carpet, orange for the ball and white for the field lines. Football player equipment is red and dark blue. The filtering is carried out by defining a top and bottom limit for the Y, U and V colour components. A pixel will successfully pass a filter only if all its components are between these limits. Fig. 1 shows a filtering example for blue.

Fig. 1. Colour filtering for the blue goal

Object recognition has to be carried out during a football match. The environment includes objects that are partially hidden behind others, so the frames taken in a football match will not always show the complete object we want to recognize. Scan-line-based methods present a lot of problems in these situations, whereas our system works properly, as will be shown in the results section.

---

[2] http://www.aldebaran-robotics.com/eng/Nao.php

3   Vision System

Genetic algorithms use individuals that represent potential solutions to problems. For our vision system, individuals have to represent the detection of the object o placed at distance d with the orientation or. This information (object o at distance d with orientation or) is contrasted with the one extracted from the last frame captured by the robot's camera. The fitness will be high for individuals with information that is plausible with respect to the last image. On the other hand, the fitness will be low if the object o does not appear in the image.

3.1   General Processing Scheme

The processing starts with the arrival of new images at the robot's camera. A new image will evolve a new population for each object to be recognized. In this work, three distinct objects are considered, the blue goal, the yellow goal and the orange ball, so three different populations will be kept. After taking a new image, the colour filtering allows the robot to know the objects likely to appear in the image. The populations of the non-plausible objects will not be evolved. Fig.2 shows the general processing scheme.

```
Capture a new image and filter it with colour filters
for each object to recognize
    if we have obtained enough pixels
        Evolve a new population
        Apply local search over the best individual
        Return the estimated distance to the object
    end if
end for
```

Fig. 2. General system processing scheme

In order to avoid local optimums, the population will be restarted after a given number of iterations failing to improve the best individual. An iteration will increase the value of a counter if the best fitness of the iteration is not greater than the best global fitness. The counter value will be set to zero if the iteration obtains the best fitness. The population will be restarted if the counter reaches a limit value.

3.2   Genetic Representation

In addition to the distance between the camera and the object to be recognized, we also need to estimate the orientation between both elements. This information is not only needed for self-location tasks[8], but also for the application of the fitness function. The shape of an object in an image will depend on the distance and the orientation between object and camera.

Fig.3 presents graphically the three parameters to be estimated: d is the distance between camera and object, α is the difference of orientation in the

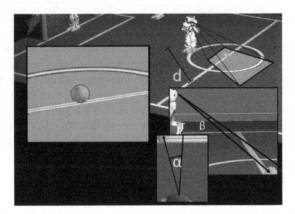

Fig. 3. Image taken with specific distance and orientation between object and camera

x-axis and β in the y-axis. With the same distance d and different α or β values, captured images will show the same ball but located at a different position within the image. The image will not show the ball with big α or β variations. A third component for the orientation difference in the z-axis is not needed, because using horizon detection techniques[9], the image can be processed to show all the objects parallel to the floor.

Each individual stores the following information (genes):

- Distance to the robot: d
- Orientation difference in the x-axis: α
- Orientation difference in the y-axis: β

All the genes are represented by a numerical value, limited by the maximum distance detection for d, and by the field of view for α and β. An additional gene is needed to perform goal detection. This gene (θ) represents the goal orientation when the frame is taken. Two frames taken with the same $< d, \alpha, \beta >$ parameters will be different if the goal orientation varies, as can be observed in Fig.4.

3.3 Obtaining the β Parameter

We can avoid modelling β if we know the angle between the camera and the floor in the y-axis, γ. Thus β can be calculated using γ, the distance d, and the

Fig. 4. Images taken varying the θ parameter

orientation in the x-axis α. With this approach, the areas of the search space that represent unreal solutions will not be explored. Using γ and the camera's field of view, we can obtain the minimum and maximum distances at which we can detect elements. For instance, if γ is close to 90 degrees, the robot will be able to recognize distant objects, but not a nearby ball.

The main problem of calculating β instead of modelling it is that our algorithm will heavily depend on γ estimation and its performance will decrease if γ is not correctly estimated. For legged robots, the movement of the robot causes an enormous variation in the camera angle, which makes it difficult to obtain a precise value for γ. For wheeled robots, the movement will not affect the camera angle as much as for legged ones and γ can be accurately calculated.

3.4   Fitness Function

The fitness function returns numeric values, according to the goodness of the projection obtained with the parameters $< d, \alpha, \beta >$ of each individual. To evaluate an individual, its genes are translated into a projection of the object that the individual represents. The projection needs a start position $< x, y >$, obtained from α and β. The size of the object depends on d.

An object projection is evaluated by comparing between it with the information obtained from the filtering process. A pixel $< x, y >$ of the projection will be valid only if the pixel $< x, y >$ of the image captured by the robot's camera successfully passes the colour filter. This evaluation is illustrated in Fig. 5, where the left image shows the original image after an orange filter. The right one shows the result of evaluating 12 different individuals, where red pixels are invalid (they have not passed the colour filter) and green pixels are valid.

After this processing we obtain the number of valid and invalid pixels for each individual. Using the percentage of pixels that pass the filter as a fitness function has a serious drawback: individuals representing distant objects obtain better fitness values. Those individuals correspond to smaller projections resulting in a higher probability of having a bigger percentage of valid pixels (few right pixels mean high percentage).

Due to this problem, and using the information obtained with the filtering, we define the fitness function as the minimum value of:

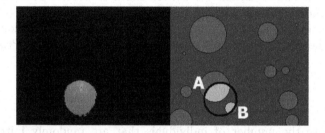

Fig. 5. Filtered image (left) and evaluation of 12 individuals to detect the ball (right)

- % of pixels of the projection that have passed the colour filter.
- % of pixels that passed the colour filter and belong to the valid projection pixels.

In order to illustrate the behaviour of the function, let's study individuals A and B in Fig. 5. Individual B has a higher percentage of pixels that passed the filter (70 versus 45). On the other hand, only 5% of the pixels that passed the orange filter belong to individual B. For A, this percentage rises to 35%. The fitness value will be 0.35 for individual A and 0.05 for B.

3.5 Population Initialization

The population is usually randomly initialized for GAs. Our approach uses additional information to initialize the first individuals. Firstly, we could use individuals from populations of previous captures. In addition to this, the information extracted from the filtering process could also be used. Such information is the number of pixels of each colour, and the x and y component of the centroid of the distribution of pixels obtained with the colour filter. According to this information, an individual can be initialized in 3 different ways:

- Randomly.
- Using the information from the filtering process.
- Cloning an individual from a previous population.

The first two ways of generating a new individual can always be used. The third one can only be used when a valid population is available. Such population must have been evolved to recognize the same object o that we want to recognize. The number of frames between the current one and the last one that evolved a population to recognize o has to be small. If the frame number difference is large, the individuals of the population will not be valid for cloning, because these individuals were evolved to solve a situation different from the current one. A draw is carried out to select the way in which an individual is initialized. All the ways have a probability that depends on the number of frames from the last frame that recognized the object we are studying. We need two parameters to obtain these probabilities:

- MW: Max probability of cloning an individual from a previous population.
- MNF: Max number of frames possible between the present frame and the last one that recognized the object we are studying.

The sum of the three parameters is normalized to be 1.0. The probability of initializing individuals by cloning them from other populations ($CloneProb$) will decrease if the number of frames without updating the population ($NFWU$) increases. The other two probabilities are calculated using $CloneProb$.

$$CloneProb : MW - MW * (NFWU/MNF)$$
$$InitialInfoProb : (1 - (CloneProb)) * 0.66$$
$$RandomlyProb : (1 - (CloneProb)) * 0.34$$

If we increase the number of individuals that are randomly initialized, the variety of the initial population will be greater. Using the initial information,

Fig. 6. Object partially captured

the algorithm's elitism will increase (with the problem of local optimums). With individuals cloned from other populations, the algorithm will converge faster with small variations between frames. The balance between elitism and generality can be obtained through a correct combination of these three ways. We selected 0.66 and 0.34 as values to obtain a heterogeneous initial population, based on preliminary empirical tests.

3.6   Partial Object Occlusion

Vision systems must cope with hard environments. For instance, the objects to recognize can be partially hidden behind other robots, or the images captured by the robot's camera may show only parts of the desired object, due to the camera's orientation. Our proposal performs the individual's evaluation using the entire object's projection and not partial characteristics. This is the reason that our system works properly with occlusions.

4   Experiments and Results

The experiments were carried out on a RoboCup Standard Platform football field, with the official goals, a 6 x 4 metre carpet and a ball. We used a Nao robot, taking 2 images per second. The format of the images is YUV and their size is 320 x 240 pixels. While the experiments were being carried out, the absolute difference between the real and estimated distance to the object we wanted to detect was stored per frame. The estimated distance was the value of the d gene of the individual with the best fitness. Lighting conditions were stable throughout the experiments, and the colour filters were optimal.

The execution time for each image was variable. We decided to use two frames per second because the maximum execution time was never greater than 450 milliseconds.

After the filtering process ($\approx$ 80 msec), the execution time was never greater than 370 milliseconds (183 for the goal and 187 for the ball).

4.1   Genetic Algorithm Parameters

The experiments were carried out with the following parameters:

– Individual number: 12 and Iteration number: 24

- Mutation probability: 5% and Crossover type: point
- Replacement: generational algorithm
- Restart after 25% iterations without improving the global optimum
- MW: 0.5 and MNF: 10

The algorithm uses a limited number of individuals and iterations. The mutation probability and the crossover type are standard, and the entire population is replaced with the offspring at the end of the iteration. Due to this, the quality of the population can decrease while the search progresses. Evolution is performed without taking into account robot's odometry.

After evolving the population, a simple local search process (Hill Climbing) is applied to the best individual. This processing will allow us to improve the best fitness. The local search is applied by evaluating positive and negative variations for the genes of the individual. The algorithms that combine concepts and strategies from different metaheuristics are called memetic algorithms [10].

4.2   Experiment 1 - Hypothesis Validation

The objective of the first experiment was to prove that the system is able to work in the given time-frame, recognizing the environment elements and estimating the distance to them. We used the standard parameters described above and we executed the same tour over the football field 6 times. 30 frames were taken per tour (15 seconds). The frames captured the yellow goal placed between 360 and 300 cm, and the orange ball placed at distances between 260 and 200 cm.

The experiment consisted of 180 different frames (6 x 30). We stored the absolute difference between real and estimated distance (denoted $DBRED$) and the fitness of the best individual of the population by frame. These fitness values were used to generate different data sets. Each one of these data sets had only the detections carried out with individuals whose fitness values were greater than certain thresholds. Table 1 shows, taking the ball and yellow goal separately, and with four different threshold values (0, 0.25, 0.5 and 0.75), the average of the $DBRED$. It also gives the percentage of frames that obtained an individual with a fitness value greater than the threshold.

Table 1. Average $DBRED$ and % of frames with a fitness value over certain thresholds

Fitness		> 0.0	> 0.25	> 0.5	> 0.75		> 0.0	> 0.25	> 0.5	> 0.75
Average (cm)	Ball	42.62	40.57	31.77	22.75	Yellow	40.03	37.88	33.1	32.69
Frames (%)		68.89	68.33	57.78	8.89	Goal	99.44	93.33	44.44	8.89

It can be seen that the fitness function properly represents the goodness of the individuals. This is because using individuals with higher fitness values reduced the average of the differences between real and estimated distances. Table 2 shows the percentage of frames that obtained a difference between estimated and real distance lower than certain thresholds.

Table 2. Percentage of frames that obtained a *DBRED* lower than certain thresholds

	Percentage of frames under 100 cm 75 cm 50 cm 30 cm		Percentage of frames under 100 cm 75 cm 50 cm 30 cm
Ball	63.63 56.11 44.44 35.55	Yellow Goal	92.77 87.78 72.22 51.67

The results obtained show a high degree of robustness, especially for the yellow goal. In an environment with a maximum distance of 721 cm, a high (37.37% and 51.67%) percentage of frames obtained differences for the distance estimation under 30 centimetres.

Ball recognition (with our genetic algorithm) was more complicated than goal recognition, because only individuals which are very close to the solution (perfect detection) obtain fitness values different from zero. Due to the small size of the ball in the frames captured, only the projections of individuals close to the solution have pixels in common with the image obtained after the colour filtering process. The convergence of a GA with this kind of individuals will not be constant. 83.83% of correct ball recognitions (fitness> 0) were carried out with fitness values greater than 0.5. For the goal, this percentage descends to 44.69%.

4.3   Experiment 2 - β Study

The main objective of the second experiment was to test whether the β parameter can be calculated using the other parameters. The performance of the algorithm obtaining β instead of modelling it was studied. The robot made the same tour as in experiment 1.

For this experiment, the individuals did not use the β gene, but the parameter is needed for the fitness function and has to be calculated. This was done using the parameters d (distance to the object), α (orientation difference in the x-axis) and γ (orientation difference between the robot's camera and the floor in the y-axis). γ is obtained using the robot's sensors. The experiment consisted of 180 frames again and a summary of the results obtained is shown in table 3.

Table 3. Average *DBRED* and % of frames with a fitness value over certain thresholds

Fitness		> 0.0 > 0.25 > 0.5 > 0.75		> 0.0 > 0.25 > 0.5 > 0.75
Average (cm)	Ball	18.70 18.05 16.89 27.7	Yellow	33.66 32.81 34.31 27.5
Frames (%)		69.44 68.33 57.78 5.55	Goal	100.0 95.00 40.56 1.11

The first conclusion drawn from the results is that the number of correct detections (frames that obtained fitness values greater than 0) has increased. However, the percentage of frames with a fitness value greater than 0.5 and 0.75 decreased. This is because modelling β instead of obtaining it from the other parameters lets the algorithm to reach situations that are not right according to the model, but which are valid due to noise or the difference between the real and estimated γ value.

The average difference between the real and estimated distance($DBRED$) decreased considerably. With lower gene numbers and the same iterations, GAs converge faster to better solutions. In order to establish a complete comparison between modelling β and calculating it with other parameters, table 4 provides the percentage of frames that obtained a $DBRED$ lower than certain thresholds.

Table 4. Percentage of frames that obtained a $DBRED$ lower than certain thresholds

	Percentage of frames under 100cm 75cm 50cm 30cm		Percentage of frames under 100cm 75cm 50cm 30cm
Ball	68.89 68.89 65.56 54.44	Yellow Goal	96.67 93.33 76.67 48.89

If we compare table 4 and 2, we can see that the robustness of the algorithm has improved. The faster convergence of the algorithm with fewer genes makes it possible to obtain a higher percentage of frames with a small $DBRED$ to the object.

The main conclusion drawn from the data is that the number of genes should always be as small as possible. If one of the parameters that are modelled can be obtained from other parameters, this parameter should be removed. In order to use fewer genes, we have to use all the possible information retrieved from the environment, the platform and the elements to recognize. This information allows us to include our knowledge about the problem in the algorithm, and with such information the algorithm will only reach individuals representing real situations (according to the robot and the environment).

4.4   Experiment 3 - MW Study

The third experiment shows how MW affects the vision system. This parameter defines the maximun probability of cloning an individual for initialization from previous populations. MW defines the weight of previous frames for the process. If the value of this parameter increases a higher number of individuals from the initial population will represent solutions reached for previous frames.

The robot captured 20 different images from a static position. While the frames were being captured, the robot's camera orientation was quickly varied. All the frames show the blue goal placed at 250 cm and the orange ball situated at 150 cm. Most of the frames only partially show these elements due to the camera movements (only the orientation changed). We used the standard parameters for the genetic algorithm, and β was modelled as a gene. The variations in MW defined the different configurations. The experiment was repeated 9 times with each different configuration to obtain a final set of 180 frames (20 * 9). 4 different configurations were tested, with MW values of 0, 25, 50 and 75%. Table 5 shows the results obtained for the experiment.

Table 5. Average $DBRED$ and % of frames with a fitness value over certain thresholds

	MW	Fit>0	Fit>0.25	Fit>0.5	Fit>0.75	Fit>0	Fit>0.25	Fit>0.5	Fit>0.75
Ball	0.00	47.37	47.37	36.93	31.75	93.59	93.59	78.84	35.26
	0.25	43.10	41.43	34.26	34.27	91.66	91.02	80.12	44.87
	0.50	41.37	41.26	33.63	33.67	89.74	89.10	75.64	29.49
	0.75	43.48	42.08	32.72	33.49	89.10	87.18	75.00	32.69
Blue	0.00	58.02	49.48	27.15	12.78	100.0	82.68	47.49	12.85
Goal	0.25	53.64	42.63	26.71	19.72	98.32	83.80	56.42	13.97
	0.50	51.22	43.54	21.76	14.16	98.88	87.71	55.87	13.97
	0.75	44.16	37.60	24.45	15.39	98.88	89.94	64.25	12.85
			Average $DBRED$				Percentage of frames		

We can observe how the changes applied to MW do not produce big variations in the difference between the real and estimated distance. Table 5 shows how the percentage of frames that obtained better fitness values increases with greater MW values. For the blue goal, this happens for all the MW values. For the ball, the optimum point for the MW value is 0.25. The performance of the algorithm gets worse if MW is greater than 0.25.

Table 6. Percentage of frames that obtained a $DBRED$ below certain thresholds

		Percentage of frames under						Percentage of frames under			
	MW	100cm	75cm	50cm	30cm		MW	100cm	75cm	50cm	30cm
Ball	0.00	82.69	72.44	62.18	33.33		0.00	77.09	68.71	55.31	32.96
	0.25	85.90	79.49	71.79	31.41	Blue	0.25	81.00	73.74	62.57	34.08
	0.50	85.90	75.00	67.31	35.30	Goal	0.50	81.56	75.41	61.45	43.01
	0.75	80.77	73.72	64.10	32.69		0.75	87.15	78.77	72.07	48.60

Finally, table 6 presents the percentage of frames that obtained differences between the real and estimated distance below certain thresholds.

The robustness of the algorithm noticeably improved when the value of MW increased. For the ball, the best results were obtained again for a MW value of 0.25. The behaviour of the algorithm varies for the different objects to be detected when MW increases.

The ball is always captured as a small round orange object and very few frames show the ball partially hidden behind other objects. Because of this, the filtering process gives us useful information for the initialization of the new individuals. The $< x, y >$ position of the ball inside a frame will be close to the centroid $< x, y >$ obtained for the orange pixels after the filtering process. If we excessively increase the number of individuals cloned from previous iterations, the number of individuals initialized with the filtering information will be lower than the number needed for optimal convergence.

In spite of these drawbacks, a small percentage of individuals from previous iterations improves the system's convergence, because the algorithm will have a more diverse initial population. The offspring obtained by crossing individuals initialized in different ways will be able to obtain better fitness values. The individuals from previous iterations will be very useful if the initial information (obtained via the filtering process) was noisy.

The situation is completely different for goal detection. The shape of the goals in the frame depends on the position and orientation between camera and goal. The size of a goal's projection is bigger than that obtained for the ball, as can be observed in Fig.6. Individuals that are far from the solution can obtain fitness values greater than zero, due to the useful information stored in their genes. The risk of falling into local optimums is much greater for goal detection and the filtering information is less useful. Initializing individuals in different ways will help the algorithm to escape from local optimums. The solution represented by individuals from previous iterations will usually be closer to the global optimum than the one represented by the individuals initialized with the filtering information, especially for minor changes between frames.

5   Conclusions and Future Work

According to the results obtained from the first experiment, our system is a robust alternative to traditional systems for object recognition. It uses the principles of genetic algorithms with a short execution time, which allows the system to be used in the RoboCup environment. The system works properly in the presence of occlusions, without the necessity of a case-based approach.

The β parameter should always be obtained from the other parameters. This parameter can be correctly obtained if the robot's angles are measured without error. The number of genes for the individuals should be as small as possible.

Based on the results obtained in the third experiment, the similarity between consecutive frames can be used to improve the performance of our system.

The system was originally developed for goals and ball recognition, but in view of the results obtained and the available alternatives, the main application for the system should be that of goal detection. This is because goal recognition is much more difficult than ball detection, which can be done by using other techniques.

For future work, we aim to integrate the system developed with a localization method, such as Montecarlo[11] or Kalman Filters[12]. The selected localization method should use the estimated distances and orientations to the goals and the fitness of the best individual, and in order to integrate the visual and the odometry information in an optimal way[13], the fitness of the best individual could be used to represent the goodness of the visual information.

Adding some restrictions to the initialization of the new individuals by taking into account the robot's estimated pose could also be considered.

Acknowledgements

The authors acknowledge the financial support provided by the Spanish "Junta de Comunidades de Castilla-La Mancha (Consejería de Educación y Ciencia)" under PCI08-0048-8577 and PBI-0210-7127 Projects and FEDER funds.

References

1. Rofer, T., Brunn, R., Dahm, I., Hebbel, M., Hoffmann, J., Jungel, M., Laue, T., Lotzsch, M., Nistico, W., Spranger, M.: GermanTeam 2004. Team Report RoboCup (2004)
2. Wasik, Z., Saffiotti, A.: Robust color segmentation for the robocup domain. In: Pattern Recognition, Proc. of the Int. Conf. on Pattern Recognition (ICPR), vol. 2, pp. 651–654 (2002)
3. Jüngel, M., Hoffmann, J., Lötzsch, M.: A real-time auto-adjusting vision system for robotic soccer. In: Polani, D., Browning, B., Bonarini, A., Yoshida, K. (eds.) RoboCup 2003. LNCS (LNAI), vol. 3020, pp. 214–225. Springer, Heidelberg (2004)
4. Coath, G., Musumeci, P.: Adaptive arc fitting for ball detection in robocup. In: APRS Workshop on Digital Image Analysing, pp. 63–68 (2003)
5. Mitchell, M.: An Introduction to Genetic Algorithms (1996)
6. Whitley, L.: Cellular Genetic Algorithms. In: Proceedings of the 5th International Conference on Genetic Algorithms table of contents. Morgan Kaufmann Publishers Inc., San Francisco (1993)
7. Foley, J.D., van Dam, A., Feiner, S.K., Hughes, J.F.: Computer graphics: principles and practice. Addison-Wesley Longman Publishing Co., Inc., Amsterdam (1990)
8. Borenstein, J., Everestt, H., Feng, L.: Where am I? Sensors and Methods for Mobile Robot Positioning (1996)
9. Bach, J., Jungel, M.: Using pattern matching on a flexible, horizon-aligned grid for robotic vision. Concurrency, Specification and Programming-CSP 1(2002), 11–19 (2002)
10. Moscato, P.: Memetic algorithms: a short introduction. Mcgraw-Hill'S Advanced Topics In Computer Science Series, pp. 219–234 (1999)
11. Fox, D., Burgard, W., Thrun, S.: Active markov localization for mobile robots (1998)
12. Negenborn, R.: Robot localization and kalman filters (2003)
13. Martínez-Gómez, J., José, A., Gámez, I.G.V.: An improved markov-based localization approach by using image quality evaluation. In: Proceedings of the 10th International Conference on Control, Automation, Robotics and Vision (ICARCV), pp. 1236–1241 (2008)

An Approximate Computation of the Dominant Region Diagram for the Real-Time Analysis of Group Behaviors

Ryota Nakanishi, Junya Maeno, Kazuhito Murakami, and Tadashi Naruse

Graduate School of Information Science and Technology,
Aichi Prefectural University, Nagakute-cho, Aichi, 480-1198 Japan

Abstract. This paper describes a method for a real-time calculation of a dominant region diagram (simply, a dominant region). The dominant region is proposed to analyze the features of group behaviors. It draws spheres of influence and is used to analyze a teamwork in the team sports such as soccer and handball. In RoboCup Soccer, particularly in small size league(SSL), the dominant region takes an important role to analyze the current situation in the game, and it is useful for evaluating the suitability of the current strategy. Another advantage of its real-time calculation is that it makes possible to predict a success or failure of passing. To let it work in a real environment, a real-time calculation of the dominant region is necessary. However, it takes 10 to 40 seconds to calculate the dominant region of the SSL's field by using the algorithm proposed in [3]. Therefore, this paper proposes a real-time calculation algorithm of the dominant region. The proposing algorithm compute an approximate dominant region. The basic idea is (1) to make a reachable polygonal region for each time $t_1, t_2, ..., t_n$, and (2) to synthesize it incrementally. Experimental result shows that this algorithm achieves about 1/1000 times shorter in computation time and 90% or more approximate accuracy compared with the algorithm proposed in [3]. Moreover, this technique can predict the success or failure of passing in 95% accuracy.

1 Introduction

In RoboCup Soccer, the cooperative plays such as passing and shooting are the important basic skills. Particularly in RoboCup Small Size League (SSL), high level cooperative plays are developed so far. Since the strategies based on them are growing year after year [1], it is important to analyze the actions of opponent team in real time and then to change team's strategy dynamically in order to overcome the opponent. For such analysis, the *voronoi diagram* [2] and the *dominant region diagram* [3] are useful. They are used to analyze the sphere of influence. The voronoi diagram divides the region based on the distance between robots, while the dominant region diagram divides the region based on the arrival time of robots. It is considered that the dominant region diagram shows an adequate sphere of influence under the dynamically changing environment such as a soccer game.

J. Baltes et al. (Eds.): RoboCup 2009, LNAI 5949, pp. 228–239, 2010.

In the SSL, the dominant region diagram has been used for arranging team-mate robots to perform the cooperative play such as passing and shooting [4][5]. However, the existing algorithm takes much time to compute the dominant region diagram, the use of the algorithm is restricted to the case that the computation time can keep, i.e. a typical case is a restart of play. If the dominant region diagram can be computed in real time, we can apply it any time.

In this paper, we propose an algorithm that computes the dominant region diagram in real time. In the SSL, it is required to compute the dominant region diagram within 5 *msec*. So, we put this time to be our present goal. Proposed algorithm is an approximate computation of the dominant region diagram so that we discuss the computation time and the approximation accuracy through the experiment. It is shown that proposed algorithm achieves 1/1000 times shorter in computing time compared with the algorithm proposed in literature [3] and over 90% accuracy. Moreover, 5 *msec* computation time can be possible under the parallel computers. We also show that the dominant region diagram is useful for the prediction of success for passing.

2   Dominant Region Diagram

Our main purpose is to discuss a real-time computation of the dominant region diagram. At first, we briefly describe it and compare it with the voronoi diagram in this section.

2.1   Computation of Dominant Region

A dominant region of an agent[1] is defined as "a region where the agent can reach faster than any other agents". A dominant region diagram, simply a dominant region, shows the dominant region of every agent [3]. The dominant region diagram is one of the generalized voronoi diagrams. Though the dominant region diagram is an n dimensional diagram in general, we discuss a two dimensional diagram here because we consider a soccer field.

The dominant region is calculated as follows. Assume that an agent i is at the point $\mathbf{P}^i(= (P_x^i, P_y^i))$ and is moving at a velocity $\mathbf{v}^i(= (v_x^i, v_y^i))$. Assume also that the agent can move to any direction and its maximum acceleration is $\mathbf{a}_\theta^i(= (a_{\theta x}^i, a_{\theta y}^i))$ for a θ-direction. The position that the agent will be after t seconds is given by[2],

$$\begin{pmatrix} x_\theta^i \\ y_\theta^i \end{pmatrix} = \begin{pmatrix} \frac{1}{2}a_{\theta x}^i t^2 + v_x^i t + P_x^i \\ \frac{1}{2}a_{\theta y}^i t^2 + v_y^i t + P_y^i \end{pmatrix}. \tag{1}$$

For given t, the set of above points makes a closed curve with respect to θ. Conversely, for given point $\mathbf{x} = (x, y)$, we can compute the time which each

---

[1] We call a considering object (such as a player) an agent.

[2] These equation do not consider the maximum velocity of the agent. If the maximum velocity must be considered, the equations should be replaced to the non-accelerated motion equations after reaching the maximum velocity.

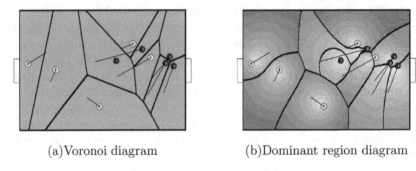

(a)Voronoi diagram   (b)Dominant region diagram

Fig. 1. Voronoi diagram vs. dominant region diagram

agent takes[3]. Therefore, for each point **x** in a region (or a soccer field), we can get the dominant region by computing the following equation,

$$I_x = \underset{i}{\operatorname{argmin}}\{t^i(\mathbf{x})\}, \tag{2}$$

where, I_x is an agent's number which comes at first to the point **x**.

Preliminary experiment using the algorithm proposed in [3] shows that the computation time takes 10 to 40 seconds when the soccer field is digitized by 610×420 grid points.

2.2 Comparisons between Voronoi Diagram and Dominant Region Diagram

The voronoi diagram divides the region based on the distance between agents while the dominant region divides the region based on the arrival time of agents. The voronoi diagram is used to analyze the spheres of influence and is shown that it is useful in RoboCup Soccer Simulation [6]. However, we think that arrival time should be considered when analyzing the sphere of influence in RoboCup Soccer, since the robots are moving in various speeds. Figure 1 shows an example of the voronoi and dominant region diagram of a scene in the game. In the figure above, small circles are agents and the straight line originated from each agent is a current velocity vector of the agent. Note that the shape of the border lines of the regions are quite different between two diagrams. The dominant region becomes a powerful tool when deciding strategy/tactics under the consideration of the motion model of the robots.

3   Approximated Dominant Region

To achieve a real-time computation of the dominant region, where the real time means a few milliseconds here, we propose an *approximated dominant region*. It

---

[3] If more than one arrival time are obtained at point **x** for the agent i, the minimal arrival time is taken.

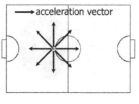

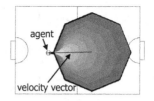

(a)Acceleration vectors   (b)Reachable polygonal region

Fig. 2. Acceleration vectors and reachable polygonal region

can be obtained as a union of *reachable polygonal regions*. A reachable polygonal region is a polygon which is uniquely calculated when the motion model and time are given.

3.1   Motion Model of Robots

In this paper, we define a motion model as a set of maximum acceleration vectors of a robot. Figure 2(a) shows an example of a motion model. Each maximum acceleration vector shows that the robot can move to that direction with the given maximum acceleration. This is an example of an omni-directional robot. Eight vectors are given. The number of vectors depends on the accuracy of obtaining the dominant region.

3.2   Computation of Reachable Polygonal Region

The reachable polygonal region is a region that is included in the polygon made by connecting the points, where each point is given as a point that an agent arrives at after t seconds when it moves toward the given direction of maximum acceleration vector in maximum acceleration. Eq. (1) is used to compute the point. Figure 2(b) shows an example of reachable polygonal region (shaded area) after 1 second passed when the acceleration vectors of figure 2(a) is given. We assume the reachable polygonal region is convex[4]. The reachable polygonal region is calculated by the following algorithm.

[**Reachable polygonal region**]
Step 1 Give a motion model of each agent (figure 2(a)).
Step 2 Give time t. Calculate each arrival point (x_θ^i, y_θ^i) according to the equation (1) using the corresponding maximum acceleration vector in Step1.
Step 3 Connect points calculated in Step2 (figure 2(b)).

3.3   Calculation of Approximated Dominant Region

The approximated dominant region is obtained from the reachable polygonal regions for every agent. When some of reachable polygonal regions are overlapped,

---

[4] If it is concave, we consider a convex hull of it.

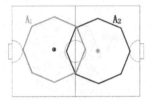
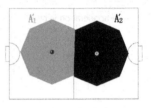

(a)Overlapped reachable polygonal
regions

(b)Divided reachable polygonal
regions

Fig. 3. Division of two overlapped Reachable polygonal regions

we have to decide which agent, the point belongs to the overlapped region. Figure
3(a) shows two overlapped reachable polygonal regions (A_1, A_2) of two agents.
In this case, it is natural to divide overlapped region into two by the line connecting the points of intersection of two polygons. Figure 3(b) shows a result for
the reachable polygonal regions. However, since the number of points of intersection between two polygons (with n vertices) varies from 0 to $2n$, we have to
clarify the method of division for each case. Moreover, we need to consider the
method of division when many reachable polygonal regions are overlapped. We
describe these methods in the following algorithm. We call this an algorithm of
the approximated dominant region.

[Approximated dominant region]
Step 1 For given time t, make a reachable polygonal region of each agent. (Figure 2(b)).
Step 2 For two reachable polygonal regions, if they are overlapped, divide the
overlapped region in the following way. Generally, a number of points of intersection between two polygons with n vertices varies from 0 to $2n$. If a
vertex of one polygon is on the other polygon, move the vertex infinitesi-
mally to the direction where the number of points of intersection does not
increase. (There is no side effect with respect to this movement.) Therefore,
the number of points of intersection is even. We show the way to divide in
case of 0, 2 and $2k$ intersections.
  1. No points of intersection: There are two cases.
     (a) Disjoint: As two reachable polygonal regions are disjoint, there is no
         need to divide.
     (b) Properly included: One includes the other. Figure 4(a) shows an
         example($A_1 \supset A_2$). In this case, $A_1 - A_2$ is a dominant region of
         agent 1 (Fig. 4(b)) and A_2 is a dominant region in agent 2 (Fig.
         4(c))[5].
  2. 2 points of intersection: The overlapped regions of A_1 and A_2 is divided
     into two region by the line connecting the points of intersection between
     two polygons to create dominant regions A_1' and A_2' (Figure 3).

---

[5] This is not correct definition, but we adopt this to perform the real time computation.

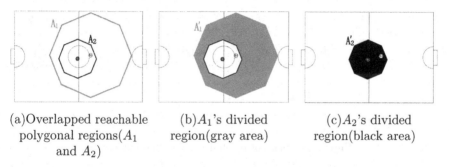

(a)Overlapped reachable   (b)A_1's divided   (c)A_2's divided
polygonal regions(A_1 region(gray area)   region(black area)
and A_2)

Fig. 4. Division of overlapped reachable polygonal regions(one-contains - other case)

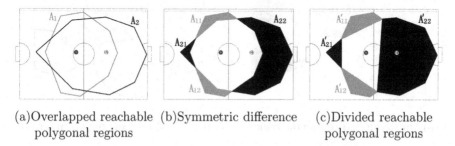

(a)Overlapped reachable   (b)Symmetric difference   (c)Divided reachable
polygonal regions   polygonal regions

Fig. 5. Division of overlapped reachable polygonal regions(4 intersecting case)

3. $2k$ points of intersection: Let A_1 and A_2 be two reachable polygonal regions and I be a set of points of intersection between the polygons of A_1 and A_2. Then, compute $A_1 - A_2$ and $A_2 - A_1$. Figure 5(a) shows an example. In this example, there are 4 points of intersection. Figure 5(b) shows a difference between two regions, where $A_1 - A_2 (= \{A_{11}, A_{12}\})$ is shaded in grey and $A_2 - A_1 (= \{A_{21}, A_{22}\})$ is shaded in black. Make convex hulls of subregions. Figure 5(c) shows the result ($A'_{11}, A'_{12}, A'_{21}, A'_{22}$). Thus, we have *partial dominant regions* ($A'_1 = A'_{11} \cup A'_{12}$ and $A'_2 = A'_{21} \cup A'_{22}$) of the agent 1 and 2. A white area in the overlapped region in figure 5(c) doesn't belong to either of two partial dominant regions.

Step 3 If n reachable polygonal regions ($A_1, A_2, \cdots, A_n$) are overlapped, we process as follows. First, for A_1 and A_2, we take partial dominant regions A'_1 and A'_2 by using the procedure in step 2. Replace A_1 and A_2 with A'_1 and A'_2. Then, for A_1 and A_3, and A_2 and A_3, do the same computation. Repeat this until A_n is computed. As a result, we get new reachable polygonal regions ($A_1, A_2, \cdots, A_n$) where any two A_is are disjoint. These are the partial dominant regions of agents at given time t. Figure 6 shows three examples of partial dominant regions of 10 agents at time $t = 0.5, 0.7$ and 0.9 seconds.

Step 4 Synthesize the partial dominant regions incrementally. For given times $t_1, t_2, \cdots, t_n$ ($t_1 < t_2 \cdots < t_n$), compute the partial dominant regions. Let them be $B_1, B_2, \cdots B_n$. Then, compute $B_1 + (B_2 - B_1) + \cdots + (B_n - \sum_{i=1}^{n-1} B_i)$.

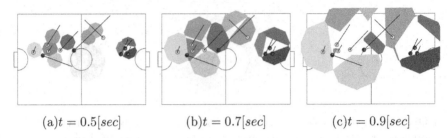

(a)$t = 0.5[sec]$ (b)$t = 0.7[sec]$ (c)$t = 0.9[sec]$

Fig. 6. Synthesis of reachable polygonal regions

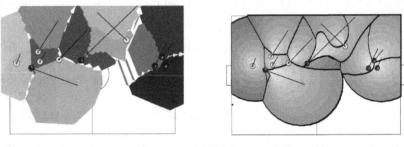

(a)Approximated dominant region   (b)Precise dominant region diagram
diagram

Fig. 7. Approximated dominant region diagram vs. dominant region diagram

This makes an approximated dominant region diagram. Figure 7(a) shows an example constructed from the examples shown in figure 6, but using 10 partial dominant regions computed by every 0.1 seconds.

3.4   Feature of Proposed Algorithm

The proposed algorithm computes the approximated dominant region and it makes the great reduction of the computation time. However, there are some small regions that are not in the approximated dominant region, but in some of the dominant regions of agents. We allowed to remain such small regions because we strongly pursue the real-time computation. We evaluate the negative effect of such small regions through the evaluation experiments.

4   Experimental Evaluation for Algorithm of Approximated Dominant Region

In this section, we apply our algorithm to the SSL and evaluate how well it works.

Table 1. Computation time and accuracy of proposed algorithm

acc. vectors	arrival-time steps	computation time[$msec$]	accuracy[%]
8	10	17.5	91.8
8	20	33.8	92.9
16	10	31.4	95.3
64	200	2.4×10^3	99.9
2048	400	5.0×10^5	100

Table 2. Computation time on various computers (parameters: max acc. vectors: 8, arrival-time steps: 10)

	CPU	proposed method(A)	existing method(B)	rate(B/A)
1	3.16GHz	17.5[msec]	13.3[sec]	760
2	3.2GHz	19.3[msec]	24.2[sec]	1254
3	2.2GHz	38.2[msec]	40.5[sec]	1060
4	2.2GHz	38.3[msec]	40.4[sec]	1055

4.1 Experiment

In the SSL, since the ball moves very fast, the standard processing cycle is 60 processings per second. One processing includes an image processing, a decision making, action planning, command sending and so on. Therefore, the allowed time for the computation of the dominant region is at most 5 milli-seconds[6]. Our purpose is to make the computation of the approximated dominant region within 5 msec.

4.2 Experimental Result

We digitize the SSL's field into 610×420 grid points (1 grid represents the area of about 1 cm^2) and, for 10 agents (5 teammates and 5 opponents), compute the approximated dominant region that can arrive within 1 second. The reason why we set arrival time to 1 sec is that almost all of the whole field can be covered by the dominant region as shown in Fig. 7. We measure the computation time and the accuracy of the approximated dominant region. We define the accuracy by the following equation,

$$\texttt{Accuracy}[\%] = \frac{\text{Total grid points that } I_x^d \text{ and } I_x^a \text{ coincide}}{\text{All grid points}} * 100 \qquad (3)$$

where, I_x^d and I_x^a are given by Eq. (2) for the precise dominant region and the approximated dominant region, respectively.

---

[6] This time constraint is sufficient when our algorithm will be applied for the other leagues in RoboCup soccer and human soccer.

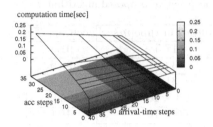
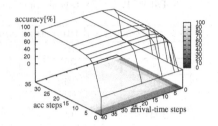

(a)Computation time vs. acceleration   (b)Accuracy vs. acceleration vector
   vector and arrival-time step pairs   and arrival-time step pairs

Fig. 8. Computation time and accuracy

We used the computer with Xeon X5460 as CPU, 8GB main memory and FreeBSD operating system for this experiment. We measured the computation time by running the program in a single thread.

The computation time and the accuracy of the approximated dominant region depend on the number of maximum acceleration vectors and the number of partial dominant regions (i.e. the number of time-divisions). We measured the computation time and the accuracy for the various values by using two parameters above. Figure 8 shows the results of the measure. Table 1 shows the computation time and the accuracy for some typical values of the parameters. The resulting approximated dominant region of the first row of the Table 1 is shown in Figure 7(a).

In comparison with Fig. 7(b), it is considered that Fig. 7(a) is a good approximation of the dominant region diagram. The accuracy is ranging from 91.8% to 100% from Table 1. These numbers show that our algorithm gives a good approximated dominant region.

Table 2 shows the computation time of the approximated dominant region diagram measured on the various computers. (The parameters on this experiment are fixed as the number of maximum acceleration vectors takes 8 and the number of time-divisions takes 10.) From the table, it is shown that the computation time can be reduced about 1/1000 times shorter compared with the algorithm shown in [3] and the accuracy keeps a little more over 90%. In addition, it is shown that, from Fig. 8(a), the computation time increases in proportion to the number of maximum acceleration vectors and the number of time-divisions, and from Fig. 8(b), the accuracy goes up rapidly to be 90% according to the increase of the number of maximum acceleration vectors and/or time-divisions, and then still slightly increases.

4.3   Discussion

From Table 1, the approximated dominant region with parameters of 8 maximum acceleration vectors and 10 time-divisions achieves the accuracy of 92%. However, its computation time takes 17.5 *msec*. It is a little bit far from our

Table 3. Computation time necessary to make and to synthesize reachable polygonal regions

acc. vectors	8	16	24	32
average computation time [$msec$]	1.71	3.10	4.37	5.46
standard deviation[$msec$]	0.016	0.017	0.014	0.014

goal, which is computation time to be 5 $msec$. To achieve this goal, we discuss a parallel computation here. Another issue whether the accuracy of 92% is enough in our purpose will be discussed in the next section.

In our algorithm, we create an approximated dominant region by synthesizing the partial dominant regions incrementally. Each partial dominant region can calculate independently and its computation time is almost equal for all the partial regions. The latter is supported by the fact that the computation time doubles when the partial dominant regions double. (See Fig. 8(a).) And also Table 3 shows an average time to compute a partial dominant region. For synthesis of 10 partial dominant regions, it takes $0.5msec$ on average. Therefore, it is expected in parallel computation that the computation time will be about $1.71 + 0.5 + \alpha$ $msec$ when 10 partial dominant regions are synthesized, where α is an overhead of the parallel computation and is considered as a constant. In the multi-core parallel computer, the α is small enough. We consider that it is possible to make the computation time within 5 $msec$, which is our goal.

5   Prediction of Success for Passing

One application of the dominant region is the prediction of success for passing. In this section, we introduce a dominant region of a ball. The approximated dominant region takes a significant role to predict the success for passing. If we can predict the success of passing accurately in real time, we can choose a defense or an offense strategy appropriately.

5.1   Approximated Dominant Region of Ball

We consider a dominant region of a ball. The motion of the kicked ball on the SSL's field can be considered as a uniform decelerated motion, since the ball receives the force by the friction of the field only, and the friction is constant over the field. In addition, the ball moves on a straight line unless it meets with an object. Thus, the dominant region of a ball is defined as a line segment that the ball does not meet with any agents. By using the following way, it is possible to find an agent who can get the ball first: 1) compute a partial dominant region for time t_i and draw the position of ball at time t_i on it. 2) If the ball is in the dominant region of an agent, then the agent can get the ball. if not, repeat computation for next time t_{i+1} until the ball is in the dominant region of an agent. By this way, we can predict the agent who gets the ball first, and also we can find a dominant region of a ball.

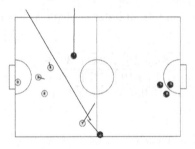

(a)An example of soccer game (at present)

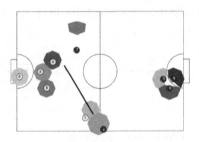

(b) A synthesized partial dominant region until $t = 0.45[sec]$

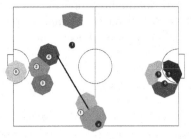

(c) A synthesized partial dominant region until $t = 0.5[sec]$

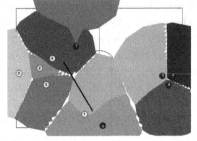

(d)Approximated dominant region diagram with ball ($t = 1sec$)

Fig. 9. Example of approximated dominant region diagram with ball

Figure 9 shows an example. Fig. 9 (a) is a current situation of the game. The ball is at the lower part of the left side from the center line. The lines in front of the agents and the ball show the velocity of them. Figs. 9 (b) and (c) are the synthesized partial dominant region until $t = 0.45sec$ and $t = 0.5sec$, respectively. In Fig. 9 (b), there is no agent who can get the ball, but in Fig. 9 (c), the agent No. 4 can get the ball, since the ball is in the dominant region of the agent 4. Fig. 9 (d) shows the approximated dominant region until $t = 1sec$ and the ball's dominant region. To make this diagram, it takes 0.5 $msec$ more time than the computation time of the diagram without the ball.

5.2   Discussion

In this experiment, we use the approximated dominant region with parameters of 8 maximum acceleration vectors and 20 time-divisions for 1 second interval[7]. We used the logged data of the third-place match in 2007 RoboCup competition to analyze the prediction of success for passing. By using the proposed algorithm, we predict the robot who gets the ball first. 60 passings are predicted correctly out of 63 total passings in the game (95% accuracy), i.e. the predicted agent and the agent that gets the ball in the game coincide.

---

[7] It is possible to obtain the approximated dominant region within 5 $msec$ under the parallel computation environment even if these parameter values are used.

In the results, 3 passings are failed to predict correct agents. The detailed analysis shows that the cause of mis-prediction is not due to the accuracy of the approximated dominant region but due to the strategy of the team. That is, the mis-predicted agent acts to achieve an other goal like moving the goal area to defend the goal by the team's strategy instead of getting the ball. Therefore, we think the approximated dominant region is very useful to judge the prediction of success for passing as well as to evaluate the team's strategy.

6   Conclusion

In this paper, we discussed the real-time computation of the dominant region. We proposed an approximated dominant region algorithm which can calculate an approximated dominant region in real time with accuracy over 90 % under the parallel computation environment.

Moreover, we proposed an approximated dominant region including a ball's dominant region. Experimental results show that it predicts the agent who will get the ball correctly over the accuracy of 95%. This also shows that the approximated dominant region is useful for the analysis of the team's strategy.

In this paper, we also discussed the application of the algorithm to the RoboCup, but it is possible to apply the algorithm to the other objects by establishing the moving model of the agents correctly. In this case, it is necessary to choose appropriate values of parameters, because the calculation cost depends on the required accuracy of approximation.

The future problems are to reduce the computation time further with keeping the approximation accuracy, and to exploit new applicable fields of the algorithm, not just for sports games.

References

1. Murakami, K., Hibino, S., Kodama, Y., Iida, T., Kato, K., Naruse, T.: Cooperative Soccer Play by Real Small-Size Robot. In: Polani, D., Browning, B., Bonarini, A., Yoshida, K. (eds.) RoboCup 2003. LNCS (LNAI), vol. 3020, pp. 410–421. Springer, Heidelberg (2004)
2. Preparata, F.P., Shamos, M.I.: Computational Geometry. Springer, Heidelberg (1988)
3. Taki, T., Hasegawa, J.: Dominant Region: A Basic Feature for Group Motion Analysis and Its Application to Teamwork Evaluation in Soccer Games. In: Proc. SPIE Conference on Videometrics VI, January 1999, vol. 3641, pp. 48–57 (1999)
4. Nakanishi, R., Bruce, J., Murakami, K., Naruse, T.: Cooperative 3-robot passing and shooting in the RoboCup Small Size League. In: Lakemeyer, G., Sklar, E., Sorrenti, D.G., Takahashi, T. (eds.) RoboCup 2006. LNCS (LNAI), vol. 4434, pp. 418–425. Springer, Heidelberg (2007)
5. Nakanishi, R., Murakami, K., Naruse, T.: Dynamic Positioning Method Based on Dominant Region Diagram to Realize Successful Cooperative Play. In: Visser, U., Ribeiro, F., Ohashi, T., Dellaert, F. (eds.) RoboCup 2007. LNCS (LNAI), vol. 5001, pp. 488–495. Springer, Heidelberg (2008)
6. Dashti, H.T., et al.: Dynamic Positioning Based on Voronoi Cells (DPVC). In: Bredenfeld, A., Jacoff, A., Noda, I., Takahashi, Y. (eds.) RoboCup 2005. LNCS (LNAI), vol. 4020, pp. 219–229. Springer, Heidelberg (2006)

A Lua-based Behavior Engine
for Controlling the Humanoid Robot Nao

Tim Niemüller[1], Alexander Ferrein[1,2], and Gerhard Lakemeyer[1]

[1] Knowledge-based Systems Group
RWTH Aachen University, Aachen, Germany
{niemueller,gerhard}@kbsg.rwth-aachen.de
[2] Robotics and Agents Research Lab
University of Cape Town, Cape Town, South Africa
alexander.ferrein@uct.ac.za

Abstract. The high-level decision making process of an autonomous robot can be seen as an hierarchically organised entity, where strategical decisions are made on the topmost layer, while the bottom layer serves as driver for the hardware. In between is a layer with monitoring and reporting functionality. In this paper we propose a behaviour engine for this middle layer which, based on formalism of hybrid state machines (HSMs), bridges the gap between high-level strategic decision making and low-level actuator control. The behaviour engine has to execute and monitor behaviours and reports status information back to the higher level. To be able to call the behaviours or skills hierarchically, we extend the model of HSMs with dependencies and sub-skills. These Skill-HSMs are implemented in the lightweight but expressive Lua scripting language which is well-suited to implement the behaviour engine on our target platform, the humanoid robot Nao.

1 Introduction

Typically, the control software of an autonomous robot is hierarchically organised with software modules that communicate directly to the hardware on the lowest level, some more elaborated entities for, say, localisation or the object detection on a middle layer, and an action selection mechanism on top. On each of these levels the time constraints are different, meaning that modules on lower levels have shorter decision cycles and need to be more reactive than those on the higher levels. The reason is that usually, the level of abstraction increases with each layer of the software. (See e.g. [1,2] for textbooks on "classical" 3-tier architectures). The same holds for the high-level action selection. From that viewpoint basic or primitive actions are selected in coordination with the teammates and the team strategy; these are broken down to actuator commands on the lowest level over several levels of software abstraction. The term basic or primitive action hides the fact that these actions are usually on a high level of abstraction. Examples for those basic actions are dribble or attack-over-the-left-wing. Many different approaches exist for how these primitive actions will be selected. These range

J. Baltes et al. (Eds.): RoboCup 2009, LNAI 5949, pp. 240–251, 2010.

from full AI planning to simply reactively couple sensor values to these actions or behaviours. Between the high-level action selection and the driver modules for the servo motors of the robot, a middle layer is required to formulate complex actions and report success or failure to the high-level control.

In this paper, we address this middle-layer and show a possibility how the gap between high-level control and the rest of the robot system can be bridged. Independent from the high-level scheme, the middle control layer for behaviours, which we call *behaviour engine*, needs to be expressive enough to hide details from the high-level control while having all the information needed to order and monitor the execution of basic actions in the low-level execution layer. This behaviour engine must thus provide control structures for monitoring the execution and facilities to hierarchically call sub-tasks etc. Moreover, it needs to be lightweight enough not to waste resources which should either be spent for the high-level decision making or for tasks like localisation. Hence we need a computationally inexpensive framework which allows for the needed kind of expressiveness. As our target platform is the standard platform Nao which has limited computational resources, the lightweight of the behaviour engine is even more important. We propose a behaviour engine which matches these criteria. The building block for our behaviour engine is a *skill*. We formalise our skills as extended Hybrid State Machines (HSMs) [3] which allow for using state machines hierarchically. Our implementation of these *Skill-HSMs* is based on the scripting language Lua [4], which is a lightweight interpreter language that was successfully used before for numerous applications ranging from hazardous gas detection systems for the NASA space shuttle to specifying opponent behaviour in computer games. We show that the combination of Lua and HSMs provide a powerful system for the specification, execution and monitoring of skills. However, by choosing a general purpose language like Lua, we do not preclude the possibility to later extend the behaviour engine. It is therefore also possible to use other skill specifications in parallel to our hybrid state machines.

The paper is organised as follows. In Sect. 2 we present the software framework FAWKES which we use for our Nao robot. In particular, we show the different control modules and how the communication between the different sub-systems takes place. One important issue is that the original motion patterns from NaoQi can be easily integrated, if desired. In Sect. 3 we define skills in terms of extended HSMs, called Skill Hybrid State Machines (SHSMs), to yield a hierarchy of skills, and distinguish between actions, behaviours, and motion patterns. Section 4 addresses the implementation of SHSMs in Lua. In Section 5 we show an example state machine for the stand-up motion. We conclude with Section 6.

2   Fawkes and the Nao

In this section we briefly introduce the ideas behind and key components of the FAWKES software framework and describe our target platform, the Nao. We then go over to describing our instantiation of FAWKES on the Nao platform.

2.1   The Humanoid Robot Nao

The Nao [5] is the new biped platform for the STANDARD PLATFORM LEAGUE by the French company Aldebaran. The league is the successor of the Sony Aibo league. The 58 cm tall robot has 21 degrees of freedom, and is equipped with two CMOS cameras providing 30 frames per second in VGA resolution, it has some force sensitive resistors, a gyroscope and an accelerometer as well as sonar sensors. The CPU is an AMD Geode running at 500 MHz accompanied by 256 MB of RAM. To communicate with the platform, Wi-Fi (IEEE 802.11b/g) and Ethernet are available. It comes with a closed source software framework called NaoQi which is the only way to access the robots actuators and sensory. It also provides basic components for image acquisition, walking, and actuator pattern execution.

2.2   Fawkes in a Nutshell

The FAWKES robot software framework [6] provides the infrastructure to run a number of plug-ins which fulfil specific tasks. Each plug-in consist of one or more threads. The application runs a main loop which is subdivided into certain stages. Threads can be executed either concurrently or synchronised with a central main loop to operate in one of the stages. All threads registered for the same stage are woken up and run concurrently. Unlike other frameworks such as Player or Carmen we pursue a more integrated approach where plugins employ threads running in a single process exchanging data via shared memory instead of message passing. Currently, we use the software in the ROBOCUP@HOME LEAGUE for domestic service robots as well as in the MIDDLE SIZE LEAGUE and STANDARD PLATFORM LEAGUE for soccer robots. The framework will soon be released as Open Source Software.

Blackboard. All data extracted and produced by the system and marked for sharing is stored in a central blackboard. It contains specified groups of values which are accessed with a unified interface. An example is an object position interface, which provides access to position information of an object, like the position of the robot itself or the ball on the field. These interfaces can be read by any other plug-in to get access to this information. Commands are sent to the writer via messages. Message passing eliminates a possible writer conflict. Opposed to IPC (see http://www.cs.cmu.edu/~{}ipc/) as used by Carmen data is provided via shared memory, while messaging is used only for commands.

Component-based Design. FAWKES follows a component-based approach for defining different functional blocks. A component is defined as a binary unit of deployment that implements one or more well-defined interfaces to provide access to an inter-related set of functionality configurable without access to the source code [7,8]. With the blackboard as a communication infrastructure, system components can be defined by a set of input and output interfaces. This

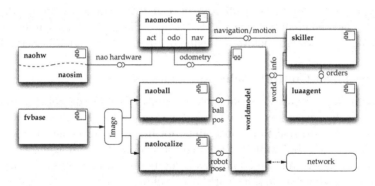

Fig. 1. Component configuration for FAWKES on the Nao robot

allows for easily replacing a component as long as the replacement component requires and provides the same sets of interfaces.

2.3 Running Fawkes on the Nao

Figure 1 shows the component configuration that has been implemented for the Nao robot platform. On the left-hand side one can find the *naohw/naosim* component. The *naohw* and *naosim* plug-ins both provide access to the underlying robot hardware, on the real robot and in a simulation environment. The *fvbase* plug-in provides access to the camera and distributes the acquired image via a shared memory segment. The *naoball* and *naolocalize* plug-ins use this image to extract information about the robot and ball position. This data together with other acquired information is processed in the *worldmodel* component. This component merges different sources of information (local extraction components and information received from other robots via Wi-Fi) to provide a unified world model to all higher level components. The *naomotion* component deals with robot locomotion and odometry calculation. It also includes a navigation component for path planning. The unified world model is used by the skill execution run-time (*skiller*) and the *luaagent* component which we are going to describe in more detail below.

NaoQi Integration. Given the current closed nature of the Nao robot platform it is essential to integrate FAWKES with NaoQi to gain access to the hardware. Beyond plain hardware access it is desirable to provide access to other NaoQi functionality. An integration module exists for this purpose that integrates FAWKES with NaoQi. For instance the actuator sub-component of *naomotion* can be provided via the integration of Aldebaran's NaoQi motion engine.

3 The Behaviour Engine and Skill Hybrid State Machines

In this section we define the behaviour engine. As stated in the introduction, the whole behaviour system can be seen as a layered system, just like a hierarchically structured system for the overall control of a robot. In Sect. 3.1 we therefore distinguish between *low-level control, behaviours,* and the *agent* as different levels of the behaviour system, before we define our behaviour engine in Sect. 3.2.

3.1   Low-Level Control, Behaviours and Agents

To model a robot's behaviour multiple different levels of control can be distinguished. On the lowest level we have tight control loops which have to run under real-time or close-to-real-time constraints, for instance, for generating joint patterns to make the robot walk. On the highest level we have an agent which takes decisions on the overall game-play or on the team strategy, possibly communicating and coordinating actions with other robots.

Especially when more elaborated approaches for designing an agent like planning or learning are used, it is beneficial to not only have the very low-level actions like "walk-a-step", but also more elaborate reactive behaviours like "search-the-ball" or "tackle-over-the-right-wing". This reduces the computational burden for the agent tremendously. Additionally it is easier to develop and debug small behaviour components. In the following we will clarify what we understand with skills and show the different levels of behaviours as three tiers.

Definition 1 (Behaviour levels)

Level 0: Low-level Control Loops *On this level modules run real-time or close-to-real-time control loops for tasks like motion pattern generation or path-planning and driving.*

Level 1: Skills *Skills are used as reactive basic behaviours that can be used by the agent as primitive actions.*

Level 2: Agent *An agent is the top-most decision-making component of the robot that makes the global decisions about what the robot is doing and the strategic direction.*

At each level, behaviours can only be called by other behaviours which are from the same or a higher level.

According to this understanding of a tiered behaviour system, the FAWKES software framework was designed. Figure 2 shows the organisation of the FAWKES software stack. On the lowest level modules for sensor acquisition, data extraction, and other low-level control programs are located. On the top, the agent is making the overall decision on the behaviour. In between lies a reactive layer which provides basic actions to the agent. For this it uses information of the low-level modules to make local decisions for a specific behaviour, e.g. when approaching a ball the walking direction might need to be adjusted for ball movements. From that it creates commands for the low-level actuator driving components like locomotion.

3.2   Behaviour Engine

Against the background of a deliberative approach, one has specific expectations what the behaviour engine has to provide, which nevertheless can be applied for

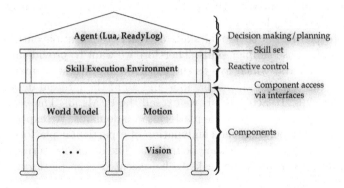

Fig. 2. The FAWKES Software Stack

reactive decision making. On the higher level strategic planning, we need primitive actions that we can model and use for plan generation. For efficiency reasons these primitive actions – skills – are written following a reactive paradigm. The skills are small reactive execution entities which have a well-defined purpose, like "go-to-position (x, y)" or "intercept-ball". When a skill is called it will try to fulfil its purpose, or report a failure if that is not possible, e.g. the ball is no longer visible while intercepting it. A skill cannot deliberately switch to another skill. This decision is the task of the higher level agent program. However, it can call another skill as part of itself, for instance the *intercept* skill has to call the *goto* skill for the movement towards the ball. But the intercept would not decide to change the behaviour to *search-ball* when the ball is not visible. While changing the active skill could make sense for a field player, a defender might better go back to the defending position. Therefore the decision to switch skills should be made by the agent controlling the overall behaviour. Skills can thus be seen as execution entities which make only local decisions, not global game-play decisions. These skills need a particular programming and run-time environment – the behaviour engine. According to Definition 1 the behaviour engine is located at level 1 in our behaviour hierarchy. From the initial proposition that skills are reactive execution entities which accomplish simple task, i.e. the primitive actions from a higher level perspective, state machines are an obvious choice for modelling the behaviour. As we need continuous transitions between states, we selected hybrid state machines (HSMs) as the model of our choice. A HSM is a finite state machine which allows for state transitions following differential equations (flow conditions) on the one side, and logical expressions (jump conditions), on the other side. Jump conditions are represented by the inequalities, while the flow conditions are stated by the differential equations (see e.g. [3]).

UML state charts combined with hybrid automata have been used by Stolzenburg et al. [9,10] to model higher level reactive behaviours. The focus of their work, however, was on level 2 of our behaviour hierarchy, where they coordinated the multi-agent behaviour with HSMs. Here, we concentrate on a formal model for hierarchical skills on level 1.

We want to use HSMs to model skills as reactive execution entities. More complex skills often use simpler skills. Therefore we want to provide an efficient way for re-using a skill, which avoids the construction of a state machine that includes both, the complex skill behaviour and all of the internal details of the included simple skill. To achieve this we extend the HSMs [3]. The behavior engine should be applicable to multiple platforms and domains. We expect that the available skills will depend on the combination of a particular platform and domain. Therefore we give the following definitions. The domain describes the area of operation and determines the tasks to be executed. The platform describes the used robot system.

Definition 2 (Skill Space). *The combination of a platform $\mathcal{P}$ and a domain $\mathcal{D}$ with regard to skills is called* skill space $(\mathcal{P}, \mathcal{D})$ *for platform $\mathcal{P}$ and domain $\mathcal{D}$.*

Definition 3 (Set of Skills). *The set $\mathcal{K}_{(\mathcal{P},\mathcal{D})}$ is called the* set of skills *for the skill space $(\mathcal{P}, \mathcal{D})$. It denotes the set of available skills for a particular skill space.*

Definition 4 (Skill Hybrid State Machine (SHSM))

$$S = (G, X, D, A, jump, flow, exec, \mathcal{K}_{(\mathcal{P},\mathcal{D})}) \tag{1}$$

Final and failure state. *The graph $G = (Q, E)$ has only two valid exit states $Q_{\text{exit}} = \{q_{\text{final}}, q_{\text{failure}}\}$.*

Control graph. *A finite directed multi-graph $G = (Q, T)$, where $Q = Q_U \cup Q_{\text{exit}}$ are the states (vertices) with Q_U being the user defined states and T are the transitions (edges).*

Dependencies. *For hierarchical definition of a skill existing skills can be re-used. These used skills are called* dependencies *of skill S. Skills that are used in the current skill are called* direct dependencies, *skills that are used in direct dependencies or their dependencies are called* indirect dependencies. *A skill may not depend directly or indirectly on itself. For this we define a set $D \subseteq \mathcal{K}_{(\mathcal{P},\mathcal{D})} \setminus S$ of dependencies. Let $D_S \subseteq D$ be the set of skills that the skill S directly depends on. Then the function $\delta : \mathcal{K}_{(\mathcal{P},\mathcal{D})} \to \wp(\mathcal{K}_{(\mathcal{P},\mathcal{D})} \setminus S)$ with $\delta(S) = D_S \cup \{\delta(d) \mid d \in D_S\}$ gives a set of all direct and indirect dependencies of S and $S \notin \delta(S)$. This can be represented as a dependency graph.*

Execution Function. *A skill is executed with the* exec *function. It assigns values to some variables $x \in X$ and runs the state machine by evaluation of the jump conditions of the current state, possibly leading to a state change. It is defined as $\text{exec}(x_1, \ldots, x_n) \to \{\text{final}, \text{running}, \text{failure}\}$ with $x_i \in X$. The return value depends on the current state after the evaluation.*

Actions. *For the execution of lower-level behaviors and other SHSMs we define a set A of actions. An action $a \in A'$ is a function $a(x_1, \ldots, x_n) \to \{\text{running}, \text{final}, \text{failure}\}$ with $x_i \in X$ that executes a lower-level system behavior (on a lower behavior level). The set $K = \{\text{exec}_d \mid d \in D\}$ is the set of execution functions of dependency skills (on the same behavior level). The set of actions is then defined as $A = A' \cup K$.*

Action Execution. *For each state $q \in Q$ we define a set $E_q \subseteq A$ of actions. Each action is executed when the state is evaluated. The set E_q may be empty.*

Before we illustrate the definition of skill hybrid state machines in Sect. 5 with an example skill from the humanoid robot Nao in detail, we address the implementation of SHSMs in Lua in the next section. For now, it is sufficient to note that skills are special hybrid state machines that can hierarchically be called by other skills. We hence model actions and sub-skills as a way to interact with the actuators of the robot and to easily call other skills. To avoid specification overhead sub-skills are defined as functions. Rather then integrating another skill's state machine when it is called into the current caller's state machine it is sufficient to call the encapsulating k-function and define appropriate jump conditions based on the outcome. Dependencies are defined as a way to avoid cyclic call graphs when a skill calls another skill.

4   Implementing Skill Hybrid State Machines in Lua

4.1   Lua

Lua [4] is a scripting language designed to be fast, lightweight, and embeddable into other applications. These features make it particularly interesting for the Nao platform. The whole binary package takes less then 200 KB of storage. When loaded, it takes only a very small amount of RAM. This is particularly important on the constrained Nao platform and the reason Lua was chosen over other scripting languages, that are usually more than an order of magnitude larger [11]. In an independent comparison Lua has turned out to be one of the fastest interpreted programming languages [11,12]. Besides that Lua is an elegant, easy-to-learn language [13] that should allow newcomers to start developing behaviours quickly. Another advantage of Lua is that it can interact easily with C/C++. As most robot software is written in C/C++, there exists an easy way to make Lua available for a particular control software.

Other approaches like XABSL [14] mandate a domain specific language designed for instance for the specification of state machines. This solves to some extent the same problems, easy integration, low memory foot print and easy development. But it also imposes restrictions in terms of the expressiveness of the language. By using a general purpose language one can easily experiment with different approaches for behaviour formulations. Using parsing expression grammars implemented in Lua (LPEG [15]), XABSL files could even be read and executed within the Lua behaviour engine.

Lua is popular in many applications. Especially in the computer game sector Lua has found a niche where it is the dominant scripting language. It has been used to develop game AI and extension modules for games. And even in RoboCup applications, Lua showed its strength as programming language before [16,17].

Integration of Lua into Fawkes. As FAWKES uses a plug-in strategy for integrating software modules, it was particularly easy to develop a Lua plug-in for FAWKES making use of the C/C++ interface. As on the level of the behaviour engine required information from the low-level control system are stored in the blackboard, access to the blackboard from Lua needed to be guaranteed. Wrappers for accessing C++ code from Lua can be generated automatically via tolua++ (cf. http://www.codenix.com/tolua/ for the reference manual). Since interfaces are generated from XML descriptions input for automated wrapper generation by means of tolua++ can easily be created. With this, data can be read from and written to the blackboard, and messages can be created and sent from Lua. With this access to all the robot's information about the current world situation is accessible and commands to any component in the framework can be sent.

The agent calls skills by forming a Lua string which calls the skills as functions. The Lua integration plug-in will create a sandbox and execute the Lua string in that sandbox. The sandbox is an environment with a limited set of functions to be able to apply policies on the executed string. This could be preventing access to system functions that could damage the robot or only providing certain behaviours.

4.2   Implementing Skill Hybrid State Machines

We chose SHSMs to model the robot's different behaviours. Each skill is designed as an HSM. The core of the implementation is the Skill-HSM to which states are added. Each state has a number of transitions, each with a target state and a jump condition. If the jump condition holds the transition is executed and the target state becomes the active state. The HSM has a start state which is the first active state. When a skill is finished or stopped by the agent, the state machine is reset to the start state.

The state machine is executed interleaved. That means that in an iteration of the main loop all transitions of the active state are checked. If a jump condition holds the appropriate transition is executed. In this case transitions of the successor state are immediately checked and possibly further transitions are executed. A maximum number of transitions is executed after which the execution is stopped for this iteration. If either a state is reached where no jump condition fires or the maximum number of transitions is reached, the execution is stopped and continued only in the next iteration, usually with fresh sensor data and thus with possibly different decisions of the jump conditions.

4.3   Tools for Developing Behaviours

An often underestimated aspect is the need for tool support when developing behaviours. Lua has the great advantage that it comes with automatic memory management and debugging utilities. Hence, we can focus on the development of skills in the behaviour engine. Instead of going along the lines of programming the behaviours graphically (as RoboCupers usually are experienced programmers), skills need to be coded. Nevertheless, we support the debugging process with visual tools, displaying the state transitions and execution traces on-line.

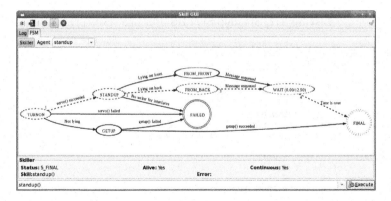

Fig. 3. SkillGUI: a GUI to execute, visualise and debug skills

The behaviour is visualised as a graph with states as nodes and transitions as edges. The developed implementation in Lua allows to create the HSM graph at run-time. From the formulation of the behaviour as HSM in Lua the graph can be directly generated. The Graphviz [18] library is used to generate the graphical representation and display to the user. In Figure 3 the visualisation of the state machine of a stand-up behaviour is shown. The dotted node marks the currently active node, the dashed lines are states and transitions which have been passed in the current run of the state machine. The numbers give the order of the transition sequence, the trace. By this trace one can follow what happened in the state machine even if the transitions happen very fast.

5   The Stand-Up State Machine in Lua

To illustrate the definition of SHSM, we show an example state machine of a stand-up behaviour of the Nao. In Listing 1.1 an excerpt from the code of the stand-up skill is given. The complete skill is visualised in Figure 3. The code shows the state machine for standing up from lying on the back. Of particular interest are line 2, where the dependencies are defined, and the following lines, which define the required blackboard interfaces. Lines 8-10 instantiate the states, line 10 specifically adds a sub-skill state which will execute and monitor the getup skill (which makes the robot stand from a sitting position). In Lines 12-13 a transition is added which is executed when the robot is standing or sitting on its feet. Line 15 adds a transition that is executed if the robot is lying on its back. Lines 17-20 show the code to order an execution in the low-level base system via the blackboard. Parts like the execution of stand-up from lying on the front and waiting for the action to finish have been omitted for space reasons.

In our experiments, besides verifying the correctness of the state machine and its monitoring, we analysed the run-time of the system. As we stated in the introduction, a behaviour engine must be lightweight, not wasting any resources. On average, the state machine takes 1 ms till the next command is executed, 3 ms with the debugging graph visualised.

Listing 1.1. Standup Skill HSM in Lua (excerpt)

```
1   fsm                = SkillHSM:new{name=name, start="STANDUP"}
2   depends_skills     = {"servo", "getup"}
3   depends_interfaces = {
4      {v = "naomotion", type = "HumanoidMotionInterface"},
5      {v = "naohw",     type = "NaoHardwareInterface"}
6   }
7
8   fsm:new_jump_state("STANDUP")
9   fsm:new_jump_state("FROM_BACK")
10  fsm:new_jump_state("GETUP", getup, FINAL, FAILED)
11
12  STANDUP:add_transition(GETUP,
13            "naohw:accel_x() >= -35 and naohw:accel_x() <= 35")
14
15  STANDUP:add_transition(FROM_BACK, "naohw:accel_x() < -35")
16
17  function FROM_BACK:init()
18     naomotion:msgq_enqueue_copy(
19        naomotion.StandupMessage:new(naomotion.STANDUP_BACK))
20  end
```

Other skills that have been implemented include walking to a certain global or relative position, controlling servos (e.g. to move the head while walking), searching for the ball, tracking the ball with the head or intercepting the ball.

In addition to the behaviour engine we have used extended HSMs and Lua to implement a simple agent program to control the overall game-play (*luaagent* component in Figure 1). In this configuration the basic actions for the HSM are the skills provided by the behaviour engine (*skiller* component). By keeping these two layers separate and not writing the agent as a top-level skill the agent could easily be replaced, for instance by a more powerful planning component.

6   Conclusion

In this paper we proposed a behaviour engine for the humanoid standard platform Nao. For this purpose, we introduced a three tier architecture for the behaviour system. The behaviour engine we propose here is located in the middle layer and interfaces between the high-level decision component and the execution of behaviour patterns. This means that the behaviour engine particularly has the task to monitor the execution of behaviours, and report success or failure to the high-level decision maker. Central for our behaviour engine is the skill, which is a piece of code which monitors the low-level execution and reports to the high-level agent. As a formal model for skills we decided for hybrid state machines. We extended these state machines to allow for skill hierarchies. A skill is therefore formalised by a skill hybrid state machine. The implementation of the behaviour engine was done in the lightweight scripting language Lua, which can be easily integrated into a robot control architecture. We use this behaviour engine successfully on the new standard biped platform Nao. In particular it is important to note that the motion patterns via NaoQi can be integrated easily. This makes our approach interesting as an extension of the Nao software architecture. For future work, we want to use the inherent features like events and multi-graphs, which come with HSMs and could be used to model multi-agent behaviour, for example for cooperative team play on the agent level as well.

Acknowledgments. This work partly supported by the German National Science Foundation (DFG) in the Priority Program 1125, and by grant no. SUA 07/031 of the International Bureau of the Federal Ministry of Education and Research (IB/BMBF). A. Ferrein is currently funded by a grant of the Alexander von Humboldt foundation. We would like to thank the anonymous reviewers for their helpful comments.

References

1. Bekey, G.A.: Autonomous Robots: From Biological Inspiration to Implementation and Control. MIT Press, Cambridge (2005)
2. Murphy, R.R.: Introduction to AI Robotics. The MIT Press, Cambridge (2000)
3. Henzinger, T.A.: The theory of hybrid automata. In: Proceedings Logic in Computer Science 1996, July 1996, pp. 278–292. IEEE, Los Alamitos (1996)
4. Ierusalimschy, R., de Figueiredo, L.H., Filho, W.C.: Lua - An Extensible Extension Language. Software: Practice and Experience 26(6), 635–652 (1999)
5. Aldebaran Robotics (2008), http://www.aldebaran-robotics.com/
6. Niemueller, T.: Developing A Behavior Engine for the FAWKESRobot-Control Software and its Adaptation to the Humanoid Platform Nao. Master's thesis, Knowledge-Based Systems Group, RWTH Aachen University (to appear, 2009)
7. Brooks, A., Kaupp, T., Makarenko, A., Williams, S., Orebäck, A.: Towards Component-Based Robotics. In: Proc. IROS 2006, August 2005, pp. 163–168 (2005)
8. Collins-Cope, M.: Component Based Development and Advanced OO Design. Whitepaper, Ratio Group Ltd. (2001)
9. Furbach, U., Murray, J., Stolzenburg, F.: Hybrid Multiagent Systems with Timed Synchronization – Specification and Model Checking. In: Dastani, M.M., El Fallah Seghrouchni, A., Ricci, A., Winikoff, M. (eds.) ProMAS 2007. LNCS (LNAI), vol. 4908, pp. 205–220. Springer, Heidelberg (2008)
10. Arai, T., Stolzenburg, F.: Multiagent systems specification by UML statecharts aiming at intelligent manufacturing. In: Proc. AAMAS 2002. ACM Press, New York (2002)
11. Ierusalimschy, R., de Figueiredo, L.H., Filho, W.C.: The Evolution of Lua. In: Proceedings of History of Programming Languages III, pp. 2-1–2-26. ACM, New York (2007)
12. The Debian Project: The Computer Language Benchmarks Game, http://shootout.alioth.debian.org/ (retrieved January 30, 2009)
13. Hirschi, A.: Traveling Light, the Lua Way. IEEE Software 24(5), 31–38 (2007)
14. Loetzsch, M., Risler, M., Jungel, M.: XABSL - A Pragmatic Approach to Behavior Engineering. In: Proc. IROS 2006, pp. 5124–5129 (2006)
15. Medeiro, S., Ierusalimschy, R.: A parsing machine for PEGs. In: Proceedings of the 2008 Symposium on Dynamic Languages, pp. 1–12. ACM, New York (2008)
16. Kobayashi, H., Ishino, A., Shinohara, A.: A framework for advanced robot programming in the robocup domain - using plug-in system and scripting language. In: Proc. IAS-9, pp. 660–667 (2006)
17. Hester, T., Quinlan, M., Stone, P.: UT Austin Villa 2008: Standing On Two Legs. Technical report, Department of Computer Sciences, The University of Texas, Austin (2008)
18. Ellson, J., Gansner, E., Koutsofios, L., North, S.C., Woodhull, G.: Graphviz – Open Source Graph Drawing Tools. In: Mutzel, P., Jünger, M., Leipert, S. (eds.) GD 2001. LNCS, vol. 2265, pp. 483–597. Springer, Heidelberg (2002)

Stable Mapping Using a Hyper Particle Filter

Johannes Pellenz and Dietrich Paulus

Active Vision Group, University of Koblenz-Landau,
Universitätsstr. 1, 56070 Koblenz, Germany
{pellenz,paulus}@uni-koblenz.de

Abstract. Often Particle Filters are used to solve the SLAM (Simultaneous Localization and Mapping) problem in robotics: The particles represent the possible poses of the robot, and their weight is determined by checking if the sensor readings are consistent with the so far acquired map. Mostly a single map is maintained during the exploration, and only with Rao-Blackwellized Particle Filters each particle carries its own map.

In this contribution, we propose a Hyper Particle Filter (HPF) – a Particle Filter of Particle Filters – for solving the SLAM problem in unstructured environments. Each particle of the HPF contains a standard Particle Filter (with a map and a set particles, that model the belief of the robot pose in this particular map). To measure the weight of a particle in the HPF, we developed two map quality measures that can be calculated automatically and do not rely on a ground truth map: The first map quality measure determines the contrast of the occupancy map. If the map has a high contrast, it is likely that the pose of the robot was always determined correctly before the map was updated, which finally leads to an overall consistent map. The second map quality measure determines the distribution of the orientation of wall pixels calculated by the Sobel operator. Using the model of a rectangular overall structure, slight but systematic errors in the map can be detected. Using the two measures, broken maps can automatically be detected. The corresponding particle is then more likely to be replaced by a particle with a better map within the HPF.

We implemented the approach on our robot "Robbie 12", which will be used in the RoboCup Rescue league in 2009. We tested the HPF using the log files from last years RoboCup Rescue autonomy final, and with new data of a larger building. The quality of the generated maps outperformed our last years (league's best) maps. With the data acquired in the larger structure, Robbie was able to close loops in the map. Due to a highly efficient implementation, the algorithm still runs online during the autonomous exploration.

1 Introduction

Many different algorithms have been proposed in the last few years to solve the SLAM (Simultaneous Localization and Mapping) problem ([2,3,4,6,8]). These solutions nowadays run in real time (for 2D maps) and produce very accurate maps, which can be used by robots for path planning and navigation. To handle

J. Baltes et al. (Eds.): RoboCup 2009, LNAI 5949, pp. 252–263, 2010.

the uncertainties that result from noisy sensor data, probabilistic approaches became very popular within the last couple of years (see [9] for a survey). Within the probabilistic approaches, Particle Filters [10] have some favorable properties: They can handle the so called "kidnapped robot problem", where the pose of the robot is unknown in the beginning. Also, multiple hypothesis of the robot pose can be tracked, and non-Gaussian distributions can be modeled.

However, the original implementation of Particle Filters suffered from the fact that only a single map (often as an occupancy grid) was maintained, and the particles only represented different assumed poses of the robot. So if the map "broke", the complete map building failed. Rao-Blackwellized Particle Filters [1] can maintain a single map for each particle. So if a map breaks, the corresponding particle is finally removed. Hähnel et al. combined Rao-Blackwellized Particle Filtering and scan matching in the FastSLAM algorithm [5]. Here each Particle Filter contains a map as a list of features that are stored within an Extended Kalman Filter.

In this work, we propose a Hyper Particle Filter (HPF) – a Particle Filter of Particle Filters – for solving the SLAM problem in unstructured environments. Each particle of the HPF contains a standard SLAM Particle Filter: The particle contains a map and a set of particles that model the belief of the robot pose in this particular map. The Particle Filter requires a measurement step. To measure the weight of a particle in the HPF, we developed two *map quality* measures: The first map quality measure mq_1 determines the contrast of the occupancy map. This is motivated by the fact that if the map has a high contrast, it is likely that the pose of the robot was determined correctly before the map was updated. This finally leads to an overall correct map. The second map quality measure mq_2 determines the distribution of the orientation of wall pixels calculated by the Sobel operator. In contrast to mq_1, the measure mq_2 can only be used if the overall structure of the mapped area is known to be rectangular. Using this measure, slight but systematic errors in the map can be detected. Both quality measures can be calculated automatically and do not rely on a ground truth map. Using mq_1 and mq_2, broken maps can automatically be identified. The corresponding particle is then more likely to be replaced by a particle with a better map within the HPF.

The paper is organized as follows: Section 2 reviews the use of Particle Filters for robotic mapping and localization, section 3 introduces the concept of the Hyper Particle Filter. Section 4 describes experiments with the HPF. Section 5 concludes the paper, and section 6 closes the paper with the topics that are addressed in the future.

2   Particle Filters

The idea of Particle Filters is adopted from the field of computer vision, where the principle is known as the Condensation Algorithm [7]. The idea is to represent the distribution function of the robot pose as a set of particles: The more likely a particular pose is, the more particles represent this area. Whenever new sensor data is available, the following steps are executed (adapted to the field of robotics):

1. **Resampling**: A new generation of particles is created from the current set of particles. The higher the weight of a particle is, the more likely it is drawn and its pose is represented in the next generation.
2. **Drift**: The particles are moved according to the control update or the odometry readings of the robot. Additionally, a scan matcher is used to improve the estimated transformation.
3. **Diffuse**: The particles are moved according to the noise of the motion model.
4. **Measure**: Using the current sensor data (e. g. readings from the laser range finder) a weight is assigned to each particle (which represents the consistency of the pose and sensor data with the so far acquired map).

For our robot "Robbie X", we used a Particle Filter with about 1,000 particles that represent possible poses $(x, y, \Theta)^T$ in 2D. Note that the map itself is *not* part of the state vector.

In our approach, the Particle Filter is used for the localization only: The acquired map is stored independently in an occupancy grid [3]. In the context of RoboCup Rescue, the grid has a size of 800×800 cells, which represents a map of an area of 40×40 meters with a grid cell size of 50×50 millimeters. The grid is stored in two planes: One plane counts how often a cell was "touched" by a laser beam. This value is increased either if the laser beam reported the cell as free or if the cell was reported as occupied. A second plane stores the information how often a cell was seen as occupied. By dividing the values of these two planes, the occupancy probability for a cell c_i is estimated by the following ratio:

$$p_{occ}(c_i) = \frac{\text{count}_{occ}(c_i)}{\text{count}_{seen}(c_i)} \tag{1}$$

To extend the map, the best pose that the Particle Filter determines is used to update the map: The current laser range scan is added to the global occupancy map at the estimated robot pose by constructing a local map and "stamping" it into the global map, incrementing the counts for *seen* and *occupied* cells. Special attention is paid to cells that are touched by the laser beams more than once during a single scan (these are cells that are close to the robot): If such a cell is seen as occupied, than other beams of this scan can not overwrite this cell as free any more.

So far only a single map was maintained, and in case of a defect in the map, there was now way to recover.

3   Hyper Particle Filter (HPF)

To overcome the problem of broken maps, we use multiple (about 30) SLAM Particle Filters concurrently. Due to the probabilistic behavior of each filter in the diffusion step, different (but similar) maps are generated in each SLAM filter. Additionally, each filter can reject a new measurement with a 20% chance. This way, some of the filters are not influenced by totally broken scans, that might be the result of scans taken while the robot was turning or tilting on top of a ramp. This happens frequently at the rough environment that is simulated in the RoboCup Rescue arena. Even though we use a gimballed laser range finder, which is actively balanced using two servo motors and the data of an dual-axis accelerometer, we get distorted scans. This frequently happens when the robot drives over the top of the ramps.

The 30 SLAM Particle Filters are organized within another Particle Filter, the so called Hyper Particle Filter. This is depicted in Fig. 1. The particular steps of this Particle Filter are implemented as follows:

1. **Resampling**: During the resampling step, SLAM Particle Filters that carry a map with a weight below 99% of the weight of the map of the best SLAM Particle Filter are replaced by the best one (including map and robot poses).
2. **Drift**: During the drift step each SLAM Particle Filter is updated with the current laser scan.
3. **Diffuse**: (No diffusion.)
4. **Measure**: For the measurement step, the map quality of each SLAM Particle Filter is analyzed using the map quality measures mq_1 and mq_2 that are described below.

3.1   Map Quality Measure mq_1

The ability to generate high quality maps is strongly related to the ability to locate the robot when the range measurement was taken: If the localization was correct (and the noise of the sensor is reasonably low), then the resulting map shows consistent assumptions for walls and other features. With incorrect pose estimations, the assumptions for obstacles are not consistent, which results in blurry regions in the occupancy grid. These situations are illustrated in Fig. 2: Two scans are take from the same position, with a different orientation (see Fig. 2(a) and 2(b)). If the rotation between the two scans is estimated correctly, then the two scans can be registered perfectly (see Fig. 2(c)). If there is an error in the estimation, then the scans are merged incorrectly in an occupancy grid (see Fig. 2(d)): The occupancy probability for parts of the map cells drops, because the location that corresponds to these cells have been seen once as free and once as occupied. It is likely that this inncorrect registration is the starting point for a broken map.

The value of mq_1 measures the contrast of the map $\mathbf{M_{occ}}$ by checking for the absence of uncertain cells for all areas that have been scanned by the range sensor so far (these cells are in the set $\mathrm{seen}(\mathbf{M_{occ}})$):

$$mq_1(\mathbf{M_{occ}}) = \frac{1}{|\mathrm{seen}(\mathbf{M_{occ}})|} \sum_{c \in \mathrm{seen}(\mathbf{M_{occ}})} \mathrm{contrast}(\mathbf{M_{occ}}(c)) \qquad (2)$$

with

$$\mathrm{contrast}(\mathbf{M_{occ}}(c)) = \left(\frac{\mathbf{M_{occ}}(c) - 0,5}{0,5} \right)^2 \qquad (3)$$

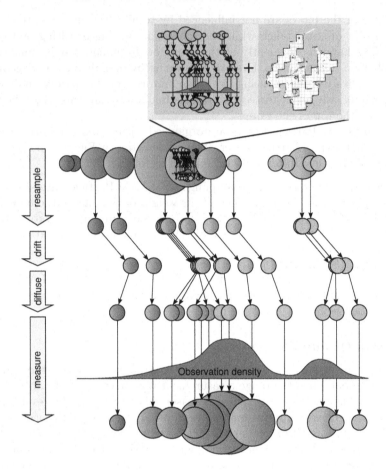

Fig. 1. The principle of the Hyper Particle Filter: The particles of the Hyper Particle Filter are standard SLAM Particle Filters (with a map and particles that represent robot poses). For the measurement step, the quality of each map is evaluated.

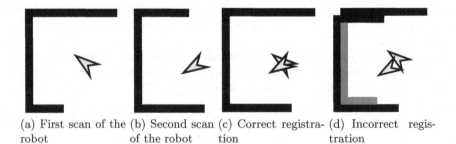

(a) First scan of the robot   (b) Second scan of the robot   (c) Correct registration   (d) Incorrect registration

Fig. 2. Correct (c) and incorrect (d) registration of the two scans (a) and (b). The incorrect registration lowers the occupancy probability of some of the grid cells. This is an indicator for the incorrect pose estimation.

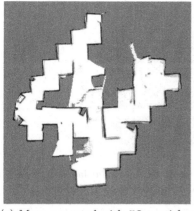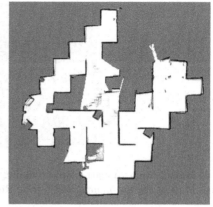

(a) Map generated with **50** particles: (b) Map generated with **1000** particles:
$mq_1 = 87.39\%$ $mq_1 = 92.55\%$

Fig. 3. Maps generated with different numbers of particles – and the corresponding map quality measure mq_1. The lower quality of map (a) compared to (b) is visually hard to detect, but sensed by the mq_1.

where $\mathbf{M_{occ}}$ is the occupancy grid representation of the map and $\mathbf{M_{occ}}(c)$ the occupancy value of a particular cell c of the grid. This way, a pixel which is clearly free ("white") or clearly occupied ("black") is counted with full contrast ($= 1.0$); a pixel which was scanned the same number of times free and occupied ("gray") is counted with no contrast ($= 0.0$). Fig. 3 gives an example for the measure mq_1 for two different maps. For the first map (Fig. 3(a)), only 50 particles were used; for the second map (Fig. 3(b)) 1000 particles. As expected, the value of mq_1 is higher for the map with more particles.

3.2 Map Quality Measure mq_2

The mq_2 measures the consistancy of wall directions within a map. Unlike mq_1, this measure is not model free: It can be used only if the building has a rectangular structure. The map becomes inconsistent if a systematic error causes the walls of the map to bend to a certain direction, or if the map is broken and the walls suddenly point to a random direction. An example of such a broken map is given in Fig. 4.

The directions of edges in an image can be calculated in various ways; one of the most common methods is the Sobel edge detector. Let $\mathbf{M_{occ}}$ be the occupancy grid of a building structure, and $\mathbf{S_x}$ and $\mathbf{S_y}$ Sobel operators for the x and y direction, then the horizontal and vertical derivative approximations $\mathbf{G_x}$ and $\mathbf{G_y}$ can be calculated as follows:

$$\mathbf{G_x} = \mathbf{S_x} * \mathbf{M_{occ}} = \begin{bmatrix} 1 & 0 & -1 \\ 2 & 0 & -2 \\ 1 & 0 & -1 \end{bmatrix} * \mathbf{M_{occ}} \tag{4}$$

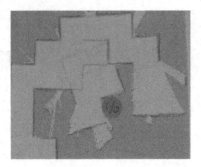

Fig. 4. Broken map: A part of the building (next to the round label) is turned and moved. Map of the autonomy final during RoboCup Rescue 2007, Atlanta, USA.

and

$$\mathbf{G_y} = \mathbf{S_y} * \mathbf{M_{occ}} = \begin{bmatrix} 1 & 2 & 1 \\ 0 & 0 & 0 \\ -1 & -2 & -1 \end{bmatrix} * \mathbf{M_{occ}} \tag{5}$$

where $*$ denotes the convolution operation. Using $\mathbf{G_x}$ and $\mathbf{G_y}$, we can calculate the strength of the edge and gradient's direction:

$$\mathbf{G(M_{occ})} = \sqrt{\mathbf{G_x}^2 + \mathbf{G_y}^2} \tag{6}$$

$$\mathbf{\Theta(M_{occ})} = \operatorname{atan2}(\mathbf{G_y}, \mathbf{G_x}) \tag{7}$$

Using the edge directions, we calculate the measure mq_2 for the consistancy of the wall direction using the following algorithm:

1. Calculate for a map $\mathbf{M_{occ}}$ the edge image $\mathbf{\Theta(M_{occ})}$.
2. Create the histogram $\operatorname{hist}(\mathbf{\Theta(M_{occ})})$ over the angles (bin size: $3°$).
3. Smooth the histogram using a mean filter of size 5.
4. For the four largest peaks $p_i, i \in [1..4]$ in the histogram:
   Calculate μ_i, σ_i and $|p_i|$ (number of angles that belong to this peak). Use all angles within $\pm 30°$ around the peak.
5. Calculate mq_2: $mq_2 = \frac{\sum_{i=1}^{4} |p_i| \sigma_i}{\sum_{i=1}^{4} |p_i|}$

The value of mq_2 is small for significant peaks, and large for non-significant peaks. Fig. 5 gives examples for two different maps: Based on the same logfiles, the parameter that determines the distance between sampling points in the laser scan is modified. With a small distance between the sampling points (100 mm) a highly accurate map is produced (with the original, rectangular structure of the building). If the distance between the sampling points is increased (e.g. to 1,000 mm), the map gets inconsistent over time; resulting in skewed (and blurry) walls. Therefore, the orientation of the walls is not consistent any more, which is clearly visible in the histogram of the wall angles: The histogram is more flat, and therefore mq_2 – which adds up the variance of the four most significant peaks – is larger.

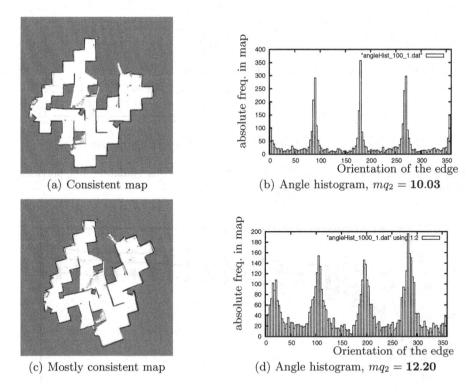

(a) Consistent map

(b) Angle histogram, $mq_2 = \mathbf{10.03}$

(c) Mostly consistent map

(d) Angle histogram, $mq_2 = \mathbf{12.20}$

Fig. 5. Map and angle histogram to determine mq_2 (particle count = 5000, cell size 50×50 mm). In (a) ten times more distance measurements from each laser scan were used compared to (c). (data set: `rescueServer_2008.07.20_13-49-13.log`)

4   Experiments

4.1   Map Quality Measures

To check how the map quality measures behave under controlled settings, we run the following experiment: Using the identical log file (from the RoboCup Rescue final in China, 2008), we produced hundreds of maps but used different parameters in the map building process. For example, we iterated the number of particles that are used for the robot localization; for each particular number we performed three complete mapping runs. From experience, we know that the more particles are used to represent the posterior of the robot pose, the more accurate the generated maps are. Typically, we used about 2,000 particles for our mapping. For the experiments, we iterated the particle count from 50 to 2,000. Fig. 6(a) shows the correlation between the number of particles and mq_1. Two conclusions can be drawn from the graph: 1. The gain in quality is large up to about 1,000 particles. 2. If the experiment is repeated, the resulting quality can vary significantly (e. g. see the three different values at particle count 1,400). So building maps concurrently definitively makes sense! Fig. 6(b) shows the

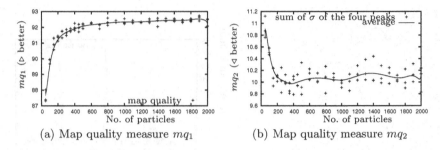

(a) Map quality measure mq_1 (b) Map quality measure mq_2

Fig. 6. Map quality measure mq_1 and mq_2 versus number of particles (data set: `rescueServer_2008.07.20_13-49-13.log`)

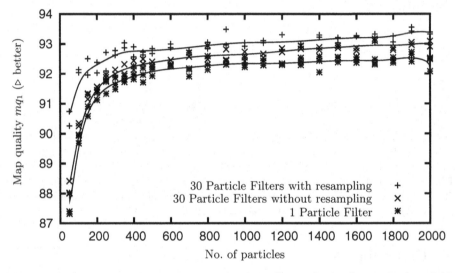

Fig. 7. Map quality measure mq_1 versus number of particles; 1 (lowest line) and 30 Particle Filters; without (middle) and with (top line) resampling (replacement of maps with low quality).

correlation between the number of particles and mq_2. Here also up to a number of about 300 particles, the increased particle count yields a gain in quality. Then again – and here more significantly than with mq_1 – the resulting quality varies a lot, even with the same number of particles. This means: Even if you choose the same parameters, the outcome of the probabilistic algorithm can be very different. But now, using mq_1 and mq_2, we can detect the resulting quality.

4.2   Hyper Particle Filter

We tested the HPF using the log files from last years RoboCup Rescue autonomy final. Using mq_1, the HPF always picks the best map out of these 30 generated maps after each mapping step. This way, the probability of a highly accurate map

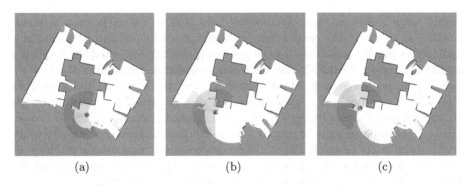

(a)   (b)   (c)

Fig. 8. Successful loop closing using a Hyper Particle Filter (HPF) with 30 particles. (a) Situation right before the loop closing. (b) The robot scans areas that it has visited before; first errors in the map are visible. (c) The HPF closes the loop by switching to a consistent map.

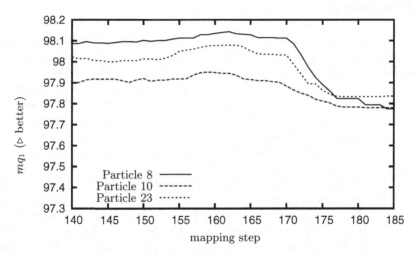

Fig. 9. Map quality measure mq_1 versus the mapping steps while the loop closing happens between steps 170 and 180: The quality of the map of particle 8 drops, but the map of particle 23 can cope with the loop closing.

was dramatically increased. Fig. 7 shows how the map quality increases with the number of particles, for one (bottom line) and for 30 concurrent Particle Filters (middle), and also the influence of the resampling (top line). Within the resampling step, SLAM particle filters with low quality maps are replaced by better particles, so that the average map quality for all particle counts can be improved further.

4.3   Loop Closing

Using mq_1 the problem of loop closing can be solved. The loop closing happens when the robot drives around a block and returns to a spot that it has visited before using a different way. Accumulated errors often cause the map to break in these areas due to the chain error. For the experiments, we recorded data in a building in which a RoboCup Rescue arena is placed. We collected the odometry and laser range data while driving the robot around this arena. Immediately after the loop closing, the map was broken for a moment; this situation is illustrated in Fig. 8(b). Using mq_1, the broken map was detected. Then, the algorithm switch to another map (with a higher mq_1 value). This map is shown in Fig. 8(c). Note that in this map the walls fit exactly. The weights of the involved particles (only 3 out of the 30) is shown in Fig. 4.2. The quality of all maps drop after the loop is closed, but some particles can cope the situation better than others. So the HPF switches to a particle with a clearer and more consistent map.

The performance of the loop closing depends on the number of particles in the HPF: The more particles (and maps) are maintained concurrently in the HPF, the more likely a loop can be closed successfully.

5   Conclusion

In this work, we presented the concept of a Hyper Particle Filter for robust online mapping: A Particle Filter that is composed of a number of conventional SLAM Particle Filters. Two new map quality measures for the evaluation can be used for the measurement function of the Hyper Particle Filter: mq_1 determines the contrast of the map, while mq_2 calculates the distribution of the orientation of wall pixels calculated by the Sobel operator. While mq_1 is model free, mq_2 relies on a orthogonal building structure assumption. Using these measures, we can detect broken maps and replace them by maps that are consistent.

We showed that the overall map quality is increased compared to the standard "single" SLAM Particle Filter by just using multiple Particle Filters concurrently. Using a simple resampling strategy, the overall quality could once again be increased. Because loop closings also influences the map quality, loop closing situations can implicitly be detected and a another map (that is consistent – even after the loop closing) is chosen automatically. We will use the new filter during RoboCup 2009 on our robot "Robbie 12" in the RoboCup Rescue league. In the test with last years log files, the quality of the maps generated by the HPF outperformed our last years (league's best) maps. With the data acquired in the larger structure, Robbie was able to close loops in the map. Due to a highly efficient implementation, the algorithm still runs online during the autonomous exploration.

6   Future Work

So far the diffusion step of the Hyper Particle Filter is not implemented. We plan to modify parameters of the SLAM Particle Filters in this step: For example, the scan matcher that reduces the error of the odometry might be switched on or off. This way, the scan matcher is turned on automatically when it increases the accuracy of the mapping, but is switched off (and therefore saves computation time) when it is not needed. So the software can adapt itself depending on the current environment.

References

1. Doucet, A., de Freitas, N., Murphy, K., Russel, S.: Rao-blackwellised particle filtering for dynamic bayesian networks (2000)
2. Durrant-Whyte, H.: Localisation, mapping and the slam problem. Technical report, The University of Sydney, Summer School (2002)
3. Elfes, A.: Using occupancy grids for mobile robot perception and navigation. Computer 22(6), 46–57 (1989)
4. Gutmann, J., Konolige, K.: Incremental mapping of large cyclic environments. In: Proceedings of the IEEE International Symposium on Computational Intelligence in Robotics and Automation (CIRA), Monterey, California, pp. 318–325 (1999)
5. Hähnel, D., Burgard, W., Fox, D.: Sebastian Thrun. An efficient fastslam algorithm for generating maps of large-scale cyclic environments from raw laser range measurements. In: Proceedings of 2003 IEEE/RSJ International Conference on Intelligent Robots and Systems, 2003 (IROS 2003), October 2003, vol. 1, pp. 206–211 (2003)
6. Hähnel, D., Burgard, W., Fox, D., Thrun, S.: A highly efficient fastslam algorithm for generating cyclic maps of large-scale environments from raw laser range measurements. In: Proceedings of the IEEE/RSJ International Conference on Intelligent Robots and Systems, IROS (2003)
7. Isard, M., Blake, A.: Condensation - conditional density propagation for visual tracking. International Journal of Computer Vision 29(1), 5–28 (1998)
8. Thrun, S.: Learning metric-topological maps for indoor mobile robot navigation. Artificial Intelligence 99(1), 21–71 (1998)
9. Thrun, S.: Robotic mapping: A survey, 1 (2002)
10. Thrun, S., Burgard, W., Fox, D.: A real-time algorithm for mobile robot mapping with applications to multi-robot and 3d mapping. In: ICRA, pp. 321–328. IEEE, Los Alamitos (2000)

A Characterization of 3D Sensors for Response Robots

Jann Poppinga, Andreas Birk, and Kaustubh Pathak

Jacobs University Bremen*
Campus Ring 1, 28759 Bremen, Germany

Abstract. Sensors that measure range information not only in a single plane are becoming more and more important for mobile robots, especially for applications in unstructured environments like response missions where 3D perception and 3D mapping is of interest. Three such sensors are characterized here, namely a Hokuyo URG-04LX laser scanner actuated with a servo in a pitching motion, a Videre STOC stereo camera and a Swissranger SR-3000. The three devices serve as prototypical examples of the according technologies, i.e., 3D laser scanners, stereo vision and time-of-flight cameras.

1 Introduction

Sensors providing 3D range data are getting more and more important for mobile robotics in general and for Safety, Security and Rescue Robotics (SSRR) in particular. 3D data allows response robots for example to estimates the size of gaps, to construct realistic maps of unstructured disaster environments, or to detect human victims from shape. Concrete examples of research with relevance for SSRR where 3D sensors are used include 3D mapping [1][2][3][4][5], semantic environment classification [6] or the detection of drivable terrain [7].

The main purposes of this paper are twofold. First, updated information concerning the state of the art of according sensors compared to previous discussions in the literature are provided. Second, the focus is on devices that can be directly used on mobile robots, especially in the context of Safety, Security and Rescue Robotics.

The rest of this paper is structured as follows. In section 2, 3, and 4 a concrete 3D laser scanner, stereo camera, respectively time-of-flight camera are introduced and their general properties are discussed. A direct comparison is presented in section 5. Section 6 concludes the paper.

2 3D Laser Range Finder

Laser Range Finders (LRF) in their standard form are the predominant sensor for mapping on mobile robots. But as the interest on 3D mapping increases, the

---

* Formerly International University Bremen.

J. Baltes et al. (Eds.): RoboCup 2009, LNAI 5949, pp. 264–275, 2010.

limitations of the standard systems that only scan a horizontal plane become more and more apparent. There are meanwhile quite some 3D laser scanners available off the shelf. A very recent but coarse overview is given in [8]. A detailed discussion of an example sensor can be found in [9]. These systems are designed for geometric applications and hence not really suited for mobile robots, mainly due to their weight and power consumption.

It is hence very popular within the 3D mapping community to take a standard 2D laser scanner and to actuate it to get data from an additional dimension. One option is to mount two scanners perpendicular to each other and to exploit the movement of the robot itself [10]. But most commonly, the sensor is directly driven with some servo-mechanism to also get significant amounts of 3D data on a stationary or slow moving robot. In doing so, different motions of the scanner with respect to the robot's frame are possible. Examples are rolling [11], pitching [12] or yawing movements [13]. The prototypical system presented here uses a pitching motion.

This system is based on a Hokuyo URG-04LX [14]. It has a 240 degrees field of view, which is scanned in 683 steps, i.e., the angular resolution is 0.36 degree. It can cover 0.2 to $4m$ with a resolution of $\pm10mm$. It is based on a near infrared laser-diode with $\lambda = 785nm$. The URG-04LX is interfaced via USB to the mobile robot. For the servo that moves the sensor, a small board based on a PIC18f2410 micro-controller is used, which is interfaced via RS232. There are of course other 2D laser scanners that could be used as basis for a 3D sensor. Example characterizations of other 2D laser scanners are [15] for the Hokuyo PBS03 and [16] for the popular Sick LMS 200.

The main advantage of the Hokuyo URG-04LX is its compactness (l:50mm, w:50mm, h:70mm), small weight (160 g) and low power consumption (2.5 W). These advantages are traded in with the relative short range of 4 m. Our system uses a standard servo for a pitching motion of the scanner (figure 1(a)). The main difference of this system compared to other 3D scanners, for which larger scanners like the Sick LMS 200 are very popular as sensor basis, is its compactness in size, weight and power combined at the cost of a shorter range. But the characteristic aspects that serve as basis for the evaluation and comparison to other 3D laser range finders are very generic.

Like laser scanners in general, the URG-04LX has a relatively slow update rate of 100 $msec/scan$. The overall time for a 3D scan depends on the range and

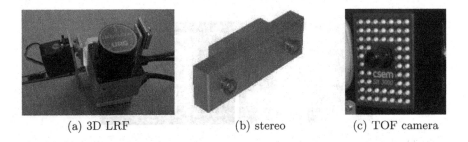

(a) 3D LRF   (b) stereo   (c) TOF camera

Fig. 1. The three different 3D sensors that are analyzed here

the resolution of the actuation of the sensor. The low cost servo in our system allows precise steps with a minimum angular resolution of 3 deg. It hence takes about 3 seconds for the 30 scans that cover a pitching motion of 90 deg, i.e., 683×90 data points. This relative slow rate for a total 3D scan is typical. The system described in [12] for example takes 3.4 seconds to produce a scan with 256×181 data points.

3   Stereo Vision

Stereo vision is a well known technique for 3D measurements in general [17] as well as for mobile robots in particular [18,19]. A common criticism for stereo vision, especially when compared to laser range finders, are its high computational requirements. This true if the generation of the disparity image is done in software. But alternatives exist like the stereo-on-chip (STOC) camera (figure 1(b)) from Videre Design [20], which has an embedded processor.

The device connects to a PC a using IEEE 1394 (Firewire) interface and consumes 2.4 Watts. It produces a 3D point cloud at a resolution of 640×480 at 30 frames per second. The cameras are CMOS imagers, rigidly mounted on an anodized aluminum chassis; the base-line is 9 cm. They have a global shutter, i.e., all pixels are exposed simultaneously. The left and right pixels are interleaved in the video stream. The device needs to be calibrated and this information is then stored on it. Both monochrome and color images can be obtained. The device board runs a version of the SRI Small Vision System (SVS) [21] stereo algorithm, which is based on area correlation.

Fig. 2. The left camera image, and the corresponding disparity image of a flat white surface with almost no visual features

Fig. 3. The left camera image, and the corresponding disparity image of a flat surface with visual features

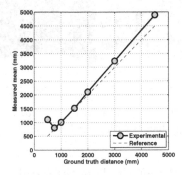
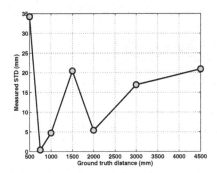

Fig. 4. The mean and the standard deviation of distance measured with a Videre STOC using the points lying within the central 150×150 sq. mm. The results are based on averaging 20 measurements at each distance.

Figures 2 and 3, illustrate the main disadvantage of stereo systems, namely their dependency on sufficiently distinctive regions for matching. As the disparity image in figure 2 shows, a flat featureless board is not detected by the device. Only the boundary of the board is detected. This should be compared to the disparity image in figure 3 where the board has some text on it. This time the interior of the board is captured quite well.

The dependence on environment conditions is also illustrated by an experiment to estimate the discrepancy from the ground truth for the points lying within the central 150×150 sq. mm. This experiment was performed on a freshly calibrated device. The results are shown in figure 4. The ranges of objects too close to the cameras, namely at a distance less than the focal length, are incorrect. As the distance of the object increases, the range error steadily increases as one can expect. Nevertheless, note the irregular behavior in the standard deviation, which can be related to environment conditions.

4   Time of Flight Cameras

The Swissranger SR-3000 (figure 1(c)) is a time-of-flight camera, i.e., a technology that is much less established than laser scanners or stereo cameras. The device is produced and marketed by the "Centre Suisse d'Electronique et de Microtechnique" (CSEM). An earlier version of this sensor, namely the SwissRanger-2, is characterized in some detail in [22]. The technological principles on which this sensor is based are described in [23]. Roughly speaking, this type of sensor uses an array of cells similar to an imager of a camera to measure the phase-shift of emitted modulated infrared light. By this, a time-of-flight based distance measurement can be done simultaneously in each cell of the array.

The sensor connects to a Computer via USB. It is delivered with an API and example applications for Windows, Linux and Mac OS X. The sensor generates color encoded distance images as well as intensity images. The first correspond

Fig. 5. At the left, the underlying scene shot with a normal camera; next to it the effects of the Amplitude Threshold (AT) on range images with a SR-3000

to the measured phase-shift, the second to the amplitude of the signal. The measured distances are encoded by hue, range from red (near) to violet (far). Black pixel indicate that no useful information was measured, mostly because too little modulated light was received. The underlying technology of the SR-3000 is relatively young and far less established than laser range finders or stereo cameras. Though promising, there are still many drawbacks.

The first fundamental problem is the *wrap-around* error. As phase-shift is measured, $c = c + k \cdot 2\pi$ for arbitrary k. This means if the phase of 2π corresponds to for example 8 m then two points in a distance of 0.2 m and 8.2 m lead to the same measurement.

In theory, there is the option to use the so-called *Amplitude Threshold* parameter to remedy this problem. When used with this parameter, the camera only returns distance values for pixels with a certain minimal brightness. Note that the range for the unused pixels is set to zero. This parameter is potentially also useful for eliminating other kinds of errors as discussed later on. Roughly speaking, a higher amplitude threshold leads to less data points but the data points are of higher quality.

The main problem is to find a suited amplitude threshold (AT) for the current conditions in a particular environment. This is illustrated in figure 5. First of all, there is some random noise on some objects. This most prominently occurs on dark objects and around the edges of objects. Second, the amplitude depends on distance and it hence can remedy wrap-around errors. But the amplitude also depends on the reflectance properties of the objects. If the AT is chosen too low, correct data points are discarded.

An other important parameter is the integration time, which is also called exposure time. It is the time used for acquiring each frame. The perceivable brightness of the illumination unit increases with the integration time. The integration time of course determines the frame rate and also the power consumption as it influences the brightness of the illumination LEDs. The quality of the images can increase with the exposure time, but also the minimal distance increases. An autoillumination feature for the device can be used to automatically determine integration time and illumination intensity. The user can supply a certain range for the integration time and the best value within the boundaries is chosen. This setting yields good results, so it should be used unless there are reasons not to do so, e.g., the urgent need to save energy or to achieve a certain frame rate.

Fig. 6. Though the SR-3000 already has an extremely limited field of view, artifacts from the focused modulated light sources are apparent. When using a short exposure time and AT=0, there are is a significant amount of noise in the corners of the range image (left). When increasing the AT to 500, any information from these regions is completely discarded and a almost perfect circular shape caused by the narrow cone of the modulated illumination can be seen (right).

There are two possible ways around the errors caused by wrap-around of the phase-shift. The simpler one as mentioned before is to apply an amplitude threshold. This works quite often, as the amount of light reflected by an obstacle is proportional to its distance to the source of light. But when dealing with obstacles with different reflective properties, this fails. The second possibility is to use two modulation frequencies in an alternating way. As each modulation frequency entails a specific non-ambiguity range, the pixels of alternating frames wrap around at different distances. Thus, the errors in one frame can be evened out by comparing it to another frame taken at a different modulation frequency.

While frames retrieved by this method still contain some errors and are also not totally dense, this method is in general advisable. The down side is that the frame rate is reduced, especially as two frames have to be dropped immediately after the change of the modulation frequency.

An other significant drawback of the sensor is its extremely small field of view. Even with its already very limited standard view of $47° × 39°$, artifacts from the LEDs as quite focused illumination sources are apparent. The main portion of the modulated light emitted by the LEDs is centered in the middle of the sensor. So when using a relatively low exposure time, the corners of the picture do not get enough light for decent measurements. An example is given in figure 6(a). Please note that the color scale for this image is extremely short; the distance between red and violet is 40 cm. This is done for illustration purposes. When the only the brightest pixels are used, i.e., AT is accordingly adjusted, the circular shape of the lighting becomes visible (figure 6(b)).

Bright light. The SR-3000 is very sensitive to ambient light conditions. Especially sunlight requires some manual tuning of AT (see figure 7).

A further problem is that objects close to the camera can make objects near them on the picture appear closer to the camera than they are; especially if the far objects are dark and the material of the near object is bright, i.e., highly reflective. An other form of irregular distortion appears when an object is close to the camera. Then, a faint ghost image of it will often appear on the opposite

(a) Photograph of a scene with some sunlight. (b) Random noise due to bright light, AT=0 (c) Corrected image with AT=300

Fig. 7. The SR-3000 is very sensitive to ambient light conditions. When confronted with a scene including a bit of sunlight (left), a significant amount of noise occurs with AT=0 (center). Only when tuning AT, here to 300, the problem is solved (right).

 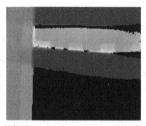 

(a) Distance picture of a static rod. (b) The rod moving down fast (about 10 m/s). The large blue and green bars are movement errors. (c) The same rod moving up slowly (about 0.25 m/s). The errors become smaller but are still present.

Fig. 8. Errors caused by a moving object (AT=500)

of the frame. Yet an other source for distortions are moving objects (figure 8). When sampling them, there are always distortions at their edges, even with low exposure times. Many of these problems are also mentioned in the manual of the sensor [24].

Last but not least, the SR-3000 behaves much like a regular camera and it is hence can be mislead by reflections. When a reflected surface appears in its view, the SR-3000 will not measure the distance to where the reflection occurs, but the perceived distance to the objects visible on the reflective surface. The measured distance is then the distance from the camera to the reflective surface plus the distance from the surface to the object. An example is shown in figure 9.

As a rough quantitative comparison between the SwissRanger to the stereo camera, the same measurements concerning accuracy were performed: the discrepancy from the ground truth for the points lying within the central 150×150

(a) Photograph of a scene (b) Distance information
with reflections due to a with AT 500
glossy floor.

Fig. 9. Reflections, here on a floor, cause the measurement of false distances

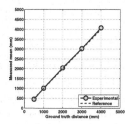
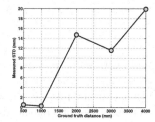

Fig. 10. The mean and the standard deviation of distance measured with a Swis-sRanger SR-3000 using the points lying within the central 150×150 sq. mm. The results are based on averaging 20 measurements at each distance.

sq. mm was determined. The results are shown in figure 10. As can be seen, the measurements are quite accurate, especially for longer distances.

5   Comparison of the Results

The three sensors are quite similar in several aspects. First of all, all three can be relatively easily incorporated on a mobile robot as they have similar low demands in terms of space, payload and power (table 1). They also deliver roughly the same amount of data points (table 2). Note that the stereo system uses an embedded hardware to do the matching of regions to produce the disparity image. Hence all three sensors directly deliver 3D data points via their interface. No additional computation is needed for the raw data, which is of interest here. Obviously, the representations differ, but any robotics application will require some coordinate transformation of this raw data anyway. The systems are also quite similar in terms of cost, which is a few thousand Euros each.

The main differences are the update frequency and the quality of the data. The stereo and the time-of-flight camera beat the 3D laser scanner by far in terms of sampling frequency. The 3D laser scanner in contrast delivers much higher quality data. This does unfortunately not only hold with respect to Gaussian noise, which could be easily compensated by averaging, which would be supported by

Table 1. General physical properties

	size	weight	power
3D URG40-LX	l:50mm, w:50mm, h:70mm	425 g	2.5 W
STOC	l:132mm, w:39mm, h:44mm	261 g	2.4 W
SR-3000	l:42.3, w:50, h:67mm	162 g	12 W[1]

Table 2. Data acquisition

	number of data points	sampling rate	field of view	range
3D URG40-LX	683×90	0.3 Hz	$240° \times 90°$	0.2 - 4 m
STOC	640×480	30 Hz	$70° \times 52°$	0.75 - 3 m
SR-3000	176×144	$\leq$ 50 Hz[2]	$47° \times 39°$	0.6 - 8 m[3]

the higher sampling rates of the two camera sensors. But in addition to high Gaussian noise, the stereo and the time-of-flight camera suffer from fundamental drawbacks that can not just be remedied by higher update frequencies. The differences in terms of quality of the data of the three sensors can not easily be presented in detailed quantitative terms as they strongly depend on environmental conditions. An according exhaustive discussion would by far exceed the limits of this paper. Hence a qualitative analysis, which should be at least as useful, is given here.

Stereo requires the matching of identifiable regions. Featureless objects are simply not detected, no matter how often or from which position they are viewed. Hence, a significant amount of regions is not coped with. This holds especially for plain walls, which can be found in many corridors, offices or other "structured" environments. At least, stereo does not provide any completely false depth information. If no match can be performed, this is indicated accordingly. The error of depth estimates significantly increases with range. But near objects, if detected, are well measured. Stereo hence can serve as a fast 3D sensor on a mobile robot.

The basic idea of a time-of-flight camera as a solid state, 3D range finder is at first glance very promising. But the technology is still in its infancy. To get useful data out of the SR-3000, some parameters have to be tuned by hand. Some problems are of a fundamental nature due to the underlying technology, e.g., the wrap-around error or wrong measurements due to reflections. Others, like the strong sensitivity to ambient light conditions or the extremely narrow field of view, may be remedied by software or future hardware generations. This sensor can nicely supplement a stereo vision system as their strengths are complementary. A time-of-flight camera performs best on homogeneous flat surfaces,

---

[1] The exact power consumption depends on the integration time.

[2] The exact sampling frequency depends on the integration time.

[3] The range depends on the frequency setting for the modulation. Here the values for 20 MHz are given for which the best results were achieved.

where stereo tends to work badly. The accuracy of data points, which are not corrupted by structural errors is very high as indicated in figure 10. Nevertheless, a usage of just this sensor for 3D data acquisition on a mobile robot in an arbitrary environment is at least non-trivial.

Laser scanners are a very mature, well established technology. The currently available off the shelf 3D scanners are mainly targeted for geometric applications. They are hence not really suited for mobile robots, mainly due to their weight and power consumption. Also, the cost is typically much higher than for the sensors presented here. But a 3D sensor can be easily constructed from any of the popular 2D sensors supplemented with an actuator. The main disadvantage in general, no matter whether off the shelf or based on an own design, is that the acquisition of a single scan takes several seconds. This is a time period where the motion of the robot usually can not be ignored anymore, i.e., either the robot stops for taking a full scan or the motion is compensated for in the acquisition process. The quality of the data points is much higher than compared to stereo and time of flight cameras. This holds with respect to two aspects. First, the mean and standard deviation of measurements compared to ground truth is much smaller. Second, the amount of completely false or non-classified points is very small. This is mainly due to the fact that laser scanners sample single points, which in turn causes their main disadvantage, namely the relatively long time it takes to generate a single scan.

6 Conclusion

The acquisition of 3D data is increasingly important for mobile robots, especially for systems operating in unstructured environments like response robots . Three different technologies are characterized here based on three prototypical devices, namely a 3D laser scanner, a stereo camera and a time-of-flight camera. The results and their discussion provides general guidelines for system developers as well as potential end users.

Laser scanners are the most mature and reliable technology. Off the shelf 3D scanners are mainly targeted for geometric applications and not really suited for mobile robots. Turning a common 2D device into a 3D scanner is relatively easy. Here, an example based on a Hokuyo URG04-LX and a servo for a pitching motion was presented. Such systems can deliver very high quality data, but at the cost of relatively slow update rates. Stereo cameras in contrast have very high sampling frequencies, especially when using embedded hardware like the Videre STOC presented here. In addition to the typically higher error at larger distances, stereo systems suffer from the drawback that they require regions that can be matched. Featureless objects, especially plain walls, are not detected. Last but not least, the CSEM Swissranger SR-3000 as a time-of-flight camera combines in theory all the advantages of a laser scanner and a stereo camera. But in reality, the technology is still in its infancy. The update rate is high, but the quality of the data is poor. The device requires a high amount of parameter tuning. Ideally, all three sensors are simply used together for 3D data acquisition if space, payload, power and budget constraints permit. To quite some extent, they supplement each other.

Acknowledgments

We gratefully acknowledge financial support by the *Deutsche Forschungsgemeinschaft* (DFG).

References

1. Howard, A., Wolf, D.F., Sukhatme, G.S.: Towards 3d mapping in large urban environments. In: Proceedings of the IEEE/RSJ International Conference on Intelligent Robots and Systems (IROS), Sendai, Japan (2004)
2. Thrun, S.D., Haehnel, D.F., Montemerlo, M., Triebel, R., Burgard, W., Baker, C., Omohundro, Z., Thayer, S., Whittaker, W.: A system for volumetric robotic mapping of abandoned mines. In: Proc. IEEE International Conference on Robotics and Automation (ICRA), Taipei, Taiwan (2003)
3. Hähnel, D., Burgard, W., Thrun, S.: Learning compact 3D models of indoor and outdoor environments with a mobile robot. Robotics and Autonomous Systems 44(1), 15–27 (2003)
4. Davison, J., Kita, N.: 3d simultaneous localisation and map-building using active vision for a robot moving on undulating terrain. In: IEEE Conference on Computer Vision and Pattern Recognition, Hawaii, December 8-14 (2001)
5. Liu, Y., Emery, R., Chakrabarti, D., Burgard, W., Thrun, S.: Using em to learn 3d models of indoor environments with mobile robots. In: 18th Conf. on Machine Learning, Williams College (2001)
6. Nuechter, A., Wulf, O., Lingemann, K., Hertzberg, J., Wagner, B., Surmann, H.: 3d mapping with semantic knowledge. In: Bredenfeld, A., Jacoff, A., Noda, I., Takahashi, Y. (eds.) RoboCup 2005. LNCS (LNAI), vol. 4020, pp. 335–346. Springer, Heidelberg (2006)
7. Poppinga, J., Birk, A., Pathak, K.: Hough based terrain classification for realtime detection of drivable ground. Journal of Field Robotics 25(1-2), 67–88 (2008)
8. Point of Beginning 2006 3D Laser Scanner Hardware Survey (2006), http://www.pobonline.com/POB/Protected/Files/PDF/POB0506/LaserScanningSurvey.pdf
9. Langer, D., Mettenleiter, M., Frohlich, C.: Imaging laser scanners for 3-d modeling and surveying applications. In: IEEE International Conference on Robotics and Automation (ICRA), vol. 1, pp. 116–121 (2000)
10. Thrun, S., Burgard, W., Fox, D.: A real-time algorithm for mobile robot mapping with applications to multi-robot and 3d mapping. In: ICRA, pp. 321–328 (2000)
11. Wulf, O., Wagner, B.: Fast 3d-scanning methods for laser measurement systems. In: International Conference on Control Systems and Computer Science, CSCS14 (2003)
12. Surmann, H., Nuechter, A., Hertzberg, J.: An autonomous mobile robot with a 3d laser range finder for 3d exploration and digitalization of indoor environments. Robotics and Autonomous Systems 45(3-4), 181–198 (2003)
13. Wulf, O., Brenneke, C., Wagner, B.: Colored 2d maps for robot navigation with 3d sensor data. In: IEEE/RSJ International Conference on Intelligent Robots and Systems (IROS), vol. 3, pp. 2991–2996. IEEE Press, Los Alamitos (2004)

14. Automatic, H.: URG-04LX Scanning Laser Range Finder (2006),
    http://www.hokuyo-aut.jp/products/urg/urg.htm
15. Alwan, M., Wagner, M., Wasson, G., Sheth, P.: Characterization of infrared range-
    finder pbs-03jn for 2-d mapping. In: IEEE International Conference on Robotics
    and Automation, ICRA, pp. 3936–3941 (2005)
16. Ye, C., Borenstein, J.: Characterization of a 2d laser scanner for mobile robot ob-
    stacle negotiation. In: IEEE International Conference on Robotics and Automation
    (ICRA), vol. 3, pp. 2512–2518 (2002)
17. Barnard, S.T., Fischler, M.A.: Computational stereo. ACM Comput. Surv. 14(4),
    553–572 (1982)
18. Murray, D., Little, J.J.: Using real-time stereo vision for mobile robot navigation.
    Autonomous Robots 8(2), 161–171 (2000)
19. Elfes, A.: Using occupancy grids for mobile robot perception and navigation. Com-
    puter 22(6), 46–57 (1989)
20. Videre-Design (2006), http://www.videredesign.com/
21. Konolige, K.: The Small Vision System (2006),
    http://www.ai.sri.com/~konolige
22. Weingarten, J., Gruener, G., Siegwart, R.: A state-of-the-art 3d sensor for robot
    navigation. In: IEEE/RSJ International Conference on Intelligent Robots and Sys-
    tems (IROS), vol. 3, pp. 2155–2160. IEEE Press, Los Alamitos (2004)
23. Lange, R., Seitz, P.: Solid-state time-of-flight range camera. IEEE Journal of Quan-
    tum Electronics 37(3), 390–397 (2001)
24. AG, M.I.: Swissranger SR-3000, Manual V1.02. (July 2006)

Multiple Model Kalman Filters: A Localization Technique for RoboCup Soccer

Michael J. Quinlan and Richard H. Middleton

[1] Department of Computer Science, University of Texas at Austin, USA
[2] Hamilton Institute, NUI Maynooth, Maynooth, Co. Kildare, Ireland
mquinlan@cs.utexas.edu, richard.middleton@nuim.ie

Abstract. In the Standard Platform League (SPL) there are substantial sensor limitations due to the rapid motion of the camera, the limited field of view of the camera, and the limited number of unique landmarks. These limitations place high demands on the performance and robustness of localization algorithms. Most of the localization algorithms implemented in RoboCup fall broadly into the class of particle based filters or Kalman type filters including Extended and Unscented variants. Particle Filters are explicitly multi-modal and therefore deal readily with ambiguous sensor data. In this paper, we discuss multiple-model Kalman filters that also are explicitly multi-modal. Motivated by the RoboCup SPL, we show how they can be used despite the highly multi-modal nature of sensed data and give a brief comparison with a particle filter based approach to localization.

1 Introduction

Localization has been studied by many researchers for several years now. Most of the algorithms implemented in RoboCup fall broadly into the class of particle based filters (see for example [5]) or Kalman type filters (see for example [6]) including Extended and Unscented ([4]) variants. In some divisions of RoboCup, algorithms are very well established, given the rich sensor data provided by laser scanners, omni-directional cameras etc. (see for example [3]). However, in the standard platform league (formerly the four legged league) there are substantial sensor limitations particularly with the rapid motion of the camera, and the need for active perception. In addition, the league has deliberately removed beacons as unique landmarks, leaving the colored goals as the only unique landmarks on the field.

Due to the non-uniqueness of most land marks in the SPL, it is important that any localization algorithm be able to handle this ambiguous data. In particular, it is clear that in many cases, the relevant probability density functions will be multi-modal. Whilst it is generally accepted that particle filters can handle this situation, it seems less well known in the RoboCup domain that Kalman type filters can be easily adapted to handle multi-modal distributions. In other research areas, however, multiple model (also called Gaussian Mixture or Gaussian Sum) filters have been used for many decades (see for example [1]).

J. Baltes et al. (Eds.): RoboCup 2009, LNAI 5949, pp. 276–287, 2010.

In this paper, we first give a review of multiple-model Kalman filters, with particular emphasis on features and approximations relevant to real time implementation within the RoboCup framework. We then present examples and results of the mutliple model Kalman filter.

2   Multiple Model Kalman Filters

2.1   Problem Formulation

In many robotics applications, localization algorithms are concerned with the problem of estimating the 'state' of the robot, from uncertain data. For example, in the Standard Platform League, we might typically be concerned with estimating the location (in 2D cartesian coordinates) and orientation of the robot given data derived from vision of objects on the field such as goal posts and field markings. In this case, the state we wish to estimate is written as the 3 dimensional vector

$$x(t) = \begin{bmatrix} x_r(t) \\ y_r(t) \\ \theta(t) \end{bmatrix} \tag{1}$$

where (x_r, y_r) denote the robot's cartesian coordinates and θ is the robot's orientation. Often, a probabilistic or statistical representation of uncertainty is used, though more recently some 'constraint' based localization techniques have also been applied (see for example [2]).

In the probabilistic setting, adopting a Bayesian estimation framework, there are two components to the state estimation problem:

- **Time Update.** Firstly, given the pdf of $x(t-1)$ conditioned on data up to time $(t-1)$, $p(x(t-1))$ and also given odometry information at time t, we wish to make an estimate of the conditional density function of $x(t)$, given data up to time t, $p(x^-(t))$.
- **Measurement Update.** Secondly, given the conditional pdf, $p(x^-(t))$ and also given measurement data at time t, we wish to find $p(x(t))$.

In the standard Kalman filter approach, we use a multivariate (n-dimensional) Gaussian to represent the conditional pdfs of $x(t)$, for example

$$p(x(t)) = \frac{1}{(2\pi)^{n/2}|P(t)|^{1/2}} e^{-\frac{1}{2}(x(t)-\hat{x}(t))^T P^{-1}(t)(x(t)-\hat{x}(t))} \tag{2}$$

where $\hat{x}(t)$ denotes the expected value of the state at time t, and $P(t)$ represents the state covariance matrix also at time t. In the standard Kalman filter, or the extended or unscented versions there are simple expressions that allow computation of the time update equations (that relate $(\hat{x}^-(t), P^-(t))$ to $(\hat{x}(t-1), P(t-1))$ and measurement update equations that relate $(\hat{x}(t), P(t))$ to $(\hat{x}^-(t), P^-(t))$.

The Kalman filter has an extensive history and has proven very useful in a wide range of applications, and also enjoys relatively simple computations. At each time, given a scalar measurement variable, the computational complexity of the time update and measurement update equations is typically $O(n^2)$ where n is the dimension of the state variable. Unfortunately, it provides a very poor representation of multi-modal distributions, since despite the generality available in (2), this distribution is always unimodal. Fortunately, this difficulty can be overcome by the use of Gaussian mixtures.

2.2   Gaussian Mixture Background

Gaussian mixtures represent the state pdf as a sum of a number of individual multivariate Gaussians, or multiple models. Each of the N models, for $i = 1..N$, is described by 3 parameters (where for simplicity we drop the explicit dependence on time):

- $\alpha_i \in [0,1]$, the probability that model i is correct, that is, the 'weight' associated with model i,
- $\hat{x}_i \in \mathbb{R}^n$, the state estimate for model i,
- $P_i = P_i^T > 0 \in \mathbb{R}^{n \times n}$, the covariance for model i.

For each model, the multivariate normal probability distribution function (pdf) is given by:

$$p_i(x) = \alpha_i \frac{1}{(2\pi)^{n/2}|P_i|^{1/2}} e^{\left(-\frac{1}{2}(x-\hat{x}_i)^T P_i^{-1}(x-\hat{x}_i)\right)}. \tag{3}$$

The overall mixture pdf is therefore:

$$p(x) = \sum_{i=1}^{N} p_i(x). \tag{4}$$

Note that all variables, α_i, $\hat{x}_i$, P_i and N can change with time in the algorithms to be discussed below.

Some of the key features of this representation are that under certain assumptions, any pdf can be approximated to an arbitrary degree of accuracy by a Gaussian mixture of sufficient degree (see for example the discussion in [1, §II]). We first consider the simple case of updates for unambiguous measurements.

2.3   Model Update with Unambiguous Measurements

We first perform an EKF (or UKF as appropriate) update of each of the N models, for all unambiguous objects from vision (e.g. ball, known goal posts, field markings that can be uniquely identified from other visual cues). This EKF (or UKF) measurement update is identical to the regular (that is single model) update, except that we need to include update equations for the model weight. For each model, and for each measurement update, an approximate heuristic for updating the weights is:

$$\alpha_i := \alpha_i \left(\frac{R}{R + (y - \hat{y})^2} \right) \tag{5}$$

where R is the variance of the measurement considered. Note the update proposed in (5) is simple and has the right general form, that is, zero innovation keeps the α value high, whilst large innovation shrinks the value. However, assuming statistically independent normally distributed measurement errors, and allowing for vectors of m measurements[1] the weights should be updated according to:

$$\alpha_i := \alpha_i \left(\frac{1}{\sqrt{(2\pi)^m |\Sigma_\eta|}} e^{-\frac{1}{2}\eta^T (\Sigma_\eta)^{-1}\eta} \right) \tag{6}$$

where $\eta = y - \hat{y}$ is the innovations associated with the measurement, and Σ_η is the variance of the innovations. Note that the innovations variance can be computed as the sum of the measurement variance R and the variance $\hat{R}$ of the prediction, $\hat{y}$, that is, $\Sigma_\eta = R + \hat{R}$.

One of the problems with the weight update given in (6) is its lack of robustness to outliers. For example, a single, slightly bad measurement where $|\eta| = 4\sigma_\eta$ would multiply α_i by almost four orders of magnitude less then if $\eta \approx 0$. To correct this, if we assume a probability of ϵ_o that our observation is an outlier (that is a false positive from vision), then a more appropriate weight update is:

$$\alpha_i := \alpha_i \left((1 - \epsilon_o) \frac{1}{\sqrt{(2\pi)^m |\Sigma_\eta|}} e^{-\frac{1}{2}\eta^T (\Sigma_\eta)^{-1}\eta} + \epsilon_o \right). \tag{7}$$

Having processed all the unambiguous measurements, we now turn to the problem of processing ambiguous measurements, which gives rise to the problem of model splitting.

2.4 Model Splitting - Ambiguous Measurements

When considering an ambiguous measurement update, with M alternate possibilities, an initial distribution with N elements (or models) can be performed by splitting each of the N initial models into M models (to a total of $M \times N$ models) and doing a standard measurement update for each possible combination of model component with each possible measurement component. Note that it is also possible that splitting could be the results of ambiguous time updates (for example, if we are uncertain whether the ball has just been kicked). In this case, similar considerations to those below will apply during the time update portion of the extended Kalman filter. For now, we look just at the measurement update equations.

Suppose that we start with N models, and a measurement that is ambiguous, and can therefore be interpreted as M different field objects, such as M different corner points. For simplicity we consider the case where each of these is equiprobable, though there is no difficulty in generalizing the algorithms below to cases where each of the measurement ambiguities has different, but known, probabilities.

---

[1] For example, it may make sense to consider the range and bearing of a single object as a single, two dimensional vector measurement.

The processing of an ambiguous measurement is performed by executing the following actions:

```
for each active model i
    for j =1 to number of ambiguous choices
        create a new copy (child) of model i;
        update this new model with measurement type j;
        if (update is an outlier) merge² new model with model i;
    end;
    renormalize the weights for all children of model i;
end;
```

Note that the distribution of weights at the end of the inner loop respects the relative weights after the measurement updates, but renormalizes the total weight. Clearly, whilst the individual actions within this procedure are relatively computationally cheap, it can give rise to an exponential growth in the number of active models, which is clearly impractical. One of the most important problems therefore in many multiple model Kalman filters is how to control the number of models. Although pruning (that is deleting) models with very small weights may be helpful, this is not a complete solution and it is important to have procedures for merging models.

2.5 Model Merge Equations

We first consider the simpler of the problems associated with merging models, namely, given a group (often a pair) of models, how do we merge (or join) these into a single resulting model that approximates the original pdf. There are many possible algorithms that may be used for merging models, see for example [8]. The discussions here follow closely these algorithms or simplified forms of them. For simplicity, we describe merging a pair of models, however, the algorithms below generalize trivially to merging more than two models at once.

Firstly, it is clear that when merging, to preserve the total weight probability of one of the models being correct, we should have [8, (2.24)]:

$$\alpha_m = \alpha_1 + \alpha_2 \tag{8}$$

where α_m is the weight of the merged model and α_1, α_2 are the weights of the two models to be merged.

Also, we can derive the merged mean as follows[3]:

$$\hat{x}_m = \frac{1}{\alpha_m} (\alpha_1 \hat{x}_1 + \alpha_2 \hat{x}_2) \tag{9}$$

---

[2] The model merge procedure will be discussed in Section 2.5. This logic frequently causes early model merges and thereby reduces the expansion in the number of active models.

[3] Note that when merging, extra care is need to merge the orientation components of the estimates. For example, merging an orientation of 179° with −179° should not give 0°.

Note however that this merging algorithm can cause 'drift', wherein, merging of a high weight, though slightly uncertain model, and low probability model with different mean, causes a small shift in the parameter estimates. If this situation persists (for example when repeatedly observing the same ambiguous object without any other observations), then the parameter estimates can drift significantly. To avoid this problem prior to computing the merged covariance, we follow (9) by the logic:

$$\text{if } \alpha_1 > 10 * \alpha_2 \text{ then } \hat{x}_m := \hat{x}_1$$
$$\text{if } \alpha_2 > 10 * \alpha_1 \text{ then } \hat{x}_m := \hat{x}_2$$

If we wish to preserve the overall covariance of the pdf corresponding to the original pair of pdfs, then the merged covariance matrix is given by:

$$P_m = \frac{\alpha_1}{\alpha_m} \left(P_1 + (\hat{x}_1 - \hat{x}_m)(\hat{x}_1 - \hat{x}_m)^T \right) + \frac{\alpha_2}{\alpha_m} \left(P_2 + (\hat{x}_2 - \hat{x}_m)(\hat{x}_2 - \hat{x}_m)^T \right)$$

(10)

Note that it is not obvious that these equations give the 'optimal' merge. In particular, some of the main contribution of the thesis [8], is to pose the merge problem as an optimization of the difference between the resultant pdf and the original mixture pdf. In this case, a recursive algorithm for computing the optimal merge can be generated. For reasons of simplicity and numerical efficiency, we propose the simpler equations (8),(9),(10). Note however, that (for example) merging a low weight high variance pdf with a high weight low variance pdf by this procedure tends to under-estimate the probability of the 'tail' of the distribution, whilst giving better accuracy in the pdf of the main mode of the distribution.

2.6 Model Merge Decisions

Model merge decisions are complex and there seem to be a number of possible algorithms for this. The authors of [8] formulate the problem of deciding which models to merge in an optimization framework. This optimization starts with a high order mixture model and seeks to find a lower order mixture model that best fits the original mixture model in the sense of the mean square deviation of the probability density functions. The only inputs needed are the original model, and the number of elements (models) in the final mixture. However, the computations for this kind of procedure seem prohibitive in the RoboCup SPL environment.

We therefore propose a computationally simpler procedure, based on a simplified form of the optimization approach. Our approach is based on computing pairwise merge metrics, that is, a measure of how much 'information' will be lost if this pair of models is merged. One metric proposed for example in [8, pp2.66] computes the distance metric, d_{ij}, for a pair of models indexed by (i, j) as

$$d_{ij} = \left(\frac{\alpha_i \alpha_j}{\alpha_i + \alpha_j} \right) (\hat{x}_i - \hat{x}_j)^T P_{ij}^{-1} (\hat{x}_i - \hat{x}_j)$$

(11)

where P_E is the covariance that would result if the models were to be joined,

$$P_{ij} = \left(\frac{\alpha_i}{\alpha_i + \alpha_j}\right) P_i + \left(\frac{\alpha_j}{\alpha_i + \alpha_j}\right) P_j + \left(\frac{2\alpha_i\alpha_j}{\alpha_i + \alpha_j}\right) \left((\hat{x}_i - \hat{x}_j)(\hat{x}_i - \hat{x}_j)^T\right) \quad (12)$$

To simplify calculations, and avoid the matrix inverse that may be problematic in higher order systems (for example combined robot, ball and ball velocity estimation where $n = 7$), we propose a simpler approximate metric

$$D_{ij} = (\alpha_i\alpha_j)(\hat{x}_i - \hat{x}_j)^T (\alpha_i\Delta_i + \alpha_j\Delta_j)^{-1} (\hat{x}_i - \hat{x}_j) \quad (13)$$

where Δ_i denotes the matrix formed by the diagonal component of P_i.

Given a mechanism, such as (13), for computing a metric on the closeness of two models, we could repeatedly search for the closest two models to merge. Note however, that to implement this, we must first compute all $N(N-1)/2$ possible distance metrics, find the smallest, merge these, then recompute merge metrics (or at least the $N-2$ metrics associated with the new merged model) and repeat. Alternatively, it may be simpler to merge based on a threshold using an algorithm such as that following:

```
for each active model i
  for each active model j
    if (mergeMetric(i,j) < threshold) then
      model i := mergeTwoModels(i,j);
    end;
end;
```

If this threshold based merge does not achieve sufficient reduction in the number of models, it can be repeated with larger thresholds.

2.7   Algorithm Implementation

To avoid the overhead of dynamic memory allocation and deletion, (as well as the potential to inadvertently create memory leaks), we implement a fixed size array of size MAX_MODELS of models. Each of these models is a normal KF model (that is, includes the states estimates and the state covariance) and in addition includes parameters for the weight alpha and a Boolean, active, denoting whether or not the model is in use.

The execution has main steps as follows:

1. **Time Update** For each of the active models, a call is made to the regular KF time update on this model, that is, it incorporates locomotion data and updates the filter covariances.
2. **Measurement Update for all Unambiguous Objects** For each of the active models, a regular measurement update is performed as suggested in Section 2.3.
3. **For each Ambiguous Object, split models**. Ambiguous objects are: (i) Unknown Intersections; (ii) Unknown Lines; and (iii) Ambiguous Goal

Posts. For each of these situations, each active model is split according to the various possibilities for the unknown object. This splitting uses algorithms as discussed in Section 2.5. After this process, the model weights, α_i, are normalized so that they sum to 1.

4. **Merge Models.** Since splitting models can leave us with a large number of possible models, after each ambiguous object update, we merge models to try to eliminate redundant models.

5. **Generate Localization Data for Behavior** Following the model merge, the α_i values are again normalized and the best model is selected to represent the most likely robot position, together with variances of the estimates. Note however, that we have a special segment of code, so that if there is a valid 2nd best model, the variance of the estimates reported to behavior is increased to account for any deviation between the state estimates for the best and 2nd best models. For example, with respect to orientation, instead of reporting just the variance $\sigma^2_{\theta_{i_1}}$ of the best model i_1, the overall heading variance σ^2_θ is computed as

$$\sigma^2_\theta = \sigma^2_{\theta_{i_1}} + \alpha_{i_2} \left(\theta_{i_1} - \theta_{i_2}\right)^2 \tag{14}$$

where i_2 denotes the index of the second most likely model.

3   Example and Results

In this section we run through an example of a Multiple Model Extended Kalman Filter (MM-EKF), as described in Section 2. One of the most demanding localization situations in RoboCup SPL is goal keeper localization since positioning needs to be very accurate, and in most cases, the only visible unique land marks are distant goals. In our test case we have a robot standing inside the yellow goal mouth looking directly up the field (in our coordinate system, the location is x=-290, y=0.0, θ=0.0). The robot then pans its head from side-to-side. In this example, the robot saw 46 unique observations (either the blue goal or identifiable blue goal posts) and 278 ambiguous observations (unidentifiable blue goal posts, intersections and lines). This gives an indication of the amount of information being ignored when not using ambiguous objects.

3.1   Comparison with a Single Model EKF

Firstly, lets present the accuracy results when comparing a MM-EKF to a Single Model EKF (S-EKF). As expected the MM-EKF easily out performs the S-EKF (see Figure 1). Note: the error at the start is due to the robot not intially knowing its location, once settled both versions converge to a stable location. In this case, the MM-EKF converged to a location 11.61cm from the true location with an average orientation error of $-1.6°$. While the S-EKF converged to a location 29.12cm from the real location and with an average orientation error of $-9.30°$.

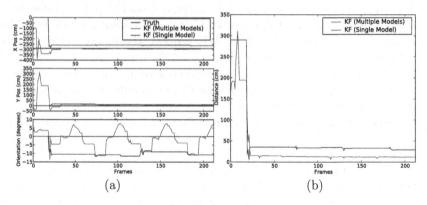

Fig. 1. Accuracy comparison between Multiple Model EKF and Single Model EKF. (a) Presents the location from the filters in each of x, y and orientation (θ). (b) Total distance error from actual location.

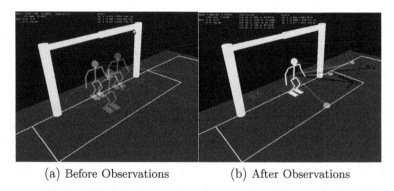

(a) Before Observations   (b) After Observations

Fig. 2. Example of ambiguous observations converging to the correct location. (a) 3 almost equal probability models (α=0.411,0.306,0.283). (b) After the observation of 3 lines and one intersection we are now left with a high certainty model (α=0.824) and two lower likelihood models (α=0.099,0.077).

3.2   Splitting and Merging

We will present two examples of splitting and merging. Firstly, an example of a merge split/merge cycle where three roughly equal models can converge to a more likely model after observering only ambiguous information. This example describes the update taking place at Frame 33 from Figure 1. Originally the MM-EKF is maintaining three models (as shown in Figure 2 (a)) with α values of 0.411,0.306 and 0.283 respectively. The opacity of the robot represents the α value of that model, with a more solid robot representing a higher α.

The first observation considered is the intersection that is 93cm away at an angle of 60°, in the ideal case this observation should be matched with the top left corner of the penalty box. Unfortunately the three models have just enough

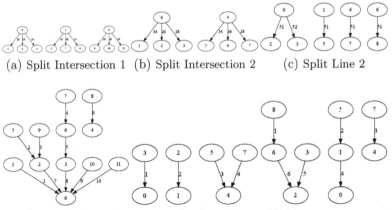

(a) Split Intersection 1   (b) Split Intersection 2   (c) Split Line 2

(d) Merge Intersection 1 (e) Merge Intersection 2 (f) Merge Lines 2 & 3

Fig. 3. (a)-(f) present the sequence of splits and merges that take place due to the observations

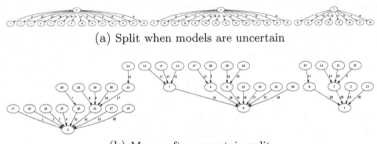

(a) Split when models are uncertain

(b) Merge after uncertain split

Fig. 4. Example of the worst case splitting/merging that can occur when seeing an ambiguous object. In this case the robot was unsure of its own location and observed an intersection. Unable to outlier many of the alternatives forced it perform a complicated merge.

uncertainty that they keep the 3 corners to the left of the robot, that is the top and bottom corners of the penalty box and the corner on the left hand side of the field. This spilt can be seen in Figure 3 (a), with the corresponding merge (d) reducing the total down to two models. Next the MM-EKF considers the second (false) intersection. Again the same three corners are considered valid options and the proceeding split/merge trees are shown in (b) and (e) respectively (Note: Model 6 is still alive at the end of the merge but was not graphed due to nothing combining with it).

The next observation considered is the line to the far left. This should be the left sideline but its has been reported with an incorrect vision distance. Because of this error all the alternatives we rejected as outliers, hence no splits or merges we undertaken. The next considered observation is the left edge of the penalty

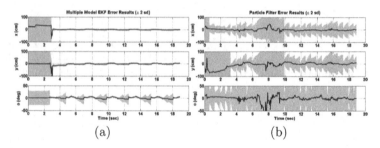

Fig. 5. Comparison of the MM-EKF (a) and a Particle Filter (b). The shaded area represents the uncertainty (two standard deviations).

box, in this case three of the four models (1,4 & 6) outliered on all but the correct line, while model 0 was sufficiently uncertain that it also considered the sideline as an alternative (shown in (c)). For space reasons we have forsaken showing the splits on the front edge of the penalty box, but rather have shown the final merge tree of the combined splits for the last two observed lines (f).

Secondly, we show the worst case scenario, that is when the robot has very little idea where it is (i.e. a high variance in location) and it observes an ambiguous object (Figure 4). In this case the MM-EKF is maintaining 3 possible models and then the robot observes an unknown intersection. The models fail to outlier on most of the possible alternatives and the merging step runs through a complicated procedure as shown in Figure 4 (b). The end results show very little improvement in accuracy as no alternative is favored. Luckily this scenario only occurs when the robot is kidnapped (or at startup) and doesn't see many unique objects before seeing an ambiguous object.

3.3   Comparison with a Particle Filter

Here we briefly compare the the MM-EKF to a Particle Filter (PF). The PF used in this comparison has been run at both the 2007 and 2008 RoboCup competitions and has been tuned for game performance. While the both filters provide similar accuracy, the MM-EKF can do so with a high certainty. This is due to the PF using a small number of particles (100) to simultaneous track position and handle noise/outliers/kidnapping. While the MM-EKF can perform optimal updates based on the observations for each model, while relying on more sophisticated approaches to outliers and kidnapping. It should also be noted that even when running so few particles the MM-EKF is substantially faster in terms of execution time, averaging less then 35% of time required to process a vision frame.

4   Conclusions

In this paper, we have discussed the background and implementation of multiple model Kalman filter based localization, with particular emphasis on the

RoboCup Soccer Standard Platform League. The MM-EKF is able to directly handle the ambiguous information, and therefore resultant multi-modal distributions common in the SPL. It shows performance that is substantially better than standard EKF implementations, and at least in a preliminary test, outperforms a particle filter applied to the same problem. The main complexity with the MM-EKF is the merge decisions required to keep the number of active models limited to a fairly low number. We have given some simple algorithms designed to achieve this with low average, and moderate peak CPU demands. Further work on a more detailed comparison with a wider variety of particle filters is required to give a more accurate picture of the relative merits of the different approaches.

References

1. Alspach, D., Sorenson, H.: Nonlinear Bayesian estimation using Gaussian sum approximations. IEEE Transactions on Automatic Control 17(4), 439–448 (1972)
2. Goehring, D., Mellmann, H., Burkhard, H.-D.: Constraint Based Belief Modeling. In: Iocchi, L., Matsubara, H., Weitzenfeld, A., Zhou, C. (eds.) RoboCup 2008. LNCS, vol. 5399, pp. 73–84. Springer, Heidelberg (2009)
3. Gutmann, J.-S., Weigel, T., Nebel, B.: Fast Accurate and Robust Self Localization in the RoboCup Environment. In: Veloso, M.M., Pagello, E., Kitano, H. (eds.) RoboCup 1999. LNCS (LNAI), vol. 1856, pp. 304–317. Springer, Heidelberg (2000)
4. Julier, S., Uhlmann, J., Durrant-Whyte, H.F.: A new method for the nonlinear transformation of means and covariances in filters and estimators. IEEE Transactions on Automatic Control 45(3), 477–482 (2000)
5. Lenser, S., Velosa, M.: Sensor resetting localization for poorly modeled mobile robots. In: Proc. of ICRA 2000, San Francisco (April 2000)
6. Middleton, R.H., Freeston, M., McNeill, L.: An Application of the Extended Kalman Filter to Robot Soccer Localisation and World Modelling. In: Proc. IFAC Symposium on Mechatronic Systems, Sydney (September 2004)
7. Nistico, W., Hebbel, M.: Temporal Smoothing Particle Filter for Vision Based Autonomous Mobile Robot Localization. In: Proc. 5th International Conference on Informatics in Control, Automation and Robotics, Funchal (May 2008)
8. Williams, J.L.: Gaussian Mixture Reduction For Tracking Multiple Maneuvering Targets In Clutter, PhD Thesis, AFIT/GE/ENG/03-19, Wright-Patterson Air Force Base, Ohio (2003)

Integrated Genetic Algorithmic and Fuzzy Logic Approach for Decision Making of Police Force Agents in Rescue Simulation Environment

Ashkan Radmand, Eslam Nazemi, and Mohammad Goodarzi

Electrical and Computer Engineering Department, Shahid Beheshti University,
Tehran, Iran
{a.radmand,nazemi,mo.goodarzi}@sbu.ac.ir

Abstract. The major task of police force agents in rescue simulation environment is to connect the isolated parts of the city. To achieve this goal, the best blocked roads should be chosen to clear. This selection is based on some issues such as number of burning buildings and victims existing in the mentioned parts. A linear combination of these factors is essential to determine a priority for each road. In this paper we propose an integrated Genetic Algorithm (GA) and Fuzzy Logic approach to optimize the combination statement. The parameters are learned via GA for some training maps. Then, because of differences between test and train maps, the agent should decide which parameters to choose according to the new map. The agents' decision is based on similarity measures between characteristics of maps using Fuzzy Logic. After utilizing this method, the simulation score increased between 2% and 7% in 20 test maps.

Keywords: Rescue Simulation, Police Force Agent, Decision Making, Genetic Algorithm, Fuzzy Logic.

1 Introduction

In rescue simulation environment, a simulated earthquake happens and the city goes in an emergency state. Some buildings start to burn, some others collapse, some civilians get damaged and blocked in collapsed buildings and some city roads close by debris caused by disaster. These blocked roads divide the city roads graph into isolated city parts. The major task of police force agents [1] is to connect the isolated parts of the city that causes the easier transportation of other types of agents (Fire brigades and Ambulances) to rescue the city and its civilians. To achieve this goal, the best blocked roads should be chosen to clear. This selection is based on some issues such as number of refuges, stuck agents, burning buildings and victim civilians existing in the mentioned parts.

Therefore, the police force agents should decide which two city parts are more important to get connected first. A linear combination of these factors is essential to determine a weight (priority) for each road.

J. Baltes et al. (Eds.): RoboCup 2009, LNAI 5949, pp. 288–295, 2010.

In section 2, we will explain major strategy of police force agents to build a linear combination of decision factors which were mentioned above. Section 3 is about description of the GA approach which has been used to learn the optimum parameters of police force agents' decision making in some fixed training maps. Section 4 explains the method of combination of the trained parameters to achieve an efficient solution for unknown city maps using Fuzzy Logic. We utilize fuzzy logic to determine the measure of similarity between new maps and learnt maps based on some characteristics of them such as state of fires, blockades, victim civilians and etc. Finally in section 5, some experimental results are reported to show the effect of proposed method on agents work efficiency.

2 Police Force Agents Main Strategy to Choose a Target Road

Police force agents should connect all city parts together as soon as possible and in a manner that leads to a higher final score which determines the performance of agents work. To achieve this goal the agents have to assign a weight to each boundary road (means any road that disconnects a city part from another) and start to open them based on these weights or priorities.

The police agents consider some conditions which determine the worth of cleaning each boundary road. The final weight of that road equals to the sum of these conditions values. The value of a condition is calculated based on its premise parameters. Each condition value should get a weight in summation step of calculating final weight of a boundary road. These weights determine the importance of each condition comparing to the other conditions.

In order to better understand, assume that we have a boundary road l which separates two city parts $cp1$ and $cp2$. Considering this situation, we have conditions described in Table1.

Table 1. Conditions to assign a weight to a boundary road

Condition	If part #1	If part#2	If part#3	Then part
Cond. 1	#(cp1.BB)>0	#(cp1.RF)=0	#(cp2.RF)>0	#(cp1.BB)* W1
Cond. 2	#(cp1.BB)>0	#(cp1.FB)=0	#(cp2.FB)>0	#(cp1.BB)* W2
Cond. 3	#(cp1.DV)>0	#(cp1.AT)=0	#(cp2.AT)>0	#(cp1.DV)* W3
Cond. 4	#(cp1.DV)>0	#(cp1.RF)=0	#(cp2.RF)>0	#(cp1.DV)* W4
Cond. 5	#(cp1.BA)>0	#(cp1.AT)=0	#(cp2.AT)>0	#(cp1.BA)* W5
Cond. 6	#(cp1.BF)>0	#(cp1.AT)=0	#(cp2.AT)>0	#(cp1.BF)* W6
Cond. 7	#(cp1.BP)>0	#(cp1.AT)=0	#(cp2.AT)>0	#(cp1.BP)* W7

BB: Burning Buildings   RF: Refuges   DV: Dying Victims
AT: Ambulance Teams   FB: Fire Brigades   PF: Police Forces
BA: Buried* AT   BF: Buried FB   BP: Buried PF
* A buried agent is a victim agent that needs ambulance team help.

Each condition consists of three "if parts" combined with AND operator and one "then part" affected by a coefficient W_i which is the weight of the condition in summation step. For example, the first condition says that "if the number of burning buildings in $cp1$ is more than zero AND the number of refuges in that city part is equal to zero AND the number of refuges in *cp2* is more than zero", then the condition 1 has a value equal to number of burning buildings multiplied by *W1*. The coefficient *Wi* is the weight of i-th condition in calculating the total weight of *l*.

The importance of connecting two city parts, one with burning buildings and other with refuges, referring to the first condition, is determined by the fact that fire brigades need to go to the refuges to fill their water tanks in order to extinguish burning buildings.

Consequently, decision making of police force agents depends on *Wi*. To achieve good weights for conditions we have used a GA approach which will explain in section 3.

3   Genetic Algorithm Approach to Determine Weights

In section 2, we described the police force agents' general strategy to choose target roads. To optimize the performance, they should assign good values to conditions weights. To achieve this goal, we utilize GA approach [3-5] that will be described in this section.

3.1   Chromosome Structure

In our method, chromosomes have a simple structure which is an array of values assigned to weights of conditions. An example of chromosomes structure has been shown in Fig1.

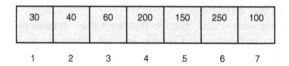

Fig. 1. An example of the chromosome structure

The size of each chromosome is equal to *n* which represents the number of conditions mentioned in section 2.

Initial population may be generated through a random or user specified process. It plays an important role in search direction. A well selected initial population increases the search procedure convergence speed and results in faster trend to optimum solution. In the proposed method, to generate initial population, values assigned to chromosomes are the same as values used before by experiment. Only one element or two of each individual chromosome take a random value.

After constructing initial population, the fitness values for all individuals should be calculated. The number of individuals in the population is constant in all generations. Some individuals that have most fitness values are gone forward to next generation. If the crossover rate is called *Pc* and number of individuals is called *Ps*, number of individuals that are passed to next generation is equal to *Ps- Ps*Pc*. Therefore, the number of new generated individuals in each generation is *Ps*Pc*. These processes are performed while the terminating condition is not satisfied. Other parts of the proposed genetic algorithm will be described in sections 3.2, 3.3 and 3.4.

3.2 Fitness Function

In the presented method, the total score of simulation is used as fitness value. This score is calculated based on work efficiency of all agents. All factors of simulation including city map, initial fires, victim civilians and decision making algorithm of other agent types (fire brigades and ambulances) should be fixed in GA training iterations. The only exception is police forces work which differs in the various training iterations. This difference is because of the changes occurred in the weights in the decision making section. So, any change in the total score is because of the change in work of police force team.

3.3 Selection Algorithm

The selection of the individuals is based on the fitness value of the solutions. The probability of selection of an individual is directly or inversely proportional to its fitness value. The roulette wheel selection [6] is used in our proposed GA. The main idea of this method is to select individuals stochastically from one generation to create the next generation. In this process, the more appropriate individuals have more chance to survive and go forward to the next generation. However, the weaker individuals will also have a little probability to select.

In selection process, $Ps*Pc$ individuals are selected to create the same number individuals from them using crossover and mutation operators.

3.4 Crossover and Mutation Operators

Since individual chromosomes based on a simple structure, complex cross over operators are not necessary. In the proposed method, two point crossovers are used. Therefore, $Ps*Pc$ individuals are selected using our selection process where Pc is crossover rate. As $Ps*Pc$ new individual is needed after doing crossover, two parents are selected and two new child are produced from them. Points in each parent are selected randomly and segments between these two points are substituted by parents to produce new individual children.

We produce a random number for each element of individual chromosomes. If it is lower than Pm, mutation will be done for that element. Note that Pm is mutation constant. Three elements of each individual chromosome can be chosen for mutation at most. Terminating criteria is the number of generations which is determined.

4 Decision Combination Using Fuzzy Logic

In rescue simulation environment, each map has some initial states such as start points of fire spread, victim civilians' positions, blocked roads and etc that affect the decision making strategy of police force agents. Therefore, it is not applicable to learn conditions weights in a certain map and use them in any other map. On the other hand, we cannot learn the weights for all possible maps because the number of maps is infinite.

To overcome this problem, we designed some training maps with specific characteristics. We have used some characteristics that are more important in classifying the maps including number of burning buildings, number of victim civilians, number of blocked roads and number of buried agents.

Each of these maps has different values of above characteristics. In fact, training maps are representatives of all maps. Applying GA approach mentioned in section 3 to all training maps gives us an array of optimum weights for each map. It is required to have a method to combine these arrays and have a solution for each new map. In this section, we will propose a method of decision combination using fuzzy logic [7-9].

To achieve this goal, each characteristic of map will create a fuzzy set that consists of three membership functions: *Low*, *Medium* and *High*. Fig. 2 demonstrates the membership functions of maps characteristics.

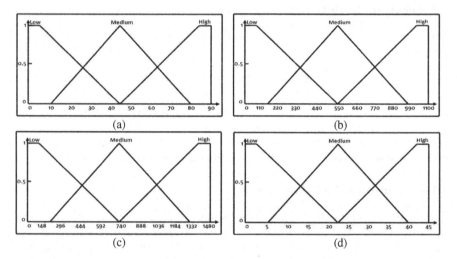

Fig. 2. "(a) Victim Civilians number (b) Burning Buildings number (c) Blocked Roads number (d) Buried Agents number" Membership Functions

4.1 Fuzzy If-Then Rules

Using defined fuzzy sets, fuzzy if-then rules will be created. Each rule has four "if parts" combined with AND operator pointing to map characteristics mentioned above and one "then part" that relates the given map to one of training maps. These rules are in Mamdani's proposed form [10]. For example, one of the rules is:

IF "number of victims" is High AND "number of burning buildings" is Low AND "number of blocked roads" is Low AND "number of buried agents" is Medium THEN map IS training map #1.

Therefore, to cover all possible conditions, we should create $3^4 = 81$ rules. So, we should design 81 training maps that each one has characteristics as same as those which are described in the center of fuzzy membership functions of corresponding rule. Having these rules, the training part is completed and we just need a fuzzy inference method that can estimate suitable weights for any new given map. This fuzzy inference method is explained in section 4.1.

4.2 Fuzzy Inference Method

Given a new map, values of map characteristics will be checked against all fuzzy If-Then rules. Each "if part" of a rule has a membership function. So, the membership value of the given map will be calculated for each of them. As a result, for a given map there will be four membership values. A rule "trigger value" (means rule weight in fuzzy inference) will be equal to minimum membership value of "if parts" of that rule.

In defuzzification step, to achieve final weights for the given map, we should combine "then part" of all rules considering their trigger values. "Weighted Average" method is used in our proposed method for defuzzification step.

For example, assume that we have just two fuzzy rules. After checking the rules for a given map, the first rule has gotten a trigger value $w1$ and refers to training map #1 in its "then part", and the second rule has gotten a trigger value $w2$ and refers to training map #2 in its "then part". So, the final weights will be calculated based on equation (1).

$$final\ weights = \frac{w1 \times weights1 + w2 \times weights2}{w1 + w2} \qquad (1)$$

weights1, weights2 and final weights are the weights arrays of training map #1, training map #2 and the given map respectively.

5 Experimental Results

In order to evaluate the presented algorithm, we implemented it on "SBCe_Saviour2008" source code which was one of the eight top teams in China2008 competitions. In this section, to demonstrate the performance of the method, we compared the results of our method with the results gained before its implementation. 81 maps with different characteristics were designed for training section. The mentioned GA-based method was applied to each map. Using try and error, GA generations consist of 10 chromosomes and the learning process continued for 100 generations for each training map. Crossover and Mutation probabilities were set to 0.8 and 0.2 respectively. Change in value of elements, which had been mutated, was equal to ±20 in first generation and reduced in a way that it reached zero in 10 last generations. Fig. 3 demonstrates changes in maximum gained scores in each generation during GA training step. All the training maps were based on Foligno city which is one of the standard maps in rescue simulation league of RoboCup competitions [2].

Maximum total score is increased about 7% as it is shown. In other 80 training maps this increase was between 5% and 8%. After completion of GA training step for all maps, 81 fuzzy if-then rules were created and each one was assigned to a training map in its "then part".

20 test maps were chosen from Robocup2009 China rescue simulation competition maps. The comparison of total scores gained using proposed method and the results gained before the training is shown in Table2.

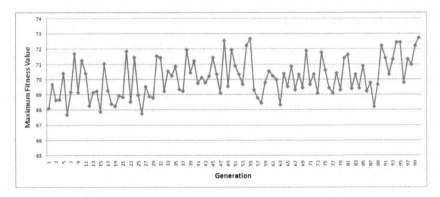

Fig. 3. Maximum gained scores in training generations

Table 2. The comparison of results gained before and after training

Map Name	Old score	New score	Map Name	Old score	New score
Map #01	65.356	67.64	Map #11	90.012	92.103
Map #02	63.165	67.265	Map #12	93.464	95.689
Map #03	76.798	78.013	Map #13	88.645	90.465
Map #04	85.679	88.625	Map #14	87.465	91.465
Map #05	72.946	75.346	Map #15	85.856	87.695
Map #06	59.689	62.345	Map #16	75.964	78.654
Map #07	75.899	78.463	Map #17	72.331	74.649
Map #08	77.689	79.341	Map #18	70.002	73.566
Map #09	74.334	76.555	Map #19	90.645	92.256
Map #10	82.645	85.756	Map #20	81.135	84.698

6   Conclusion

In this paper we proposed a GA-based approach to achieve suitable weights for decision making conditions of police force agents in rescue simulation environment. This method is applied to some training maps and the result weights will be combined to achieve suitable weights for any new given map using fuzzy inference method. The simulation results showed that the method which was presented has positive effect on decision making of police force agents and increases the total score of whole team.

References

1. Morimoto, T.: How to Develop a RoboCup Rescue Agent. Rescue Technical Committee
2. RoboCup Rescue official Website, http://www.robocuprescue.org
3. Holland, J.H.: Adaptation in Natural and Artificial Systems. University of Michigan Press, Ann Arbor (1975)
4. Goldberg, D.E.: Genetic Algorithms in Search, Optimization and Machine Learning. Kluwer Academic Publishers, Boston (1989)
5. Banzhaf, W., Nordin, P., Keller, R., Francone, F.: Genetic Programming - An Introduction. Morgan Kaufmann, San Francisco (1998)

6. Al Jadaan, O., Rajamani, L.: Improved Selection Operator for GA. Journal of Theoretical and Applied Information Technology, 269–277 (2005)
7. Kruse, R., Gebhardt, J., Klawonn, F.: Foundations of Fuzzy Systems. John Wiley & Sons, New York (1994)
8. Dubois, D., Prade, H.: Fuzzy Sets and Systems: Theory and Applications. Academic Press, New York (1980)
9. Zimmermann, H.-J.: Fuzzy Set Theory—and its Applications, 2nd edn. Kluwer Academic Publishers, Boston (1991)
10. Mamdami, E.H., Assilina, S.: An experiment in linguistic synthesis with a fuzzy logic controller. International Journal of Man-Machine Studies 7(1), 1–13 (1975)

IntellWheels MMI: A Flexible Interface for an Intelligent Wheelchair

Luis Paulo Reis[1,2], Rodrigo A. M. Braga[1,2], Márcio Sousa[1],
and Antonio Paulo Moreira[1]

[1] FEUP – Faculty of Engineering of the University of Porto, Rua Dr. Roberto Frias,
s/n 4200-465, Porto, Portugal
[2] LIACC - Artificial Intelligence and Computer Science Lab., University of Porto, Portugal
lpreis@fe.up.pt, rodrigo.braga@fe.up.pt,
marcio.mc.sousa@gmail.com, amoreira@fe.up.pt

Abstract. With the rising concern about the needs of people with physical disabilities and with the aging of the population there is a major concern of creating electronic devices that may improve the life of the physically handicapped and elderly person. One of these new solutions passes through the adaptation of electric wheelchairs in order to give them environmental perception, more intelligent capabilities and more adequate Human – Machine Interaction. This paper focuses in the development of a user-friendly multimodal interface, which is integrated in the Intellwheels project. This simple multimodal human-robot interface developed allows the connection of several input modules, enabling the wheelchair control through flexible input sequences of distinct types of inputs (voice, facial expressions, head movements, keyboard and, joystick). The system created is capable of storing user defined associations, of input's sequences and corresponding output commands. The tests performed have proved the system efficiency and the capabilities of this multimodal interface.

Keywords: Multimodal Interface, Intelligent Wheelchair, Intelligent Robotics.

1 Introduction

Physical injuries occur frequently caused by accidents affecting the mobile capabilities of individuals, among other damages. Physical injuries could also be caused by medical conditions, like brain palsy, multiple sclerosis, diseases respiratory and circulatory diseases, genetic diseases or chemical and drugs exposition. Usually, the physical deficiency result on a limited control of some muscles of the arms, legs or face. It's very difficult to generalize physical deficiencies and each person has different symptoms and uses different strategies to deal with it. An example is the cerebral palsy, which concern with injuries on some brain areas responsible for the movement control, resulting on a difficulty that could be slight or cause total incapacity of moving the arms, legs or even talk. Two persons with brain palsy are different on each one's deficiency and degree of muscle control. Cerebral palsy as no cure but the effects could change with the age.

J. Baltes et al. (Eds.): RoboCup 2009, LNAI 5949, pp. 296–307, 2010.

Nowadays, society is more and more concerned with enabling handicapped persons to have an as independent life as possible. Wheelchairs are important locomotion devices for handicapped and senior people. With the increase in the number of senior citizens and the increment of people bearing physical deficiencies in the social activities, there is a growing demand for safer and more comfortable Wheelchairs and the new Intelligent Wheelchair (IW) concept was introduced. Like many other robotic systems, the main capabilities of an intelligent wheelchair should be: Autonomous navigation with safety, flexibility and capability of avoiding obstacles; intelligent interface with the user; communication with other devices (like automatic doors and other wheelchairs). However, most of the Intelligent Wheelchairs developed by distinct research laboratories, [6][10], have hardware and software architectures too specific for the wheelchair model used/project developed and are typically very difficult to configure in order for the user to start using them.

The Intellwheels prototype includes most of the typical IW capabilities, like facial expression recognition based command, voice command, sensor base command, advanced sensorial capabilities, the use of computer vision as an aid for navigation, obstacle avoidance, intelligent planning of high-level actions and communication with other devices. However the project is based on two main innovative ideas that will tackle the abovementioned IW problems. Firstly the Intellwheels project is based on a generic IW framework that enables easy development of new intelligent wheelchairs and IW control algorithms. The framework is flexible enough to enable easy transformation of commercial wheelchairs into intelligent wheelchairs with minor hardware changes and to enable the introduction of new modules and algorithms in the intelligent wheelchair. It includes a complete IW simulation module enabling to conduct virtual reality and mixed reality experiments.

The second innovation is concerned with the Intelligent Wheelchair command methodology that is based on a flexible multimodal interface. The wheelchair is commanded at a very high-level using a high-level command language based on simple commands such as "go to bedroom", "wander", "follow wall". The commands are triggered by user selected input sequences using the multimodal interface. An input sequence may be something like "blink left eye" and then "say: go" or any given sequence of inputs coming from distinct input devices. The wheelchair enables the user to even use the same type of input sequences to select its preferred inputs/action association. In order to enable the user to start using the wheelchair, a simple patient classification module based on machine learning techniques is now under development. It will be capable of identifying the user basic capabilities and enable him to start using the wheelchair flexible multimodal interface in a straightforward manner.

This work focuses in the development of a user-friendly multimodal interface, which is integrated in the Intellwheels project. This paper presents the first prototype of the multimodal human-robot interface developed that allows the connection of several input modules, enabling the wheelchair control through flexible input sequences of distinct types of inputs (voice, facial expressions, head movements, keyboard and, joystick). The system created is capable of storing user defined associations, of input's sequences and corresponding output commands. This interface can provide an interaction between the wheelchair environment and the input method, so that at any instance the input information can be analyzed and checked if it's reliable, to assure the user safety.

The rest of this paper is organized as follows. Section 2 describes the concept of multimodal interface and indicates some of its desired characteristics. Section 3 describes the Intellwheels project and its main features and characerics. Section 4 describes the developed work reagrding the IW multimodal interface and section 5 describes the experiments performed and the results achieved. Section 6 presents the paper main conclusions and some pointers to future work.

2 Multimodal Interfaces

Generically an interface is an element that establishes a frontier between two entities. When an interface is used to assist in the Human-Computer Interaction it is called a user interface, being able to be graphical or command line based.

Most of the traditional graphical user interfaces are based in the WIMP (Window, Icon, Menu, and Pointing device) paradigm, which uses the mouse and keyboard as physical input devices to interact with the interface, for example to access information or accomplish any needed task.

An evolution to this paradigm and a way to create a more natural interaction with the user is the establishment of a multimodal interaction. This interaction contemplates a broader range of input devices such as video, voice, pen, etc, and so these interfaces are called Multimodal Interfaces.

A Multimodal Interface [1] "processes two or more user input modes – such as speech, pen, touch, manual gestures, gaze, and head and body movements – in a coordinated manner with multimedia system output. They are a new class of interfaces that aim to recognize naturally occurring forms of human language or behavior, and that incorporate one or more recognition-based technologies (e.g., speech, pen, vision)". This type of interface can be used in several fields such as, for example, navigational devices – [2] and [3] – and health care solutions – [4] and [5].

Considering the purpose of this work the main aspects to consider should be the adaptability to users, usability and safety. These factors are determinative in a Multimodal Interface design, where subjective characteristics, like user satisfaction and cognitive learning, and user interaction depend on them. The adaptability to users is necessary so that the interface can be usable and understandable by any person, independently from his informatics knowledge and cognition. With the multimodal interaction between inputs comes a wider range of output control options and a complementarily between inputs.

The output control is achieved by the combination of several inputs, only being limited by the total number of inputs. As the interaction between the inputs can differ depending on the environment, this multimodality achieves a complementarily that when any input become less recognizable, it can be compensated by another, but this must be done being in mind the interface accessibility.

Finally, having in account the project enclosure, the multimodality must enable the access to any user, despite his deficiency. This shows the Multimodal Interface accessibility importance, so that if a user as any deficiency that suppress the use of one input, there is another that compensates this handicap [4].

Since this is a Multimodal Interface, it is necessary that this project allows a transparent and intuitive control of the Wheelchair and also a flawless input interaction.

This is achieved by the understanding of the user and inputs interaction. The inputs interaction is one of the key points of a multimodal interface, since it will be this interaction that will produce the desired output to the user. It is necessary the existence of a support for integrating any kind of inputs like: video, speech, handwriting, gestures, etc, but also this support must contemplate a robust processing of the inputs to fully recognize the user intentions.

The user interaction is another key point of a multimodal interface, if not the most important, so that the user can have and pleasant experience with the interface. It is necessary to consistently verify the disposition of every component of the interface so that the visual information and content can be easily accessed. Also it is needed to assure an intuitive interaction with the system, regarding the information about the available actions and how the user can interact with them.

Other factor is the interface output, which is divided in two parts: the processes concerned with the interpretation of the user inputs and processes regarding the correct visualization of the information given to the user about the system state.

3   Intellwheels Project

This Multimodal Interface is included in the Intellwheels Project, which main objective is to provide an intelligent wheelchair development framework to aid any person with special mobility needs.

This project encloses the prototype of an intelligent wheelchair, since its hardware structure to all software needed to provide a flawless control and experience to the user, being the hardware architecture shown in figure 1.

This architecture was created with the objective of being flexible and generic, so that it does not imply considerable modifications in the wheelchair structure [6].

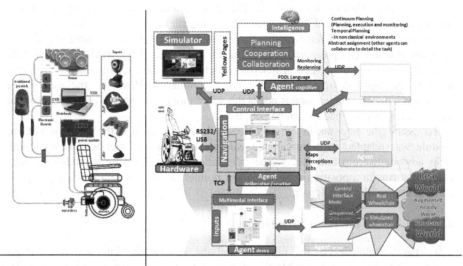

Fig. 1. Hardware architecture [6]

Fig. 2. Software architecture[6]

To enable a multimodal control of the wheelchair it is necessary to provide several inputs to user. It is also essential that these devices can map a broad kind of input methods, so that given any type of movement needs the user always has a way to control the wheelchair. With that in mind the following input devices were implemented: USB Joystick; Microphone; Head movements; Keyboard; Mouse; Video camera. With this it is possible to control the wheelchair using several types of inputs, from head movements to facial expressions [7] [8] [9]. Apart from the user inputs, the wheelchair also uses sensoring devices like: encoders, for the odometry calculation, and sonars, for obstacle detection. Several hardware interface modules are included to deal with the encoders and sensors.

One final, and important hardware device, is the laptop HP Pavillion tx1270EP, which is used to run all the developed software. In figure 2 is possible to see the global multi-agent software architecture defined for the Intelwheels project.

Focusing in the multimodal interface, it interacts with the Control Interface through a TCP socket connection, where the Control Interface will inform the Multimodal Interface of the available actions and state of any pending planning.

The user interacts with the Multimodal Interface which provides the connection, also through a TCP socket, of several independent input modules. The input modules are used for the user interaction and, therefore, create input sequences to execute the control actions assign by the Control Interface.

4   IntellWheels Multimodal Interface

The Multimodal Interface shows, in a graphical way, information about the actions, and input modules, such as kind, name or type of action or input, respectively. It also shows the defined input sequences, for the actions execution.

The joystick module works as a driver to establish a connection between an USB joystick and the Multimodal Interface. This module was adapted from [6], and it gets the information of the available buttons and analog sticks.

To enable the voice interaction it was necessary to implement a simple speech recognition module. The presented solution takes advantage of the IBM Via Voice [11] capabilities using the navigation macros, which allows the user interaction with any software through, previously recorded voice commands. However, the use of Via Voice has a disadvantage since it needs the voice module window to be active so that the voice commands macros can be perceived.

To assure the integration of the already developed inputs, the head movement module was adapted to communicate with the multimodal interface. This module takes advantage from one accelerometer installed in a cap, where it reads its values and transforms in a position type value, for pointer control, or in a percentage speed value to control the wheelchair.

4.1   Multimodal Interface Architecture

Since the wheelchair control platform and the multimodal interface are distinct agents, it was necessary to enable the multimodal interface agent to interact with the already developed control agent [6]. With that intention, data structure and information processing methods were created for the components interaction.

The system architecture, illustrated in figure 3, is a zoom in on the main architecture shown in figure 2. In this figure is possible to see the exchanged information between all the involved agents. The control interface acts as a communications server to the multimodal interface, as well as the multimodal interface acts as a communications server to the input modules.

Since the communications are totally established by the used Delphi components, as soon as the multimodal interface connects to the control agent, the control sends the information about the available high-level actions. For the input modules, as soon as one of them connects to the interface, firstly it sends its id and, number of module commands. Secondly, upon the receiving of a request from the interface, the input module sends the description of all the commands.

The interface information processing is divided in two logical parts: the server side and the client side processing. This division is derived from the need of the Multimodal Interface to act as a client to the control connection, but as a server to the inputs' connection.

For these models two data structures were created. One for storing the control actions information and other for storing the input modules commands information.

The inputs' structure is composed of six fields:
- Number: the internal number of the command;
- Name: the name of the command;
- Kind: this defines the name of the input module;
- State: for a button, this represents if its pressed – "True" – or if it was released – "False";
- Value 1 and value 2: these fields are used for transmitting the analog values of a command, for example the analog stick of the joystick.

For a digital command, like for example a button, the value fields will return a "n/a" string, being the same analogously applicable to an analog command, it returns the state field with a "n/a" string.

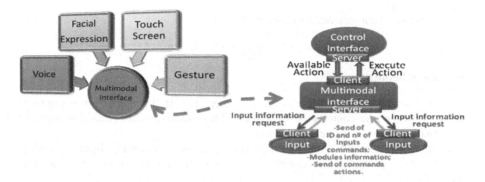

Fig. 3. Multimodal Interface architecture

The actions' structure is also composed of six fields:
- Name: the name of the action;
- Kind: the kind of action, for example movement;
- State: the availability for executing a action, returns "True" for a available action, or "False" if not available;
- Value: informs the interface about the execution of an action, returns "ON" if the action is under execution, or "OFF" when it stops its execution;
- Data: this field acts as an information about the level type of the action, being its options in table 1.

Table 1. Action structure: Data field

Data	Type name	Sent Parameters
0	Stop action	0
1	Manual action	2
2	Mid-Level action	1
3	High-Level action	0

The information is passed through one of the following messages:
- From the control interface:
  ```
  <cmd id= name="" kind="" state="" value="" data=""\>
  <cmd_state id="" state=""\>
  ```
- To the control interface
  - ➢ High level or Stop: `<action id=""\>`;
  - ➢ Mid level: `<action id="" value=""\>`;
  - ➢ Manual mode: `<manual value1="" value2=""\>`;
- From the input modules
  - ➢ Registration at the multimodal interface
    ```
    <input_info id="" mods=""\>
    <input id="">
    <module num="" name="" kind=""\>
    ...
    <module num="" name="" kind=""\>
    <input\>
    ```
  - ➢ Input event generated by user interaction
    ```
    <input_action num="" state=""\>
    <input_action num="" value1="" value2=""\>
    ```

4.2 Input Sequences

The input sequences represent how the user interacts with the interface or being more precise how the user controls the wheelchair. These sequences are created through the combination of two or more input actions.

Independently from the input module kind, or if the command is digital or analog, the associated event has two common identifiers: the module id, and the command number.

To standardize the inputs events representation was defined that if the command is digital, then its state is "True" – T – when the button is pressed, or "False" – F - when the button is released.

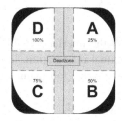

Fig. 4. Division for fixed speed values

For an analog command the state field is not used, but the input module will return the values of the analog axis. These are converted to a percentual value and are used to directly control the wheelchair, if the manual mode was activated, or are "processed" and define a fixed speed value.

This fixed speed value is achieved by logically "dividing" the cursor area in one of the analog axis of the joystick. Due to the short length of the stick it is only possible to divide in four areas without losing precision, being the division shown in figure 4. The variation assumes increments of 25% per zone, from A to D.

To simplify the sequence creation method, it was imposed a maximum number of six input command actions (fragments) to generate a sequence. Also, a minimum number of one input was imposed. Each fragment has the following format:

```
#<input_module_id>.<command_number><state>
```

The state field can be composed by one of three possibilities: "T" or "F", in case of a button, or "%" in case of one of the four pre established values (A, B, C or D).

The sequence entrance is limited by the already referred maximum number of fragments, or at any instance by the detection of an existent or nonexistent sequence. That is, for each fragment received the developed algorithm updates the input sequence under construction and, searches in the sequence list for the same occurrence. The search returns one of three available options:

- The occurrence is unique, and therefore the composed sequence can be immediately analysed;
- There are more occurrences of the same sequence fragment and thus it must be further processed;
- The occurrence does not exist in the list, meaning that the user is entering a not valid sequence and therefore the process is stopped;

If the search indicates the current sequence is not unique, the algorithm waits for a given predefined time for more input event actions in order to complete this sequence to a unique sequence. With this it is possible to evaluate if the user is still entering the sequence or, if during a pre established time interval none input action is received, if the sequence is already completed.

The use of this process turned the sequence input method more reactive to the user by providing an almost instant response to unique or wrong sequences, allowing a more effective control.

4.3 Interface Components

The interface components are all the interface visible components, since the simple buttons to images, menus and textual information. In order for the Multimodal Interface to be very simple it only contains the following components:

- List of available actions;
- Summary of the inputs connected;
- Input modules and control connections status;
- Input sequence graphical information;
- Sequence's list;
- Sequence's analysis result;
- Wheels speed information;
- Menus for programming the interface options and adding more sequences.

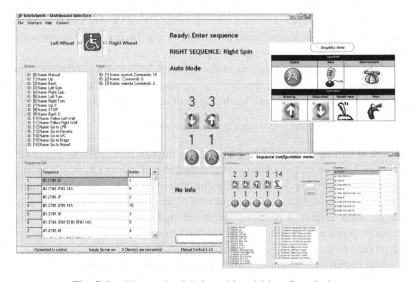

Fig. 5. Intelligent wheelchair multimodal interface design

All these components show the available information in a textual way, except the input sequence and wheels speed that show the information in a graphical way. Figure 5 displays the multimodal interface design.

5 Results

In order to evaluate the Multimodal Interface integration in the Intellwheels project several experiments were made using the Intellwheels simulator [12], with the objective of controlling the wheelchair in manual mode using distinct inputs.

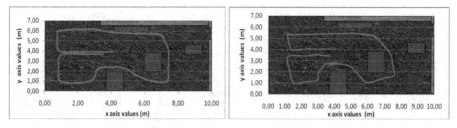

Fig. 6. Wheelchair movement in a room with obstacles, with joystick control and with head movement control

These experiments were made in a simulated room with several obstacles, where the wheelchair starts from a middle position and tries to go around the room perimeter, deviating from the objects, finishing in the start position. When using voice commands it was also tested the voice recognition software by introducing background noise during the tests. For the control with voice commands five commands (front, back, left, right and, stop) were defined to control the wheelchair. With these results of figure 6-a it is possible to see the wheelchair movement through the room, being the input method able to drive the wheelchair in the predetermined course without any problem.

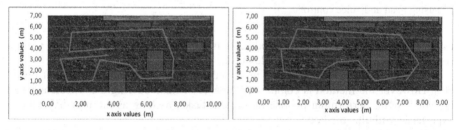

Fig. 7. Wheelchair movement in a room with obstacles, with voice control, without background noise and with background noise

Analysing the experiments of figures 7-a and 7-b it is possible to see that the implemented voice input method, for directly control the wheelchair, is a preferably input for open areas without obstacles. This is due to the delay in the response of the speech recognition software, which in emergency situations can become dangerous.

Another aspect to be considered analyzing these experiments is the sensibility of the speech recognition software to the background noise. During the experiments the microphone was approximately at 30 cm from the user, and the background noise source was a radio playing music with low volume. In these conditions it was necessary to repeat the voice commands several times, which has increased the experiment time and also made the control more difficult. Finally, figure 9 shows tests with a real wheelchair.

The results show that it is possible to drive the wheelchair just using head movements with good performance. However, these tests were made without any source of distraction. In other experiments it was confirmed that controlling the wheelchair

Fig. 8. Real wheelchair movement in a corridor with obstacles using joystick and voice control and using head and voice control

using only the head movement module, with a high sensibility, and with several sources of distraction is a very complex task.

Again, this movement method is preferred for open environments without obstacles. However the method is completely capable of manoeuvring the wheelchair in a crowded room, performing precise movement tasks, as long as the user has enough experience with this method.

Although the set of experiments performed was still very simple and separate simple experiments were performed for each input module, it is possible to take some interesting conclusions from the results. The inputs perform well and individually enable to control the wheelchair. However with distraction sources it is very complex to control the wheelchair with a single input and thus the use of high-level commands and input sequences to trigger them, seems to be an appropriate approach.

6   Conclusions and Future Work

The developed multimodal interface showed to be very flexible enabling the user to define distinct types of command sequences and associate them to the available high-level outputs.

To verify the system efficiency and the wheelchair control through the developed multimodal interface several experiments were conducted, where the wheelchair was controlled with the available inputs (joystick, voice, head movements and several inputs) in different kinds of environments (noise in the background, obstacles, etc.). The results achieved enabled to confirm the multimodal interface capabilities, except for the voice module, which proved not to be precise when there is noise in the background. However, the main capabilities of high-level commanding through input sequences of the multimodal interface need further experiments to be evaluated.

Some future directions for this project development are obvious and concern performing a set of experiments with the complete multimodal interface and the development of the yet missing input modules. One missing feature is a robust facial expressions recognizing module, needed to create a more multimodal experience to the user.

With the intention of making the Multimodal Interface more user friendly, a text to speech output and some kind of virtual user assistant could be implemented. These elements would function as an user integration process with the interface.

Acknowledgements

This work was partially supported by FCT Project PTDC/EIA/70695/2006 "ACORD: Adaptative Coordination of Robotic Teams" and LIACC/University of Porto. The second author would like to thank for CAPES for his PhD Scholarship.

References

1. Oviatt, S.: Multimodal Interfaces. In: Handbook of Human-Computer Interaction, pp. 286–304 (2002)
2. Johnston, M., Bangalore, S., Vasireddy, G.: MATCH: Multimodal Access to City Help. In: IEEE Workshop on Automatic Speech Recognition and Understanding, pp. 256–259 (2001)
3. Cohen, P.R., et al.: QuickSet: Multimodal Interaction for Distributed Applications. In: Proc. 5th ACM Int. Conf. on Multimedia, Seattle, Washington, United States, pp. 31–40 (1997)
4. Keskin, C., Balci, K., Aran, O., Sankur, B., Akarun, L.: A Multimodal 3D Healthcare Communication System. In: 3DTV Conference, May 7-9, pp. 1–4 (2007)
5. Zudilova, E.V., Sloot, P.M.A., Belleman, R.G.: A Multi-modal Interface for an Interactive Simulated Vascular Reconstruction System. In: Proceedings of Fourth IEEE International Conference on Multimodal Interfaces, pp. 313–318 (2002)
6. Braga, R.A.M., Petry, M., Moreira, A.P., Reis, L.P.: Intellwhels - A Development Platform for Intelligent Wheelchairs for Disabled People. In: 5th Int. Conf. on Informatics in Control, Automation and Robotics, Funchal, Madeira, Portugal, vol. I, pp. 115–121 (2008)
7. Martins, B., Valgôde, E., Faria, P., Reis, L.P.: Multimedia Interface with an Intelligent Wheelchair. In: Proc. of CompImage 2006 – Comp. Modelling of Objects Represented in Images: Fundamentals Methods and Applications, Coimbra, Portugal, pp. 267–274 (2006)
8. Faria, P.M., Braga, R.A.M., Valgôde, E., Reis, L.P.: Platform to Drive an Intelligent Wheelchair Using Facial Expressions. In: Proceedings 9th Int.Conf. Enterprise Information Systems - Human-Computer Interaction, Funchal, Madeira, pp. 164–169 (2007)
9. Faria, P.M., Braga, R.A.M., Valgôde, E., Reis, L.P.: Interface framework to drive an intelligent wheelchair using facial expressions. In: IEEE International Symposium on Industrial Electronics (ISIE 2007), pp. 1791–1796 (2007)
10. Simpson, R.C.: Smart Wheelchairs: A Literature Review. Journal of Rehabilitation Research & Development 42, 423–436 (2005)
11. IBM. Via Voice (June 2008),
    http://www-306.ibm.com/software/pervasive/embedded_viavoice/
    (Consulted on July 2008)
12. Braga, R.A.M., Malheiro, P., Reis, L.P.: Development of a Realistic Simulator for Robotic Intelligent Wheelchairs in a Hospital Environment. In: Baltes, J., Lagoudakis, M.G., Naruse, T., Shiry, S. (eds.) RoboCup 2009. LNCS (LNAI), vol. 5949, pp. 23–34. Springer, Heidelberg (2010)

Analyzing the Human-Robot Interaction Abilities of a General-Purpose Social Robot in Different Naturalistic Environments

J. Ruiz-del-Solar[1,2], M. Mascaró[1], M. Correa[1,2], F. Bernuy[1], R. Riquelme[1], and R. Verschae

[1] Department of Electrical Engineering, Universidad de Chile
[2] Center for Mining Technology, Universidad de Chile
jruizd@ing.uchile.cl

Abstract. The main goal of this article is to report and analyze the applicability of a general-purpose social robot, developed in the context of the RoboCup @Home league, in three different naturalistic environments: (i) home, (ii) school classroom, and (iii) public space settings. The evaluation of the robot's performance relies on its degree of social acceptance, and its abilities to express emotions and to interact with humans using human-like codes. The reported experiments show that the robot has a large acceptance from expert and non-expert human users, and that it is able to successfully interact with humans using human-like interaction mechanisms, such as speech and visual cues (particularly face information). It is remarkable that the robot can even teach children in a real classroom.

Keywords: Human-Robot Interaction, Social Robots.

1 Introduction

Social robots are becoming of increasing interest in the robotics community. A social robot is a subclass of a mobile service robot designed to interact with humans and to behave as a partner, providing entertainment, companion and communication interfaces. It is expected that the morphology and dimensions of social robots allow them to adequately operate in human environments. It is projected that social robots will play a fundamental role in the next years as companions for elderly people and as entertainment machines.

Among other abilities, social robots should be able to: (1) move in human environments, (2) interact with humans using human-like communication mechanisms (speech, face and hand gestures), (3) manipulate objects, (4) determine the identity of the human user (e.g. "owner 1", "unknown user", "Peter") and its mood (e.g. happy, sad, excited) to personalize its services, (5) store and reproduce digital multimedia material (images, videos, music, digitized books), and (6) connect humans with data or telephone networks. In addition, (7) they should be empathic (humans should like them), (8) their usage should be natural without requiring any technical or computational knowledge, and (9) they should be robust enough to operate in natural

J. Baltes et al. (Eds.): RoboCup 2009, LNAI 5949, pp. 308–319, 2010.

environments. Social robots with these abilities can assist humans in different environments such as public spaces, hospitals, home settings, and museums. Furthermore, social robots can be used for educational purposes.

Social robots should have acceptance by every kind of human user, including non-expert ones as elderly and children. We postulate that in order to have acceptance, it is far more important to be empathic and to produce sympathy in humans than to have an elaborated and elegant design. Moreover, to produce effective interaction with humans, and even enable humans to behave as if they were communicating with peers, it has been suggested that the robot body should be "based on a human's" [5] or being human-like [3]. We propose that it is important to have a somehow anthropomorphic body, but that to have a body that exactly look likes a human body is not required. Many researchers have also mentioned the importance that when interacting with humans, the robot tracks or gazes the face of the speaker [7][8][6][4]. We also believe that these attention mechanisms are important for the human user. In particular, the detection of the user's face allows the robot to keep track of it, and the recognition of the identity of the user's face allow the robot to identify the user, to personalize its services and to make the user feel important (e.g. "Sorry Peter, can you repeat this?"). In addition, it is also relevant that the interaction with the robot has to be natural, intuitive and based primarily on speech and visual cues (still some humans do not like to use standard computers, complex remote controls o even cell phones).

The question is how to achieve all these requirements. We believe that they can be achieved if the robot has a simple and anthropomorphic body design, it is able to express emotions, and it has human-like interaction capabilities, such as speech, face and hand gestures interaction. We also believe that it is important that the cost of a social robot be low, if our final goal is to introduce social robots in natural human environments, where they will be used by normal persons with limited budgets. Taking all this into consideration we have developed a general-purpose social robot that incorporates these characteristics.

The main goal of this article is to report and analyze the applicability of the developed robot in three different naturalistic environments: (i) home, (ii) school classroom and (iii) public space settings. The evaluation of the robot's performance relies in the robot's social acceptance, the ability of the robot to express emotions, and the ability of the robot to communicate with humans using human-like gestures. The article is structured as follows. In section 2, the hardware and software components of the social robot are briefly outlined. We emphasize the description of the functionalities that allow the robot to provide human-like communication capabilities and to be emphatic. Section 3 describes the robot applicability in three different naturalistic environments. Finally, in sections 4 and 5, discussion and some conclusions of this work are given.

2   Bender: A General-Purpose Social Robot

The main idea behind the design of Bender, our social robot, was to have an open, flexible, and low-cost platform that provides human-like communications capabilities, as well as empathy. Bender has an anthropomorphic upper body (head, arms, chest), and a differential-drive platform provides mobility. The electronic and mechanical hardware components of the robot are described in [12]. A detailed description of the robot as well as pictures and videos can be found in its personal website: http://bender.li2.uchile.cl/. Among Bender's most innovative hardware components to be to mention is the robot head, which incorporates the ability of expressing emotions (see figure 1).

The main components of the robot's software architecture are shown in figure 2. The *Speech Analysis & Synthesis* module provides a speech-based interface to the robot. Speech Recognition is based on the use of several grammars suitable for different situations instead of continuous speech recognition. Speech Synthesis uses Festival's Text to Speech tool, dynamically changing certain parameters between words in order to obtain a more human-like speech. This module is implemented using a control interface with a CSLU toolkit (http://cslu.cse.ogi.edu/toolkit/) custom application. Similarly, the *Vision* module provides a visual-based interface to the robot. This module is implemented using algorithms developed by our group. The *High-Level Robot Control* is in charge of providing an interface between the *Strategy* module and the low-level modules. The first task of the *Low-Level Control* module is to generate control orders to the robot's head, arm and mobile platform. The *Emotions Generator* module is in charge of generating the specific orders corresponding to each emotion. Emotions are called in response to specific situations within the finite-state machine that implements high-level behaviors. Finally, the *Strategy* module is in charge of selecting the high-level behaviors to be executed, taking into account sensorial, speech, visual and Internet information. Of special interest for this article are the capabilities for face and hand analysis included in the *Vision* module. The *Face and Hand Analysis* module incorporates the following functionalities: face detection (using boosted classifiers) [16][18], face recognition (histogram of LBP features) [1], people tracking (using face information and Kalman Filters) [14], gender classification using facial information [17], age classification using facial information, hand detection using skin information and recognition of static hand gestures [2].

Bender's most important functionalities are listed in table 1. All these functionalities have been already successfully tested as single modules. Table 2 shows quantitative evaluations of the human-robot interaction functionalities, measured in standard databases. As it can be observed in these databases, the obtained results are among the best-reported ones. This is an important issue, because we would like that our social robot has the best tools and algorithms when interacting with people. For instance, we do not want that the robot to have problems by detecting people when immersed in an environment with variable lighting conditions.

Surprised   Angry   Sad   Happy

Fig. 1. Facial expressions of Bender

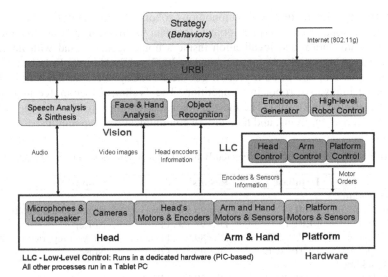

Fig. 2. Software architecture. In the bottom the hardware components: platform, head, and arm. In an upper level, low-level control processes running in dedicated hardware. All high level processes run in a tablet PC.

Table 1. Bender's main functionalities

Ability	How is achieved
Mobility	A differential-drive platform provides this ability.
Speech recognition and synthesis	CSLU toolkit (http://cslu.cse.ogi.edu/toolkit/).
Face detection and recognition	Face and hand analysis module.
Gender and age determination using facial information	Face and hand analysis module.
Hand gesture recognition	Face and hand analysis module.
General purpose object recognition	SIFT-based object recognition module
Emotions expression	Anthropomorphic 7 DOF mechatronics head.
Object manipulation	A 3 DOF arm with 3, 2 DOF fingers.
Information visualization	The robot's chest incorporates a 12 inch display
Standard computer inputs (keyboard and mouse)	The chest's display is *touch screen*. In addition, a virtual keyboard is employed in some applications.
Internet access	802.11b connectivity.

3 Applicability in Naturalistic Environments

3.1 Real Home Setting

One of the main goals behind the development of our social robot is to use it as an assistant and companion for humans in home settings. The idea is that the robot will be able to freely interact with non-expert users in those environments. Naturally, we know that we need to follow a large process until achieving this goal. In 2006 we decided that a very appropriate way to achieve this was to regularly participate in the RoboCup@Home. RoboCup@Home focuses on real-world applications and

human-machine interaction with autonomous robots in home settings. Tests are related with manipulation of typical objects that can be found in a home-like environment, with navigation and localization inside a home scenario, and with interaction
with humans. Our social robot participated in 2007 and 2008 in the RoboCup@Home
world competition, and in both years it got the RoboCup @Home Innovation Award
as the most innovative robot in competition. The Technical Committee members of
the league decide this award. The most appreciated robot's abilities were its empathy,
ability to express emotions, and human-like communications capabilities.

Table 2. Evaluation of some selected Bender's functionalities in standard databases

	Database	Results	Comments
Face Detection [1]			
- Single face	BioID	DR=95.1%, FP=1	Best reported results
- Single face	FERET	DR=98.7%, FP=0	NoRep
- Multiple faces	CMU-MIT	DR=89.9%, FP=25	4th best reported results
- Multiple faces	UCHFACE	DR=96.5%, FP=3	NoRep
Face Tracking [2]			
- Multiple faces	PETS-ICVS 2003	DR=70.7%, FP=88 (set A) DR=70.2%, FP=750 (set A)	Best reported results.
Eyes Detection [1]			
- Single Face	BioID	DR=97.8%, MEP=3.02	Best reported results
- Single Face	FERET	DR=99.7%, MEP =3.69	NoRep
- Multiple faces	UCHFACE	DR=95.2%, MEP =3.69	NoRep
Gender Classification [1]			
- Single Face	BioID	CR: 81.5%	NoRep
- Single Face	FERET	CR: 85.9%	NoRep
- Multiple faces	UCHFACE	CR: 80.1%	NoRep
Face Recognition			
- Standard test [3]	FERET *fafb*	Top-1 RR=97%	Among the best reported results
- Variable Illumination [4]	YaleB	7 individuals per class, Top-1 RR=100% 2 individuals per class, Top-1 RR=96.4%	Best reported results
- Variable Illumination [4]	PIE	2 individuals per class, Top-1 RR=99.9	Best reported results
Hand Gesture Recognition [5]			
- Variable illumination	Own Database, real-word vid-eos, 4 static gestures	RR=70.4%	NoRep

(1) Reported in [18]; (2) Reported in [14]; (3) Reported in [1]; (4) Reported in [13]; (5)
Reported in [2]. DR: Detection Rate; FP: Number of False Positives; RR= Recognition Rate;
CR= Classification Rate; MEP; Mean Error in Pixels; NoRep: No other reports in the same
dataset.

3.2 Classroom Setting

Robotics is a highly motivating activity for children. It allows them to approach technology both amusingly and intuitively, while discovering the underlying science principles. Indeed, robotics has emerged as a useful tool in education since, unlike many
others, it provides the place where fields or ideas of science and technology intersect
and overlap [11]. With the objective of using social robots as a tool for fostering the
interest of children in science and technology, we tested our social robot as lecturer
for school children in a classroom setting. The robot gave talks to schoolchildren of
10-13 years old. Altogether 228 schoolchildren participated in this activity, and at
each time one complete course assisted to the talk in a multimedia classroom (more
than 10 talks were given by the robot). The duration of each talk was 55 minutes, and
it was divided in two parts. In the first part the robot presented itself, and talked about
its experiences as a social robot. In the second part the robot explained some basic
concepts about renewable energies, and about the responsible use of energy. After the
talk students could interact freely with the robot. The talk was given using the multimedia capabilities of the robot; speech and multimedia presentation, which was projected by the robot (see pictures in figure 3).

After the robot's lecture the children, without any previous advice, answered a poll
regarding their personal appreciation of the robot and some specific contents mentioned by the robot. In the robot evaluation part, the children were asked to give an
overall evaluation of the robot. On a linear scale of grades going from 1 to 7, the robot
was given an average score of 6.4, which is about 90%. In the second part children
evaluated the robot's presentation: 59.6% rated it as excellent, 28.1% as good, 11.4%
as regular, 0.9% as bad, and 0% as very bad. The third question was, "Do you think
that it is a good idea for robots to teach some specific topics to schoolchildren in the
future?" 92% of the children answered yes. In the technical content evaluation part,
the first three questions were related to energy sources (classification of different
energy sources as renewable or non- renewable, availability of renewable sources, and
indirect pollution produced by renewable sources). The fourth question asked about
the differences between rechargeable and non-rechargeable batteries, and the fifth
question asked about the benefits of the efficient use of energy. The percentage of
correctness of the children's answers to each of the five technical content questions is
shown in Table 3. The overall percentage of correct answers was 55.4%.

In summary, we can observe that children had a very good evaluation of the robot
(6.4 over 7), and that 87.7% of them evaluated the presentation as excellent or good.
They also have a very favorable opinion about the use of robots as lecturers in a

Table 3. Percentage of correctness of the children answers to the 5 technical questions

Technical Questions	Correctness
TQ1	75.9%
TQ2	33.7%
TQ3	31.6%
TQ4	75.0%
TQ5	60.6%
Overall	55.4%

classroom environment (92%). Moreover, the children were able to learn some basic technical concepts (the overall percentage of correct answers was 55.4%), although they just heard them once from a robot. The main goal of this technical content part of the evaluation was just to see if the children could learn some basic content from the robot, and not to measure how well they learned it. Therefore, control experiments with human instructors were not carried out. This will be part of the future work. Finally, it is important to stress that the robot was able to give its talk and to interact with the children without any human assistance.

Fig. 3. Bender giving talks to schoolchildren

3.3 Public Space Setting

We tested the applicability of our social robot in a public space setting. The main idea of the experiment was to let humans interact freely with the robot, using only speech and visual cues (face, hand gestures, facial expressions, etc.). The robot did not moved by itself during the whole experience, in order to avoid any collision risks with the students, therefore it needed to catch the people's attention just using speech synthesis, visual cues and other strategies such as complaining about being alone, bored, or calling far-away detected people. The robot was placed in a few different public spaces inside our university campus (mainly building's halls), and the students passing through these public spaces could interact with the robot, if they wanted (see pictures in figure 4). When the robot detected a student in its neighborhood, it asked the student to approach and have a little conversation with him. The robot presented itself, then it asked some basic information to the student, and afterwards it asked the student to evaluate its capabilities to express emotions. Finally, after the evaluation, the robot thanked the student and the interaction finished. To evaluate the ability of the robot to express emotions, the robot randomly expressed an emotion, and it asked the student to identify the emotion. The student gave its answer using the touch screen (choosing one of the alternatives).

This process was repeated four times, to allow the student to evaluate different emotions. We decided that the human users gave their answer using the touch screen, to be sure that the speech recognition mistakes would not affect the experiment. This was the only time that the interaction between the robot and the human was not based on speech or visual cues. In all moments, no external human assistance was given to the robot's users. After the human–robot interaction finished, and the humans left the robot's surround, they were asked to evaluate its experience using a poll.

In all experiments the robot was left alone in a hall, and the laboratory team observed the situation several meters away. Our first observation was that from the total of students that passed near the robot, about 37% modified their behavior and approached the robot. 31% of them interacted with the robot, the rest just observed it. The total number of students that interacted with the robot was 83. The age range was 18 to 25 years old, and the gender distribution was 70% males and 30% females. Out of the 83 students, 74.7% finalized the interaction, and 26.3% left before finishing. The main reasons for leaving prematurely were: (i) the students were not able to interact with the robot properly (speech recognition problems, see discussion section), (ii) they did not have enough time to make the emotions' evaluation, or (iii) they were not interested in making the evaluation. The mean interaction time of the humans that finalized the interaction, including the emotions' evaluation, was 124 seconds.

In table 4 is displayed the recognition rate of the different expressions. It can be observed that the overall recognition rate was 70.6%, and that all expressions, but "happy" have a recognition rate larger than 75%. In table 5 and 6 the results of the robot's evaluation poll, made by the users after interacting with the robot are presented. It should be remembered that only the 74.7% of the users that finished the interaction with the robot, answered the poll. As it can be observed in tables 5 and 6, 83.9% of the users evaluate the robot's appearance as excellent or good, 88.5% evaluate the robot's ability to express emotions as excellent or good, and 80.7% evaluate the robot's ability to interact with humans as excellent or good. In addition, 90% of them think that it is easy to interact with the robot, 84% believe that the robot is suitable to be a receptionist, museum guide or butler, and 67% think that the robot can be used with educational purposes with children. It should be mentioned that the whole experiment was carried out inside an engineering campus, and that therefore the participants in the test were engineering students, who with a high probability enjoy technology and robots. On the other hand, we believe that as expert users in technology, they can be more critical about robots than standard users. Nevertheless, we think that the obtained results show than in general terms the social robot under evaluation has a large acceptance in humans, and that its abilities to interact with humans using speech and visual cues, as well as its ability to express emotions, are suitable for free human-robot interaction situations in naturalistic environments.

Table 4. Recognition rate of robot's facial-expressions

Expression	Correctness
Happy	51.0%
Angry	76.5%
Sad	78.4%
Surprised	76.5%
Overall	70.6%

Table 5. Human's evaluation of the robot's appearance and interaction abilities

	Excellent	Good	Regular	Bad	Very Bad
Robot appearance	30.7%	53.2%	14.5%	1.6%	0%
Ability to express emotions	31.1%	57.4%	8.2%	3.3%	0%
Ability to interact with humans	17.8%	62.9%	17.7%	1.6%	0%

Table 6. Human's evaluation of the robot's applicability and simplicity of use

	Yes	No
Do you think that it is easy to interact with the robot?	90%	10%
Do you think that the robot is suitable to be a receptionist, museum guide or butler?	84%	16%
Do you think that the robot can be useful in tasks related with children interaction?	67%	33%

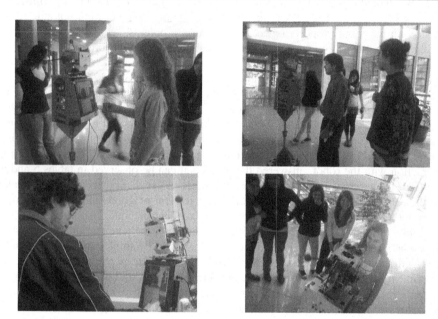

Fig. 4. Bender interaction with students in a public space inside the university

4   Discussion

Evaluation Methodology. There exist different approaches to evaluate the performance of social robots when interacting with humans. Although, isolated algorithms' performance should be measured (e.g. recognition rate of a face recognition algorithm), it is also necessary to analyze how robots affect humans. Some researchers have proposed to employ quantitative measures of the human attention (attitude [10], eye gaze [9], etc.) or body movement interaction between the human and the robot [5]. We do believe that acceptance and empathy are two of the most important factors to be measured in a human-robot interaction context, and that these factors can be measured using poll-based methods that express the user's opinion. The described social robot has been evaluated by about 300 people with different backgrounds (228 schoolchildren, 62 engineering students, and 5 international researchers in the RoboCup @Home competitions), which validates the obtained results.

Evaluation of robot capabilities. As it can be observed in table 2, the visual-based human-robot interaction functionalities of the robot, measured in standard databases are among the best-reported ones. We believe that this is very important, because the robot should have robust tools and algorithms to deal with dynamic conditions in the environment. In addition, the robot has received two innovation awards from the service-robot scientific community, which indicates that the robot theoretically is able to adequately interact with people.

Robot Evaluation when interacting with people. In our experiments with children in a real classroom setting, we observed that children gave a very good evaluation to the robot, and that 87.7% of them evaluated its presentation as excellent or good. They have also a very favorable opinion about the use of robots as lecturers in a classroom environment. We can conclude that the robot achieved the acceptance of the children (10-13 years old), who for the first time had the opportunity to interact with a robot. The robot was able to give its talk and to interact with the children without any human assistance. We conclude that the robot is robust enough to interact with non-expert users in the task of giving talks to groups of humans. In addition, the children were able to learn some basic technical concepts from the robot (55.4% correct answers to 5 technical questions). It should be stressed that the robot presentation was a standard lecture, without any repetition of contents. Besides, it should be observed that the robot, unlike a human teacher, can not detect distracted children in order to call for their attention, and also can not achieve the same level of expressivity neither in the speech or the gestures, leaving it only with his empathy and other mechanisms such as simulating breathing or moving the mouth while talking to catch the listener's attention. These results encourage us to further explore in the relevance of an appealing human robot interaction interface. Naturally, it seems necessary to carry out a comparative study of the performance of robot-teachers against human-teachers, and to analyze the dependence of the results on the specific topics that are to be taught (technical topics, foreign language, history, etc.).

In our experiments in public space settings we tested the ability of the social robot to freely interact with people. The experiments were conducted in different building's halls inside our engineering college. 37% of the students passing near the robot approached it; 31% of them interacted directly with the robot. In all cases the robot actively tried to attract the students, by talking to them. It was interesting to note that 26.3% of the students that interacted with the robot left before finishing the interaction. One of the main reasons for leaving was that the students were not able to interact properly with the robot, due to speech recognition problems. Our speech recognition module has limited capabilities, it is not able to recognize unstructured natural language, and the recognition is perturbed by the environmental noise. This is one of the main technical limitations of our robot, and in general of other service robots. Nevertheless, 74.7% of the students finished the emotion's evaluation that the robot proposed them, with a mean interaction time of 124 seconds.

Before carrying out these experiments we had the qualitative impression that, the emotions that our robot could generate were adequate, and that a human could understand them. The quantitative evaluation obtained in the experiments showed us that this perception was correct, and the humans can recognize correctly the robot's expression in 70.6% of the cases. This overall result can be improved if we design a new "happy" expression, which was recognized in only 51% of the cases. Although the mechanics of the robot head imposes some limits to the expressions that can be generated by the robot (limitation in the number of degrees of freedom in the face), we believe the current expressions are rich enough to produce empathy in the users. We have seen these in all reported experiments, and also in non-reported interactions between the robot and external visitors in our laboratory.

The acceptance of the robot by the engineering students, as in the case of the children, was high (83.9% evaluated the robot's appearance as excellent or good, 88.5% evaluated the robot's ability to express emotions as excellent or good, 80.7% evaluate the robot's ability to interact with humans as excellent or good). In addition, 90% of the students think that it is easy to interact with the robot, and 84% and 67% of the students think that the robot can be used as an assistant or with educational purposes, respectively. We believe that this favorable evaluation is due to the fact that: (i) the robot has an anthropomorphic body, (ii) it can interact using human-like interaction mechanisms (speech, face information, hand gestures), (iii) it can express emotions, and (iv) when interacting with a human user it tracks his/her face.

5 Conclusions

The main goal of this article was to report and analyze the applicability of a low-cost social robot in three different naturalistic environments: (i) home setting, (ii) school classroom, and (iii) public spaces. The evaluation of the robot's performance relied in the robot social acceptance, and its abilities to express emotions and interact with humans using human-like codes. The experiments show that the robot has a large acceptance from different groups of human users, and that the robot is able to interact successfully with humans using human-like interaction mechanisms, such as speech and visual cues (specially face information). It is remarkable that children learnt something from the robot despite its limitations.

From the technical point of view, the visual-based human-robot interaction functionalities of the robot, measured in standard databases are among the best-reported ones, and the robot has received two innovation awards from the scientific community, which indicates that the robot is able to adequately interact with people. However, one of the main technical limitations is the speech recognition module, which should be improved.

As future work we would like to further analyze the teaching abilities of our robot. In general terms, we believe that more complex methodologies should be used to measure how much the children learn with the robot, and how is this learning compared with the case when children learn with a human teacher.

Acknowledgements

This research was partially funded by FONDECYT project 1090250, Chile.

References

1. Correa, M., Ruiz-del-Solar, J., Bernuy, F.: Face Recognition for Human-Robot Interaction Applications: A Comparative Study. In: Iocchi, L., Matsubara, H., Weitzenfeld, A., Zhou, C. (eds.) RoboCup 2008: Robot Soccer World Cup XII. LNCS (LNAI), vol. 5399, pp. 473–484. Springer, Heidelberg (2009)
2. Francke, H., Ruiz-del-Solar, J., Verschae, R.: Real-time Hand Gesture Detection and Recognition using Boosted Classifiers and Active Learning. In: Mery, D., Rueda, L. (eds.) PSIVT 2007. LNCS, vol. 4872, pp. 533–547. Springer, Heidelberg (2007)
3. Hayashi, K., Sakamoto, D., Kanda, T., Shiomi, M., Koizumi, S., Ishiguro, H., Ogasawara, T., Hagita, N.: Humanoid Robots as a Passive-Social Medium – A Field Experiment at a Train Station. In: Proc. Conf. Human-Robot Interaction – HRI 2007, Virginia, March 8-11, pp. 137–144 (2007)
4. Ishiguro, H., Ono, T., Imai, M., Kanda, T.: Development of an interactive humanoid robot Robovie—An interdisciplinary approach. In: Jarvis, R.A., Zelinsky, A. (eds.) Robotics Research, pp. 179–191. Springer, New York (2003)
5. Kanda, T., Ishiguro, H., Imai, M., Ono, T.: Development and Evaluation of Interactive Humanoid Robots. Proc. IEEE 92(11), 1839–1850 (2004)
6. Kanda, T., Ishiguro, H., Ono, T., Imai, M., Nakatsu, R.: Development and evaluation of an interactive humanoid robot Robovie. In: Proc. IEEE Int. Conf. Robotics and Automation, pp. 1848–1855 (2002)
7. Matsusaka, Y., et al.: Multi-person conversation robot using multimodal interface. In: Proc. World Multiconf. Systems, Cybernetics and Informatics, vol. 7, pp. 450–455 (1999)
8. Nakadai, K., Hidai, K., Mizoguchi, H., Okuno, H.G., Kitano, H.: Real-time auditory and visual multiple-object tracking for robots. In: Proc. Int. Joint Conf. Artificial Intelligence, pp. 1425–1432 (2001)
9. Ono, T., Imai, M., Ishiguro, H.: A model of embodied communications with gestures between humans and robots. In: Proc. 23rd Annu. Meeting Cognitive Science Soc., pp. 732–737 (2001)
10. Reeves, B., Nass, C.: The Media Equation. CSLI, Stanford (1996)
11. Ruiz-del-Solar, J., Aviles, R.: Robotics Courses for Children as a Motivation Tool: The Chilean Experience. IEEE Trans. on Education 47(4), 474–480 (2004)
12. Ruiz-del-Solar, J., Correa, M., Bernuy, F., Cubillos, S., Mascaró, M., Vargas, J., Norambuena, S., Marinkovic, A., Galaz, J.: UChile HomeBreakers 2008 TDP. In: RoboCup Symposium 2008, CD Proceedings, Suzhou, China, July 15-18 (2008)
13. Ruiz-del-Solar, J., Quinteros, J.: Illumination Compensation and Normalization in Eigenspace-based Face Recognition: A comparative study of different pre-processing approaches. Pattern Recognition Letters 29(14), 1966–1979 (2008)
14. Ruiz-del-Solar, J., Verschae, R., Vallejos, P., Correa, M.: Face Analysis for Human Computer Interaction Applications. In: Proc. 2nd Int. Conf. on Computer Vision Theory and Appl. – VISAPP 2007, Special Sessions, Barcelona, Spain, pp. 23–30 (2007)
15. Sakamoto, D., Kanda, T., Ono, T., Ishiguro, H., Hagita, N.: Android as a Telecommunication Medium with a Human-like Presence. In: Proc. Conf. Human-Robot Interaction – HRI 2007, Virginia, March 8-11, pp. 193–200 (2007)
16. Verschae, R., Ruiz-del-Solar, J.: A Hybrid Face Detector based on an Asymmetrical Adaboost Cascade Detector and a Wavelet-Bayesian-Detector. In: Mira, J., Álvarez, J.R. (eds.) IWANN 2003. LNCS, vol. 2686, pp. 742–749. Springer, Heidelberg (2003)
17. Verschae, R., Ruiz-del-Solar, J., Correa, M.: Gender Classification of Faces using Adaboost. In: Martínez-Trinidad, J.F., Carrasco Ochoa, J.A., Kittler, J. (eds.) CIARP 2006. LNCS, vol. 4225, pp. 68–78. Springer, Heidelberg (2006)
18. Verschae, R., Ruiz-del-Solar, J., Correa, M.: A Unified Learning Framework for object Detection and Classification using Nested Cascades of Boosted Classifiers. Machine Vision and Applications 19(2), 85–103 (2008)

Communicating among Robots in the RoboCup Middle-Size League

Frederico Santos[1,3], Luís Almeida[2,3], Luís Seabra Lopes[3],
José Luís Azevedo[3], and M. Bernardo Cunha[3]

[1] DEE - ISEC, Inst. Politécnico de Coimbra, Portugal
fred@isec.pt
[2] DEEC - FEUP, Univ. Porto, Portugal
lda@fe.up.pt
[3] IEETA - DETI, Univ. Aveiro, Portugal
{lsl,jla}@ua.pt, mbc@det.ua.pt

Abstract. The RoboCup Middle-Size League robotic soccer competitions pose a real cooperation problem for teams of mobile autonomous robots. In the current state-of-practice cooperation is essential to overcome the opponent team and thus a wireless communication protocol and associated middleware are now fundamental components in the multi-robots system architecture. Nevertheless, the wireless communication has relatively low reliability and limited bandwidth. Since it is shared by both teams, it is a fundamental resource that must be used parsimoniously. Curiously, to the best of our knowledge, no previous study on the effective use of the wireless medium in actual game situations was done. In this paper we show how current teams use the wireless medium and we propose a set of best practices towards a more efficient utilization. Then, we present a communication protocol and middleware that follow such best practices and have been successfully used by one particular MSL team in the past four years.

1 Introduction

The RoboCup Middle-Size League (MSL) [1] has been an effective testpad for cooperative robotics. In fact, beyond all the issues associated with the construction of actual robots for operation in harsh conditions, each team now needs to develop coordinated behaviors to effectively overcome the opponent team. This cooperation is becoming more sophisticated involving the communication of team mates positions, fusion of the ball position, dynamic role assignment, formations and ball passes, among others. The cooperative behaviors are developed on top of an adequate middleware that allows the team members to exchange information. In turn, such middleware relies on a wireless communication protocol.

Despite its importance, however, the wireless communication is known to be less reliable than its wired counterpart with significantly higher bit-error rates, to have limited and variable bandwidth and to be open to the access by other stations not involved in the team, among other undesired phenomena [2]. Nevertheless, the wireless channel must be shared by both teams involved in a game, thus becoming

J. Baltes et al. (Eds.): RoboCup 2009, LNAI 5949, pp. 320–331, 2010.

a critical shared resource. Curiously, the MSL rules have had not constrains regarding the use of wireless communication. To the best of our knowledge, despite the recurrent problems with wireless communication, no study was ever done to analyze the actual use of the wireless channel in game situations.

Recently, the concern with the wireless communication has increased and the MSL rules are now already including some restrictions on the use of the wireless channel by the participating teams. As a contribution to such effort, in this paper we present an analysis of the actual use of the wireless channel by several MSL teams during the RoboCup 2008 event. We show that there is substantial difference between teams, with some making a parsimonious use of the channel while others use substantial slices of the available bandwidth, few transmiting in a sparse periodic fashion and others sending bursts of data with very short intervals. The patterns of transmission depend on the middleware layer that manages the exchange of information. From the analysis of the communication, we can also infer the kind of middleware being used.

This paper discusses issues related with the wireless communication in the MSL, shows the trend in the MSL rules with respect to the communications, and proposes a few best practices that can improve the general behavior of the wireless channel. Finally, the paper includes a brief description of a specific middleware and communication protocol that follow such best practices and which have been successfully used in MSL competitions in the last four years.

2   Wireless Communication within the MSL

For several years that the MSL rules already stipulate that the wireless technology to be used is IEEE802.11a/b. The more popular IEEE802.11g technology is not allowed simply because it uses the same band as IEEE802.11b but with fewer, despite larger, frequency channels, which increases the difficulty in channel planning and assignment per competition area to minimize cross-interference [3]. Generally, one channel is assigned to one competition field and both teams playing therein must share it. An attempt is always made to assign non-interfering channels to neighboring fields. Moreover, the communication must be infrastructured, i.e., using access points. Direct ad-hoc communication is not allowed. Other constraints have been included and this year a limitation on the bandwidth allowed to each team was introduced for the first time. Briefly, the MSL rules, concerning the wireless communication currently stipulate:

- IEEE802.11a/b technology
- Infrastructured mode (through Access Point)
- Single a + single b channels per game (each shared by both teams)
- IPv4 addressing within pre-defined networks
- Only unicasts/multicasts (broacasts are forbidden)
- Up to 2.2Mbps bandwidth utilization per team

The bandwidth limitation was calculated considering the lower bandwidth technology IEEE802.11b (11Mbit/s), which is still used by some teams due to national regulations. This is applied to both types of network, either 'b' and 'a' for fairness reasons.

2.1   Logs from the MSL at RoboCup 2008

In order to gather information on how current teams actually use the wireless channel, we monitored the communications during several games of the last edition of RoboCup, in Suzhou. We used one PC with a wireless adaptor configured in monitor mode, which disables filtering and allows receiving all IEEE802.11 packets that arrive at its antenna. The monitoring software was the Wireshark network protocol analyzer and we monitored 6 teams, during periods of approximately 1 minute, randomly taken during the third round-robin games. In all these games all communications took place in IEEE802.11a but the effective bit-rates achieve during the competitions varied widely between 6 Mbit/s and 54 Mbit/s with an approximate average of 36 Mbit/s.

Figure 1 shows a set of histograms concerning the distribution of the inter packet intervals related to each team considering the transmissions of all its members as they are effectively transmitted in the wireless medium. We can clearly identify three classes, one of teams 1 and 2 that do some level of traffic spread in the time domain, exhibiting inter-packet intervals that extend up to approximately 80ms. In the former case, the team uses multicast packets to share information in a producer-consumer fashion. On the other hand, team 2 uses unicasts, with the robots sending exchanging between them in pairs. Then, teams 3 and 4 show a clear dual mode operation with many packets sent in sequence but others sent with longer well defined intervals. Looking in more detail to their logs, we can see that all robots of team 3 transmit periodically and synchronized, with all robots transmiting in sequence and then waiting for a period of approximately 75ms. Curiosly, this team used IP broadcast frames, which are now banned by the current rules, to exchange information in a producer-consumer fashion. On the other hand, team 4 uses a middleware probably based on a centralized blackboard that resides in one particular station to which all robots send their sensing data periodically, approximate every 150ms, but often faster. Then, such station carries out some computation, probably sensor fusion, and delivers the result back to the nodes in unicast packets sent in sequence, thus generating a peak of packets sent within a very short interval. Finally, teams 5 and 6 send their traffic in an almost continuous fashion, with very short intervals, leading to numbers of packets that are an order of magnitude higher and to much higher bandwidth utilization levels.

Figure 2 shows the histograms of the packet sizes used by each team in Bytes. Clearly two situations arise, one of teams 1 through 4 that use mainly fixed size packets, in some cases with 2 different sizes, team 1 with average size packets and teams 2, 3 and 4 with relatively small packets, only, and then teams 5 and 6 that use a wide variability of packet lengths with significant use of large (1.5kB) packets. These teams were the only ones sending bursts of information, too. We detected bursts of up to six 1.5kB packets in the case of team 5 and up to twelve 1.5kB packets in the case of team 6. In the IEEE802.11a channel

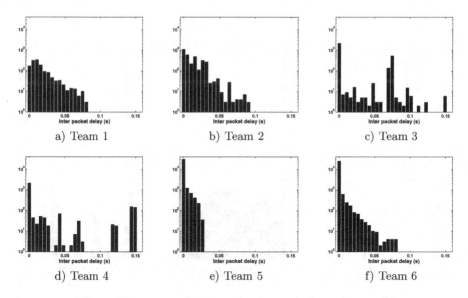

a) Team 1   b) Team 2   c) Team 3

d) Team 4   e) Team 5   f) Team 6

Fig. 1. Histograms of inter-packet intervals for each team (s)

used, these bursts, could cause interference of up to 10ms, approximately, but if an IEEE802.11b channel was used instead, these bursts could imply near 50ms delays.

Table 1 shows a summary of the main traffic statistics of the monitored teams, covering inter-packet interval in miliseconds, packet size in bytes, burst size in number of consecutive 1.5kB packets, total number of bytes transmitted in the monitoring interval and respective approximate utilization in IEEE802.11a/b channels. The traffic classes that were identified in the analysis of the histograms are naturally reflected in this table but the information on the approximate bandwidth utilization of the IEEE802.11a/b channels reveals the huge variations in channel utilization. It is curious to see that team 5 was already using approximately 25% of the IEEE802.11a channel, which corresponds to about 125% the width of an IEEE802.11b channel. The figures for team 6 are slightly better but still revealling a substantial channel overuse. The other teams use singnificantly lower bandwidths, near 2 orders of magnitude less, which allows them to play without problems among each other using any of the two kinds of channels. According to the current rules, teams 5 and 6 will have to readjust their use of the channel to meet the new 2.2Mbit/s limit. One curious detail is the fact that team 6 was using 11 different computers, substantially more that the maximum of 6 robots plus one remote station.

Figure 3 shows the impact that different opponets can have on the timeliness of the transmissions of a robot. In this particular case we used robot1 of team 2 (any other robots yielded similar results) in two games, one against team 1 that makes a relatively light use of the channel with good separations between packets and, on the other hand, against team 6 that is one of the heavy users

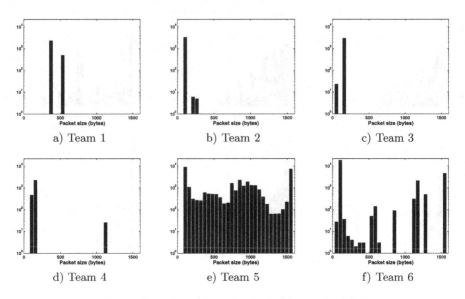

Fig. 2. Histograms of packet sizes for each team (B)

Table 1. Traffic statistics of 6 MSL teams

		Team 1	Team 2	Team 3	Team 4	Team 5	Team 6
Inter Packet	avr	17.74	15.20	20.03	21.72	1.74	1.90
(ms)	std	17.63	14.65	33.23	48.16	3.62	4.44
Packet Size	avr	412.87	139.68	160.51	187.67	787.40	497.81
(Bytes)	std	73.66	8.03	5.59	93.77	549.09	598.36
Burst Size (# 1.5kB pk)		–	–	–	–	6	12
Total kBytes		1158	460	480	517	26154	13072
% of max		4.43	1.75	1.84	1.98	100.00	49.98
Bandwidth	802.11a	1.1%	0.4%	0.5%	0.6%	25%	13%
utilization	802.11b	5.5%	2.0%	2.5%	3.0%	125%	65%

of the channel. The figures clearly illustrate the impact of playing against a heavy channel user team. When playing against team 1, the traffic pattern shows a significant regularity, indicating neglectable interference. However, the same robot playing against team 6 shows a significant change in the traffic pattern with a loss of the previous regularity and wide spread (strong jitter) of the inter-packet intervals, with a strong peak close to zero, meaning that many packets are strongly delayed and accumulated at the network interface, being then transmitted in a burst.

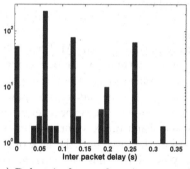
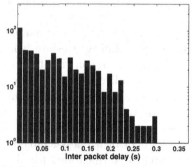

a) Robot 1 of team 2 against team 1   b) Robot 1 of team 2 against team 6

Fig. 3. Inter-packet intervals for one robot of team 2 against teams 1 and 6

Finally, it is important to refer that in these games the traffic external to the competition, including beacon frames from the AP, packets from other teams that were not playing, unknown packets, etc., was always neglectable, representing less than 1% of the channel bandwidth. Another observation, during a different game than those monitored, was the use of raw (non-IP) packets by another team, which is also in violation with the current rules.

2.2   Problems and Solutions

In MSL, and probably in other RoboCup competitions as well, the wireless communication has always been a source of concerns, given the frequent occurrence of problems. These were of diverse kinds and we could, in a simplified approach, classify them in four categories: infrastructure configuration, team communications configuration, lack of policing and channel overuse by teams.

– **Infrastructure configuration.** This category includes the cases in which the planning of the APs placement and channel assignment was non-optimal, frequently caused by constraints of the physical space in which the competitions must be layed out. Since it may be impossible to completely avoid this situaiton, we may have to live with a certain level of background interference, corresponding to an effective lower available channel bandwidth. Another problem we experienced, was the interference with pre-installed WLANs for general Internet access, which should have been switched off. This is a relevant issue that local organizers sometimes overlook.
– **Team communications configuration.** This has been one of the most common sources of problems due to frequent poor knowledge of the wireless communications technology. In fact, it is still common to find teams that bring their own APs and connect them freely in the team work area, often close to competion fields. Other times, the teams use erroneous configurations without being aware (e.g. ad-hoc mode), or send bursts of short packets overloading the network interfaces of the opponent team and causing some

device drivers to crash, disabling communications and preventing a team from playing. Without an accurate analysis of the situation, the wrong team can be disqualified due to inability to play.

- **Lack of policing**. Despite being generally permissive, the MSL rules have dictated certain constraints for some time. Unfortunately, there was always a lack of policing to verify their effective application and enforcing them. In some years, the local organization has hired a specialized company to monitor and control the use of the wireless channels. However, even in such cases it was difficult to mitigate all undesired situations, given their diversity, the number of wireless-enabled computers in the area, and the lack of rules compliance verifications for the teams. We believe that the current rules improved substantially in this aspect by including a network monitor in the games communication architecture and requesting teams to carry out an a priori communications check as a mandatory step for admission to the competitions. Moreover, the network monitor will also allow detecting situations that often occur in which teams in the work area keep their wireless interfaces open and transmitting, causing interference, as well as situations reported in the previous section of teams that use more computers than allowed, that use logical broadcasts and non-IP traffic. Nevertheless, it is still important that the organization is prepared to seek for spurious sources of interference, in case of need. This might require the use of a specific wireless channel monitoring device that provides information on the channel status, not only at the network protocol level (transmitted valid packets) but also at the physical level (bit-error rate, spurious packet fragments, medium spectral analysis, ...).

- **Channel overuse by teams**. Even without spurious interference, when the channel utilization approaches high values the channel performance deteriorates in terms of packet transmission delays and packet losses due to increased collisions and channel saturation. These delays and losses have a direct negative impact on the quality of the cooperating behaviors given their real-time character, mainly when they involve feedback control over the wireless channel. As it became clear in the previous section, in 2008 some teams were using, alone, more bandwidth than the 'b' technology can provide! This will hopefully be avoided in 2009 given the bandwidth limitation imposed by the new rules. Nevertheless, beyond the channel permanent saturation, transient saturation must also be considered and prevented, such as caused by bursts transmitted by the same station, e.g., file transfers. These can also cause a transient increase in packet delays and losses suffered by the opposing team that can harm the performance of its cooperative applications. To prevent these situations the teams must adhere to some kind of control of the consecutive amount of data that each of their robots transmits in an agreed interval of time. On the other hand, detecting such situations requires monitoring the traffic with increased temporal resolution.

As it was clear with the previous discussion, most of the problems that existed in the past can be solved or strongly attenuated with adequate restrictions on the use of the wireless channel and an effective policing of the channel utilization. Nevertheless, it is interesting to quickly analyze certains misconceptions that hindered the deployment of such solutions:

- **No need for restricting teams transmissions.** Ideally, if the channel bandwidth was infinite and there was no mutual interference between the competiting teams, restricting the teams transmissions would make no sense. However, that is not the case and finite bandwidth and mutual interference are facts that need to be considered. Then, while some teams do a parsimonious use of the channel, others exist that use substantial amount of bandwidth, often in a bursty way, with negative impact on the timeliness of the transmissions of the opposing team, as shown in the previous section, and consequently on the performance of its cooperative behaviors. Thus, some form of restriction that considers both bandwidth and bursts must be enforced.
- **Larger bandwidth solves the problem.** Unfortunately, just increasing the available bandwidth alone, as when moving from IEEE802.11b (11Mbit/s) to IEEE802.11a (54Mbit/s), is not a self-sustained solution and tends to generate wasteful patterns in bandwidth utilization. Such kind of simplistic solutions is always transitory and end up coming back to the same problem but with a larger magnitude. This trend was verified with two of the teams shown in the previous section.
- **Use a technology with QoS support.** In order to provide better support to time-sensitive traffic with respect to non-time-sensitive one in WLANs, a new standard was recently proposed, namely IEEE802.11e. Similarly to the original protocol, it includes two channel access policies, one that is distributed (EDCA) and another one that is controlled (HCCA). The former is the one that is starting to be accessible commercially while the latter has not received significant adherence by equipment manufacturers so far. Unfortunately, the latter is also the one that could bring more advantages to the RoboCup environment since it allows creating isolated channels with negotiated bandwidth, thus without mutual interference. The former just creates prioritized traffic classes, which does not help since, within a game, one team cannot be prioritized with respect to the other and rules would still be needed to guarantee fairness when sharing the same priority class. Moreover, there would be no guarantee that other external sources of interference would not transmit at the same or higher priority level, thus not avoiding the interference problem. Since it is not clear whether equipment supporting HCCA will ever be available due to market reasons, and its expected higher cost, it seems unnecessary to change the current technology and worth working on enforcing appropriate bandwidth sharing policies and mechanisms.
- **No need for technical verifications.** Ideally, teams should verify and enforce compliance of their equipment with the rules. However, in some cases, particularly with the wireless communication technology due to its idiosyncrasies, the teams often lack the knowledge to adequately enforce the needed

configurations. Without technical verifications before the actual competitions, those problems will be discovered in the game, only, and will be hard to diagnose correctly.

2.3   Further Improvements

We believe that the recent change in the MSL rules, in what concerns communications, positively addressed most of the issues discussed above and constitutes a clear step toward a reliable and efficient use of the wireless medium. Namely, the architecture with predefined IP addresses and with a network monitor, the enforcement of technical verifications specifically concerning the wireless communication as a pre-requisite to admission to the competitions and the bandwidth limitation per team will provide the needed tools to reduce the problems that have been hindering the league. However, we also believe that a further restriction is still needed to definitely increase the robustness of the communications, namely to bound the burstiness of the teams transmissions. This will enforce an adequate permeability of the traffic patterns allowing the adequate interleaving of packets from different sources resulting in lower transmission latencies. This effect is well known in the real-time communications community and can be enforced with techniques that limit the amount of traffic sent within a predefined time interval, such as the leaky bucket [4], or simply using a periodic transmission pattern with relatively small amounts of information [5].

Moreover, the bandwidth limitation of 2.2Mbit/s is hard to apply by the teams because of two reasons. On one hand, it is hard to convert actual transmitted bytes to bandwidth due to the idiosyncrasies of the wireless communication (dynamic bit-rate). On the other hand, bandwidth is a compound metric that represents an average amount of information sent per unit of time. If no interval of time is specified, it is still possible for a team to block the channel with a long burst and then compensate with some time of silence and still meet the average bandwidth stipulated in the rules. This has a significantly different impact than using the same bandwidth frequently transmitting short amounts of information. Thus, we believe the rules should not provide a limitation in bandwidth but in number of bytes per given interval of time, which is a metric that teams can easily work with. In particular, we believe that a limitation similar to 1.5kB per 20ms interval are reasonable values to work with for three reasons. Firstly, they correspond approximately to the current target of 20% channel utilization in the MSL rules for normal game situations with an IEEE802.11b channel. Secondly, they are very easy to enforce by any team, since a periodic process of 20ms is easily achieved with general purpose operating systems (i.e., no special real-time support is needed) and the 1.5kB data fits in a single packet, without need for fragmentation. Thirdly, the period of 20ms is normally adequate to the dynamics of the cooperative behaviors. Nevertheless, it is still possible to send 750B every 10ms or even 375B every 5ms if faster reactivity is needed, without violating the same constraint. Note, equally, that, as demonstrated in this paper with the logs that were carried out, such restriction can be easily policed with a common PC-based/Wireshark network monitor, or the specific monitor that the new MSL rules now refer to.

Finally, the middleware used also has a significant impact. For example, using multicasts in a producer-consumer style allows a faster dissemination of the information, with better synchronization, for four or more stations, on average. A preliminary study of the effect of using multicast/broadcast packets versus unicast ones in a multi-robot scenario is shown in [6]. Direct pair-wise exchange of information, in a peer-to-peer fashion, tends to generate much more traffic for disseminating the same information. Similarly, the use of a central blackboard used in a client-server fashion requires about twice the transmissions than a corresponding producer-consumer model.

As a summary of best practices, we suggest:

- **For the teams:**
  - using a middleware that minimizes transmissions, e.g., with multicasts,
  - using a periodic transmission pattern that is permeable to the traffic from the opponent team,
  - using low bandwidth cooperation approaches that can work well with the exchange of reasonably small amounts of data,
  - verifying the wireless communications compliance with the rules before the actual competitions,
  - not connecting APs that are not under the control of the organization,
  - not transmitting wireless traffic during competitions while in the neighborhood of the fields.
- **For the organization:**
  - carry out the adequate planning of APs and channels,
  - making sure that any pre-installed WLAN for general Internet access in the compound is switched off,
  - enforcing technical verifications of the wireless communications,
  - deploying the communications architecture in the 2009 MSL rules,
  - particularly carrying out the traffic policing using the network monitor,
  - having a specific network analyzer at hand, capable of providing information on the status on the physical channel.

3   RTDB Middleware and Reconfigurable TDMA Protocol

An example of a pair middleware / communication protocol that we consider that fulfills the best-practices referred above for the teams side, is the Real-Time DataBase (RTDB) middleware, originally developed in 2004 [7], and the reconfigurable adaptive-TDMA protocol described in [8]. The RTDB middleware targets providing an efficient and timely support for the fusion of the distributed perception and the development of coordinated behaviors by providing a seamless access to the complete team state using a distributed database that is partially replicated in all team members. This database contains images of both local and remote state variables that are accessed locally with fast non-blocking operations. The images of the remote data are updated autonomously in the background by

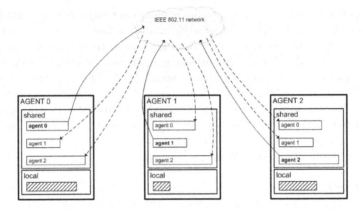

Fig. 4. Each agent transmits periodically its subset of state data that might be required by other agents

the wireless communication protocol, with each team member disseminating its shared state variables using one multicast packet periodically, at a refresh rate that is adapted to the data dynamics (Fig. 4).

The protocol works on top of IEEE802.11 with an innovative layer that enforces a TDMA-like synchronization among the team members. This synchronization aims at avoiding collisions within the team and it is based on the reception instants of the packets from the other team members, without using clock synchronization. The adaptive nature of the protocol arises from its ability to rotate the phase of the TDMA round to avoid periodic interference patterns. The TDMA round period is the only parameter set off-line and it is called *team update period (T_{tup})*, setting the responsiveness and the temporal resolution of the global communication. It is, thus, an application requirement. T_{tup} is divided equally by the number of currently active team members generating the TDMA slot structure. This structure is reconfigured dynamically everytime a node leaves (e.g., crashes) or joins the team. The stations transmit at the begining of their slots, thus maximizing the interval between their transmissions and increasing the resilience of the protocol. This middleware plus communication protocol are fully distributed and need minimal configuration facilitating its deployment.

4   Conclusions

Cooperative robotics is an exciting field that has received growing attention in recent years. RoboCup robotic soccer competitions, including the MiddleSize League (MSL), are examples of initiatives that have been fostering the research in that domain. However, the actual cooperation relies on effective wireless communication and middleware layer, but achieving the desired effective use of the wireless channel is not trivial and requires imposing rules to its fair sharing, enforcing them and requesting cooperation from the teams to comply with them.

In this paper we showed a first analysis of the wireless channel usage within MSL during actual games. We showed that there is a substantial variability in the way teams communicate and that, in several cases, the use practices are not compliant with the current rules, making it evident the need for traffic policing. We then deduced a set of recommendations that we believe help achieving the desired effective use of the wireless channel. Then the paper briefly presented one specific middleware and communication protocol that comply with the suggested recommendations and have been successfully used in MSL competitions in the last four years.

References

1. RoboCup: website, http://www.robocup.org
2. Willig, A., Matheus, K., Wolisz, A.: Wireless Technology in Industrial Networks. Proceedings of the IEEE 93, 1130–1151 (1995)
3. IEEE: Standard for Information technology-Telecommunications and information exchange between systems-Local and metropolitan area networks-Specific requirements - Part 11: Wireless LAN Medium Access Control (MAC) and Physical Layer (PHY) Specifications. IEEE Std 802.11-2007 (Revision of IEEE Std 802.11-1999) (December 2007) C1–1184
4. Carpenzano, A., Caponetto, R., Bello, L.L., Mirabella, O.: Fuzzy traffic smoothing: an approach for real-time communication over ethernet networks. In: 4th IEEE International Workshop on Factory Communication Systems, 2002, pp. 241–248 (2002)
5. Santos, F., Almeida, L., Pedreiras, P., Lopes, L.S., Facchinetti, T.: An Adaptive TDMA Protocol for Soft Real-Time Wireless Communication among Mobile Autonomous Agents. In: Proceedings of the WACERTS 2004 Workshop on Architectures for Cooperative Embedded Real-Time Systems, Lisbon, Portugal (December 2004)
6. Santos, F., Almeida, L.: On the effectiveness of IEEE802.11 broadcasts for soft real-time communication. In: Proceedings of the RTN 2005 - 4th International Workshop on Real-Time Networks, Palma de Mallorca, Spain (July 2005)
7. Almeida, L., Santos, F., Facchinetti, T., Pedreiras, P., Silva, V., Lopes, L.S.: Coordinating Distributed Autonomous Agents with a Real-Time Database: The CAMBADA Project. In: Aykanat, C., Dayar, T., Körpeoğlu, İ. (eds.) ISCIS 2004. LNCS, vol. 3280, pp. 876–886. Springer, Heidelberg (2004)
8. Santos, F., Almeida, L., Lopes, L.S.: Self-configuration of an Adaptive TDMA wireless communication protocol for teams of mobile robots. In: Proceedings of the 13th IEEE International Conference on Emerging Technologies and Factory Automation, Hamburg, Germany (2008)

Multi-robot Cooperative Object Localization

Decentralized Bayesian Approach

João Santos and Pedro Lima

Institute for Systems and Robotics, Instituto Superior Técnico, 1049-001 Lisboa, Portugal
{jsantos,pal}@isr.ist.utl.pt

Abstract. We introduce a multi-robot/sensor cooperative object detection and tracking method based on a decentralized Bayesian approach which uses particle filters to avoid simplifying assumptions about the object motion and the sensors' observation models. Our method is composed of a local filter and a team filter. The local filter receives a reduced dimension representation of its teammates' sample belief about the object location, i.e., the parameters of a Gaussian Mixture Model (GMM) approximating the other sensors' particles, and mixes the particles representing its own belief about the object location with particles sampling the received GMM. All particles are weighted by the local observation model and the best ones are re-sampled for the next local iteration. The team filter receives GMM representations of the object in the world frame, from the sensor teammates, and fuses them all performing Covariance Intersection among GMM components. The local estimate is used when the sensor sees the object, to improve its estimate from the teammates' estimates. The team estimate is used when the sensor does not see the object alone. To prevent the fusion of incorrect estimates, the disagreement between estimates is measured by a divergence measure for GMMs. Results of the method application to real RoboCup MSL robots are presented.

1 Introduction

A team of robots cooperatively tracking an object becomes a team of sensors, each making observations to build a perception of reality that can be improved by the others. Multisensor fusion addresses the problem of combining all the information from multiple sensors in order to yield a consistent and coherent description of the observed environment. The problem itself comes from the fact that the sensors information is always uncertain, usually partial, occasionally incorrect and often geographically or geometrically incomparable with other sensor views [1].

A sensor model describes the uncertainty associated with each sensor observation and location allowing to extract relevant information. The models are often nonlinear resulting in non-Gaussian posterior distributions. However, a parametric (e.g. Gaussian) approximation of sensors information is usually a better choice given the low computational power and low communications bandwidth it requires. This is achieved at the cost of a limited representation of the sensors belief. On the other hand, non parametric discrete approximations, such as Particle Filters, are able to capture arbitrarily complex uncertainty, but are intractable when it comes to communicating the state distribution due to the necessity of transmitting a large sample-based representation.

J. Baltes et al. (Eds.): RoboCup 2009, LNAI 5949, pp. 332–343, 2010.

Each sensor is part of a network node which has local computational power and is able to communicate with nearby nodes. In RoboCup, recent rules to forbid communications with exterior computers push the research towards decentralized sensor network topologies or centralized based topologies with a dynamic leader node [2]. Several teams have taken the decentralized way for a fully multi-agent approach [3] [4] [5] [6]. However, the implementations described rely mostly on parametric sensor models. We propose a decentralized approach based on a probabilistic framework from non-parametric sensors, where communication constraints must be taken into account.

This paper introduces a cooperative perception model based on particle filters and a framework for representing and measuring disagreement of sensor information based on Gaussian Mixture Models. Our soccer robots (RoboCup Middle Size League (MSL) ISocRob team) are equipped with an omnidirectional camera with limited resolution that hardly provides a global view of the field. Our main motivation is to take real advantage of this team of mobile sensors scattered across the field, in order to provide a broader view while locating and tracking the ball. We are further motivated in benefiting from a multisensor system upon the challenges constantly imposed by RoboCup MSL such as the global localization in a symmetric environment or the tracking of the (yet to come) arbitrary color ball.

The paper is organized as follows. In Section 2 we review related work. Section 3 describes the implementation of a shape-based 3D tracker for the ball using a single camera. Section 4 presents a compact sensor information representation based on Gaussian Mixture Models (GMMs) and introduces a decentralized Bayesian approach to multisensor fusion that takes advantage of distributed particle filters and GMM modeling. In Section 5 we present experimental results to validate the presented methods. Section 6 outlines our conclusions.

2   Related Work

Most of the previous work focus on merging the ball localization estimates provided by several sensors to one consistent estimate among the team of robots. Lau et al. [7] calculate the mean and standard deviation of all ball estimates for discarding outliers and then assumes the ball information of the teammate closest to it. Ferrein et al. [8] describe a weighted mean of the estimates according to the distance from the robot to the ball and a time factor denoting how long ago the robot has seen the ball for the last time. On a more probabilistic approach, Stroupe et al. [9] represent ball estimates as a two-dimensional gaussian in canonical form, allowing to merge them by multiplication, and use a Kalman filter to predict the ball position. Pinheiro and Lima [10] implemented a multi-Bayesian team of robots as a direct application of the sensor fusion method introduced by Durrant-Whyte [1]. This approach detects sensors disagreement based on the Mahalanobis distance and achieve a team consensus faster. Other approaches also accounted for merging weighted gridcells from ball occupancy maps [11], Monte Carlo (ball) localization [12] or a combination of Kalman filter with Markov localization [13]. However, although mentioned in some approaches, none of these take into consideration the robots own localization uncertainty, frequently assuming a highly accurate self-localization method. This is problematic because fusion usually takes place in the global reference frame for the team, therefore local estimates must be transformed to global estimates before fusion, and the sensor localization uncertainty plays a major role in this. Pahliani and Lima [14] proposed a new cooperative localization algorithm that reduces the uncertainty of both self-localization and object localization. This method tries to overcome the performance of two popular algorithms for fusing sensor observations: Linear Opinion Pool and Logarithmic Opinion Pool. The implementation although, is based on multi-robot Markov Localization and assumes one can distinguish and locate different team-mates, which is a complex task given the current RoboCup environment.

On other domains, Rosencrantz et al. [15] introduced a scalable Bayesian technique for decentralized state estimation with distributed particle filters using a selective communication procedure over the particle set. On the other hand, instead of selecting which particles to communicate, Upcroft et al. [16] demonstrated the validity of approximating a particle set using Gaussian mixture models or Parzen representations in Decentralized Data Fusion (DDF) systems. However, at every given network node, all sensors are treated as equals, i.e., there is one data association proccess that is impartial to whether the current node is actually tracking the target or not. This means that we are implicitly assuming that the result of the fusion process is more relevant than the local sensor observations. Therefore, we present an approach where each node builds its perception from other sensor nodes observations, and yet relys on a fusion estimation proccess for critical situations, i.e., when the the target is out of the sensor range.

3   Ball Detection and Tracking

Our ball tracking observation model is based on Taiana et al. [17]. A 3D model of the ball is used to calculate it's 2D contour projected on the image. The expected ball contour on the image is computed from its 3D shape projection on the 2D image plane. The ball has rotational symmetry which reduces the problem dimension for there is no need to consider the object orientation. Given a 3-dimensional position, the projection model tell us how the ball contour is going to look like in the image. However, to track it, one needs to estimate the ball's location with respect to the robot. For that we use a particle filter to represent the ball's state space regarding position and velocity $\mathbf{x}_t = [x, y, z, \dot{x}, \dot{y}, \dot{z}]^T$. We start by assuming a simple Markov process for the underlying dynamics of the ball specified by a transition probability, from here and henceforth denoted as motion-model, $p(\mathbf{x}_t|\mathbf{x}_{t-1})$, and that for every time step $t > 1$ a new observation z_t about the state $\mathbf{x}_t$ is made. Given the observation history at time t by $Z_t = [z_1, ..., z_t]$ our goal is to estimate the posterior distribution $p(\mathbf{x}_t|Z_t)$ for each time step. This can be done recursively over *Prediction* and *Update* steps:

$$Prediction : p(\mathbf{x}_t|Z_{t-1}) = \int p(\mathbf{x}_t|\mathbf{x}_{t-1})p(\mathbf{x}_{t-1}|Z_{t-1})d\mathbf{x}_{t-1} \qquad (1)$$

$$Update : p(\mathbf{x}_t|Z_t) \propto p(z_t|\mathbf{x}_t)p(\mathbf{x}_t|Z_{t-1}) \qquad (2)$$

where $p(\mathbf{x}_{t-1}|Z_{t-1})$ is the previous estimate and $p(z_t|\mathbf{x}_t)$ is the observation model. At a given moment in time t, the particle filter represents the probability distribution of the state as a set of N weighted samples $\{\mathbf{x}_t^{(i)}, w_t^{(i)}\}_{i=1}^{N}$, such that the posterior is approximated by an empirical estimate:

$$p(\mathbf{x}_t|Z_t) \approx \sum_{i=1}^{N} w_t^{(i)} \delta(\mathbf{x}_t - \mathbf{x}_t^{(i)}) \tag{3}$$

where $\delta(.)$ is the Dirac delta function. The estimation of the best state is computed through a discrete Monte Carlo approximation of the expectation:

$$\hat{\mathbf{x}}_t \doteq \frac{1}{N} \sum_{i=1}^{N} w_t^{(i)} \mathbf{x}_t^{(i)} \tag{4}$$

Prediction computes an approximation of $p(\mathbf{x}_t|Z_{t-1})$ by moving each particle according to the ball motion model. We assume a constant velocity model where the motion equations correspond to a uniform acceleration during one time step:

$$\mathbf{x}_t = \begin{bmatrix} I & (\Delta t)I \\ 0 & I \end{bmatrix} \mathbf{x}_{t-1} + \begin{bmatrix} (\frac{\Delta t^2}{2})I \\ (\Delta t)I \end{bmatrix} \mathbf{a}_t \tag{5}$$

where I is the 3×3 identity matrix, Δt in general represents the sampling time, and $\mathbf{a}_t$ is a 3×1 white zero mean random vector corresponding to an acceleration disturbance.

In the *Update* step, the particle's weights are updated according to the computed likelihood $p(z_t|\mathbf{x}_t^{(i)})$ for each hypothesis, from the observation model. We follow Taiana's [17] approach to compute the likelihood as a function of similarities between color histograms. We compute two YUV histograms for the inner and outer boundaries of the ball 2D projection contour and apply the Bhattacharyya [18] similarity metric. In order to track arbitrary color balls, we do not define a reference color model for the inner boundary and rely strictly on its mismatch to the outer boundary, that is the object to background dissimilarity. This is well suited given the RoboCup scenario, where the background is mostly the field color and the ball color, no matter what, will always have to contrast with it. The motion model described in Eq. (5) remains valid as long as we express the state of the ball in terms of the world reference frame W which, as opposed to the robot reference frame R, is inertial. As so, the robot pose must be taken into account in the observation model in order to project a 3D point M onto the image plane. This means that, at every time step, the coordinates expressed in the world reference frame $^W M = [^W X, ^W Y, ^W Z, 1]$ must be transformed to the robot reference frame $^R M = [^R X, ^R Y, ^R Z, 1]$ by means of a transformation matrix, which comprises a rotation matrix $^R R_W$ and a translation vector $^R t_W$:

$$^R T_W = \begin{bmatrix} ^R R_W & ^R t_W \\ 0 & 1 \end{bmatrix} \tag{6}$$

The particles that have a higher weight are replicated in the *Resampling* step, and the rest of the particle set is discarded. To prevent the loss of diversity in the particle population, we use a low variance resampling technique.

We initialize our tracker by uniformly spreading a fixed number of ball hypothesis on the ground, in a 5 meter radius circle surrounding the robot. This enable us to reduce the search state space, as we assume the ball is on the floor, and constrains the detection according to the camera resolution.

4 Cooperative Perception in Mobile Sensor Networks

4.1 Information Representation

In order to communicate the ball location and the sensor uncertatinty to other team-mates one cannot transmite the entire particle set that approximates the posterior in Eq. (3). The conversion of our sample-based non-parametric representation to a continuous distribution requires the use of methods such as kernel density estimation, but in order to achieve efficient communication a parametrization of the probability density function is, in fact, mandatory. A mixture model provides this type of representation and can also be viewed as a type of kernel method [19]. If the kernel function of the mixture model is Gaussian, the distribution is expressed as a Gaussian Mixture Model (GMM) of the form:

$$P(\mathbf{x}) = \sum_{k=1}^{N} w_k G(\mathbf{x}|\mu_k, \Sigma_k) \qquad (7)$$

where $\mathbf{x}$ are the observations of the random variable $\mathbf{X}$, w_k are positive weights such that $\sum_{i=k}^{N} w_k = 1$, G is a Gaussian probability density (Gaussian mixture component) with mean μ_k and covariance Σ_k, and N is the total number of mixture components. For the GMM to be of practical importance both for data fusion and communications, the density estimation technique, which will lead to the parametrization of the mixture model, must be computationally fast and accurate.

The Expectation Maximization (EM) algorithm is an efficient iterative method to the general approach of the maximum likelihood parameter estimation in the presence of missing data. Our main intuition while using EM is to alternate between estimating which sample from our sample-based representation belongs to which mixture component (missing data) and estimating the unknown parameters $\Theta_k = (w_k, \theta_k)$, where $\theta_k = (\mu_k, \Sigma_k)$, for each of those components. Each iteration of the EM consists of an expectation (E-step) and a maximization step (M-step). In the E-step we compute the expected likelihood for the complete data Γ (also known as Q-function) as the conditional distribution of the missing data Y, given the current settings of parameters Θ and the observed incomplete data $\mathbf{X}$. So, using Bayes's rule, for each mixture component k:

$$p(y_i = k|x_i, \theta_k) = \frac{p(y_i = k, x_i|\theta_k)}{p(x_i|\theta_k)} = \frac{p(x_i|y_i = k, \theta_k)p(y_i = k|\theta_k)}{\sum_{k=1}^{N} p(x_i|y_i = k, \theta_k)p(y_i = k|\theta_k)} \qquad (8)$$

where N is the total number of mixture components and $p(x_i|y_i = k, \theta_k)$ is, in our case, the multivariate Gaussian probability density function. One should also note that the probability of an observation being part of the kth component $p(y_i = k|\theta_k)$ is actually its relative weight w_k in the mixture model. In the M-step we re-estimate the mixtures parameters Θ by maximizing the Q-function (see [19],[20] for the in-depth derivation). From here we can compute a new approximation Θ' for each component k:

$$\mu_k' = \frac{\sum_{i=1}^{M} x_i p(y_i = k|x_i, \theta_k)}{\sum_{i=1}^{M} p(x_i|y_i = k, \theta_k)}, \Sigma_k' = \frac{\sum_{i=1}^{M} p(x_i|y_i = k, \theta_k)(x_i - \mu_k')(x_i - \mu_k')^T}{\sum_{i=1}^{M} p(x_i|y_i = k, \theta_k)}$$

$$(9)$$

where M is the number of total observations. The relative weight of each Gaussian mixture is given by:

$$w'_k = \frac{1}{M} \sum_{i=1}^{M} p(y_i = k|x_i, \theta_k). \tag{10}$$

4.2 Cooperative Sensor Model

The decentralized sensor fusion typical approach is to build one single estimate of the target, regardless of whether it's being tracked by the local sensor or not, and always assume that in the worst case we filter out our individual local estimate and use the other robot fused estimate. We propose a different approach that consists of not taking other sensors beliefs for granted, and instead use them as if they were observations gathered by the local sensor (virtual observations).

From the previously described particle filter based perception framework in Section 3, we present herein a cooperative perception model that copes both with a local sensor-distributed estimate of the object and a fused team estimate, deals with the correlation between common information and can be used to improve self-localization. The model, based on sequential Bayesian filtering representation, is illustrated in Fig. 1.

In the Local Filter, observations are made and used to compute the likelihood of the sensor model, which is then multiplied by the prior belief in the Update step. Both the *local* prior (before observation), predicted from the local posterior (after Update) over the previous state, and the *team* prior, predicted from the received posterior distributions of the teammates, are concurrently computed at each robot. This way, the other robots information will only influence the prior belief and the posterior will be determined

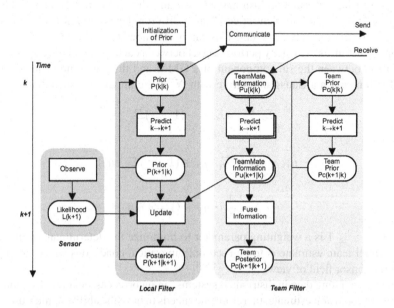

Fig. 1. Decentralized Mobile Cooperative Sensor Model (adapted from [16])

according to the local sensor measurement model. In the update step, we sample from the prior distribution, denoted in particle filters as the proposal distribution $\overline{bel}(\mathbf{x}_t)$, and our goal is that the weighted particle set approximates the posterior, denoted as the target density $bel(\mathbf{x}_t)$. Upon resampling, the particles are distributed according to the posterior:

$$bel(\mathbf{x}_t^{[m]}) = \eta p(z_t|\mathbf{x}_t^{[m]})\overline{bel}(\mathbf{x}_t) \tag{11}$$

where $p(z_t|\mathbf{x}_t^{[m]})$ is the probability of measurement z_t under the mth particle $\mathbf{x}_t^{[m]}$. The target density is then transformed in a compact GMM representation and passed on to the other robots. When it is received, new samples will be drawn from it contributing for the proposal distribution. The ability to sample is not given for arbitrary distributions, however, since our distributions can actually be decomposed in a sum of Gaussians, we can draw a random vector $\mathbf{X} = (\mathbf{x}_1, \mathbf{x}_2, ..., \mathbf{x}_n)^T$ from each bivariate component k with mean μ_k and covariance matrix Σ_k from:

$$\mathbf{x}_k^{[n]} = A_k v^{[n]} + \mu_k \tag{12}$$

where v are n independent samples drawn from $N(0, I_2)$ and A_k is the Cholesky decomposition of Σ_k, such that $\Sigma_k = AA^T$. For each new particle $\mathbf{x}^{[n]}$ we then calculate the importance factor w as described in the ball tracking Update step, Section 3. As such, samples generated from received GMMs that do not follow the local observation model will have a low likelihood and will be discarded on resampling.

In the Team Filter we receive GMM representations of the ball's posterior in the world frame. Regarding information fusion, the Covariance Intersection (CI) filter yields consistent estimates to the problem of combining different Gaussian random vectors with unknown correlation between them. This can be extended to a GMM Covariance Intersection algorithm as in [16], by performing CI between each of the mixture components. The fusion between the ith component of a GMM and the jth component of another GMM will result in a Gaussian mixture with $N \times N$ components, such that:

$$\Sigma_{ij}^{-1} = \gamma \Sigma_i^{-1} + (1-\gamma)\Sigma_j^{-1} \tag{13}$$

$$\mu_{ij} = \Sigma_{ij}(\gamma \Sigma_i^{-1}\mu_i + (1-\gamma)\Sigma_j^{-1}\mu_j) \tag{14}$$

$$w_{ij} = \frac{1}{N}(\gamma w_i + (1-\gamma)w_j) \tag{15}$$

where $0 \le \gamma \le 1$ is a weighting parameter to minimize the determinant of the result. This parallel team estimate is to be used only in critical conditions when the target is out of the sensor field of view.

When associating data in distributed systems, an incorrect association decision leads to an incorrect fusion estimate, therefore ones needs to have the ability to measure agreement among disparate sensors before fusing their observations. A distance measure between Gaussian distributions can be defined as Kullback-Leiber distance [21], Bhattacharyya distance [18] and others. However there's no analytical solution of computing these measures to evaluate the distance between Gaussian mixture models. Therefore, we take Beigi et al. [22] approach to measure distances between collections of distributions in speech recognition, and define our measure of divergence between GMMs as:

$$D(G_1, G_2) = \frac{\sum_{i=1}^{N} W_i^1 + \sum_{j=1}^{N} W_j^2}{\sum_{i=1}^{N} c_i + \sum_{j=1}^{N} c_j} \leq \xi \tag{16}$$

and assume there is agreement if $D(G_1, G_2) \leq \xi$, where ξ is a positive threshold. Consider the matrix of distances between $N \times N$ mixture componentes:

$$T = \begin{bmatrix} d_{11} & d_{12} & ... & d_{1N} \\ d_{21} & d_{22} & ... & d_{2N} \\ ... & ... & ... & ... \\ d_{N1} & d_{N2} & ... & d_{NN} \end{bmatrix} \tag{17}$$

W_i^1 is the minima of the elements in the row times the row number c_i. Likewise, W_j^2 is the minima of the elements in the column times the column number c_j. We can compute d_{ij} from the above metrics for Gaussian distributions. We choose to apply the Bhattacharyya distance for multivariate Gaussian distributions.

4.3 Improving Self-localization

Our current self-localization method is a combination of Monte Carlo Localization with gyrodometry and line points extraction. However, one of the issues that affects MCL performance is the difficulty to recover from failures. One typical recover approach consists in gradually augmenting the proposal distribution by systematically adding more and more particles until better observation likelihoods can be obtained. Two major drawbacks can put this approach at risk. One is the large amount of computational power required to draw and test samples from an augmented proposal distribution that can comprise the entire state space. The other drawback is the inability to deal with local maxima that are present in symmetric environments, such as the RoboCup field.

Instead, one can now see the problem as feature-based map localization with known correspondence, that is $p(\mathbf{r}_t | f_t^i, c_t^i, \mathbf{m})$, where $\mathbf{r}_t$ is the robot pose and f_t denotes a given feature that has a correspondence c_t in a list of landmarks $\mathbf{m}$. Let's consider the ball as a landmark $\mathbf{m}_1$. If some other robots are localized and tracking the ball, the coordinates $\mathbf{m}_{1,x}$ and $\mathbf{m}_{1,y}$ of our landmark in the world frame of the map are given by the Team Filter estimate. If the lost robot is tracking the ball relative to its local coordinate frame (Local Filter), it can make new guesses of its own whereabouts for it now knows it may be on a circle around the landmark. These new guesses represent new poses that incorporate the sensor measurement $p(f_t^i | c_t^i, \mathbf{r}_t, \mathbf{m})$. We can assume the robot is completely lost and therefore the prior $p(\mathbf{r}_t | c_t^i, \mathbf{m})$ is uniform. This assumption leads to:

$$\begin{aligned} p(\mathbf{r}_t | f_t^i, c_t^i, \mathbf{m}) &= \eta p(f_t^i | c_t^i, \mathbf{r}_t, \mathbf{m}) p(\mathbf{r}_t | c_t^i, \mathbf{m}) \\ &= \eta p(f_t^i | c_t^i, \mathbf{r}_t, \mathbf{m}) \end{aligned} \tag{18}$$

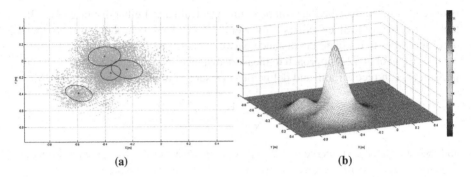

(a)   (b)

Fig. 2. Parametrization of the Ball Tracking particle set using Expectation Maximization. Particle set in green, and the respective Gaussian mixture components in black. (b) Density function.

from where we concluded that sampling from $p(\mathbf{r}_t | f_t^i, c_t^i, \mathbf{m})$ can, in this particular case, be achieved from $p(f_t^i | c_t^i, \mathbf{r}_t, \mathbf{m})$. As so, we only add a limited amount of new sample poses that derive from a common target observation to the MCL proposal distribution.

5 Experimental Results

All experiments were made online, on a Intel Centrino 1.6 GHz processor, in real game situations. To achieve maximum processor performance, the implementation was done in C++ with extensive use of Intel Performance Primitives (IPP) for optimized vector and matrices operations and Intel Math Kernel Library (MKL) for statistics procedures.

5.1 Generating Compact Information Representations

In this experiment we tested the particle set approximation with EM. The purpose was to test the algorithm efficency and determine a good number of Gaussian components that would suit an aceptable EM convergence time. We ran our EM implementation with a different number of mixture components for the same scenario and registered the average run time of the algorithm in Table 1. As so, we choose to use GMMs with 4 components for it is enough to capture a good approximation of the particle set (Fig. 2) in an aceptable time. All the processing was made online with 1200 particles.

5.2 Fusing Data

In this experiment three robots are able to localize the ball, while a fourth robot (the goalkeeper) cannot (see Fig. 3). The robots tracking the ball compute their GMM

Table 1. Average execution time while computing GMMs with our EM implementation for 12000 particles, concurrently with other modules used to play soccer

Number of mixture components	1	2	4	10
Time taken [seconds]	0.0246	0.0690	0.1131	0.1924

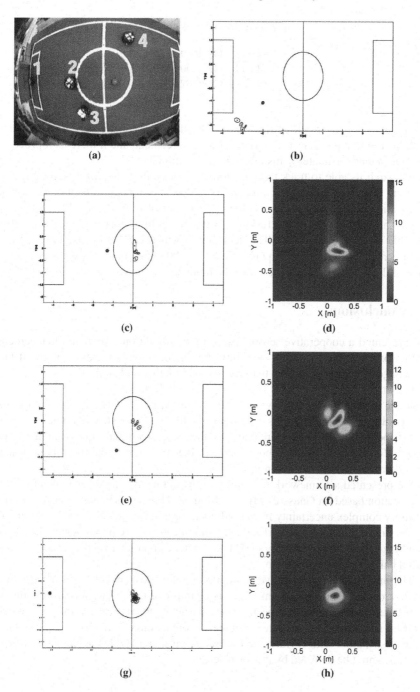

Fig. 3. GMM Data Fusion with Disagreement. (a) Top field camera view. (b) Robot4 tracks the ball and broadcasts its GMM, but is not well localized. (c-f) Robots 2 and 3 track the ball, compute their GMMs and broadcast them. (g-h) The goalkeeper (robot1) tests received GMMs for disagreement and computes GMM CI only for those that are in agreement.

Table 2. Distance measurements between the GMMs received by robot1

$D(G_2, G_3)$	9.6880
$D(G_3, G_4)$	200.5146
$D(G_2, G_4)$	162.7252

approximation and broadcast it to others. As the goalkeeper receives the teammates GMM estimates, it first tests it to see if there's agreement, and if there is, it proceeds to compute a team estimate by fusing the GMMs with CI.

Although its able to track the ball, robot4 is not able to localize itself correctly on the field (Fig. 3a). As such, it broadcasts GMM approximation of its erroneous ball localization belief, since it is corrupted by its self localization belief (Fig. 3b).

We show the results of the fusion estimate made by robot1 (Fig. 3g, 3h), which is not able to see the ball at all. The decision on which GMMs to fuse is based on the disagreement measurement Eq.(16) with $\xi = 30$. The computed distances between each of the received GMMs are shown is Table 2.

6   Conclusions

We presented a cooperative sensor fusion model based on a particle filter perception framework, for mobile robots operating in dynamic environments. We aim at taking advantage of a team of sensors to detect the ball on the field at all time.

For that we implemented a 3D shaped-based ball tracker that comprises a realistic dynamic motion model. The system is based on particle filters and also comprises an observation model that allow us to compute the likelihood of a ball hypothesis, given the ball shape model, the projection model for the omnidirectional camera and an acquired image. To acquaint for the robot motion in the tracker we take the robots pose into consideration in the observation model.

We presented a framework for representing and measuring disagreement of sensor information based on Gaussian Mixture Models. This representation allows to capture arbitrary complex uncertainty from nonlinear observation models, yet it's parametrization is simple and takes no overhead in communications. We implemented the Expectation Maximization algorithm for GMM parameter estimation to approximate the sample based ball posterior distribution.

The implemented cooperative perception model takes advantage of the GMM representation in two distinct forms. One is to improve the local ball particle filter in a distributed fashion way by injecting new particles drawn directly from the received GMMs. The other is to compute a ball team estimate directly from the received GMMs target distribution with Covariance Intersection if there's GMM agreemeant, when the the ball cannot be detected by the local sensor.

References

1. Durrant-Whyte, H.F.: Sensor models and multisensor integration. International Journal of Robotics Research 7(6), 97–113 (1988)

2. Zhao, F., Shin, J., Reich, J.: Information-driven dynamic sensor collaboration for tracking applications. IEEE Signal Processing Magazine 19(2), 61–72 (2002)
3. Azevedo, J.L., Lau, N., Corrente, G., Neves, A., Cunha, M.B., Santos, F., Pereira, A., Almeida, L., Lopes, L.S., Pedreiras, P., Vieira, J., Martins, D., Figueiredo, N., Silva, J., Filipe, N., Pinheiro, I.: Cambada 2008: Team description paper. University of Aveiro, Tech. Rep. (2008)
4. Zweigle, O., Kappeler, U.-P., Ruhr, T., Haussermann, K., Lafrenz, R., Schreiber, F., Tamke, A., Rajaie, H., Burla, A., Schanz, M., Levi, P.: Cops stuttgart team description 2007, University of Stuttgart, Tech. Rep. (2007)
5. Hafner, T., Lange, S., Lauer, M., Riedmiller, M.: Brainstormers tribots team description. University of Osnabruck, Tech. Rep. (2008)
6. Suzuki, T., Tomoyasu, N., Takafashi, M., Yoshida, K.: Eigen keio univ. team description. Keio University, Tech. Rep. (2008)
7. Lau, N., Lopes, L.S., Corrente, G.A.: Cambada: Information sharing and team coordination. In: Robótica 2008 (2008)
8. Ferrein, A., Hermanns, L., Lakemeyer, G.: Comparing sensor fusion techniques for ball position estimation. In: Bredenfeld, A., Jacoff, A., Noda, I., Takahashi, Y. (eds.) RoboCup 2005. LNCS (LNAI), vol. 4020, pp. 154–165. Springer, Heidelberg (2006)
9. Stroupe, A.W., Martin, M.C., Balch, T.: Merging probabilistic observations for mobile distributed sensing, Carnegie Mellon University, Tech. Rep. (2000)
10. Pinheiro, P., Lima, P.: Bayesian sensor fusion for cooperative object localization and world modeling. In: 8th Conference on Intelligent Autonomous Systems (2004)
11. Cai, A., Fakuda, T., Arai, F.: Information sharing among multiple robots for cooperation in cellular robotic system. In: Intelligent Robots and Systems (1997)
12. Steinbauer, G., Faschinger, M., Fraser, G., Muhlenfeld, A., Richter, S., Wober, G., Wolf, J.: Mostly harmless team description, Graz University of Technology, Tech. Rep. (2003)
13. Dietl, M., Gutmann, J.-S., Nebel, B.: Cs freiburg: Global view by cooperative sensing. In: Birk, A., Coradeschi, S., Tadokoro, S. (eds.) RoboCup 2001. LNCS (LNAI), vol. 2377, pp. 133–143. Springer, Heidelberg (2002)
14. Pahliani, A., Lima, P.: Cooperative opinion pool: a new method for sensor fusion by a robot team. In: Intelligent Robots and Systems (2007)
15. Matt Rosencrantz, G.G., Thrun, S.: Decentralized sensor fusion with distributed particle filters. In: Conference on Uncertatinty in AI, UAI (2003)
16. Upcroft, B., Ong, L., Kumar, S., Ridley, M., Bailey, T., Sukkarieh, S., Durrant-Whyte, H.: Rich probabilistic representation for bearing only decentralized data fusion. In: 7th International Conference on Information Fusion (2005)
17. Taiana, M., Santos, J., Gaspar, J., Nascimento, J., Bernardino, A., Lima, P.: Color 3d model-based tracking with arbitrary projection model. In: SIMPAR Omnidirectional Vision Workshop (2008)
18. Bhattacharyya, A.: On a measure of divergence between two statistical populations defined by their probability distributions. Bulletin Calcutta Mathematical Society (1943)
19. Hastie, T., Tibshirani, R., Friedman, J.: The Elements of Statistical Learning, pp. 165–190, 236–242. Springer, Heidelberg (2003)
20. Bilmes, J.A.: A gentle tutorial of the em algorithm and its application to parameter estimation for gaussian mixture and hidden markov models. U.C. Berkeley, Tech. Rep. (1998)
21. Kullback, S.: Information Theory and Statistics. Dover Books (1968)
22. Beiji, H., Maes, S., Sorensen, J.: A distance measure between collections of distributions and its application to speaker recognition. In: International Conference on Acoustics, Speech and Signal Processing, vol. 2, pp. 753–756 (1998)

Evolution of Biped Walking Using Truncated Fourier Series and Particle Swarm Optimization

Nima Shafii[1], Siavash Aslani[1], Omid Mohamad Nezami[1], and Saeed Shiry[2]

[1] Mechatronics Research Laboratoy (MRL), Department of Computer and Electerical Engineering, Qazvin Islamic Azad University,
Qazvin, Iran
{shafii,saslani,mohamadnezami}@mrl.ir
[2] Computer Engineering Department, Amirkabir University,
Tehran, Iran
shiry@ce.aut.ac.ir

Abstract. Controlling a biped robot with a high degree of freedom to achieve stable and straight movement patterns is a complex problem. With growing computational power of computer hardware, high resolution real time simulation of such robot models has become more and more applicable. This paper presents a novel approach to generate bipedal gait for humanoid locomotion. This approach is based on modified Truncated Fourier Series (TFS) for generating angular trajectories. It is also the first time that Particle Swarm Optimization (PSO) is used to find the best angular trajectory and optimize TFS. This method has been implemented on Simulated NAO robot in Robocup 3D soccer simulation environment (rcssserver3d). To overcome inherent noise of the simulator we applied a Resampling algorithm which could lead the robustness in nondeterministic environments. Experimental results show that PSO optimizes TFS faster and better than GA to generate straighter and faster humanoid locomotion.

Keywords: Bipedal Locomotion; Particle Swarm Optimization; Truncated Fourier series.

1 Introduction

In recent years, bipedal locomotion, especially "bipedal walking" has been one of the interesting research topics in multi disciplinary topic. Bipedal walking as a very complex motion, involves most of humanoid joints including its sensors and actuators. Many researchers have focused on this topic and a lot of approaches have been presented. But so far no method exists that can walk a robot as stable as human's do. There are two major approaches in bipedal walking researches; model-based and model free approaches. In model-based approach the designer first derives model of the robot and then builds a controller for the model. Two well known methods in this approach are "Zero Moment Point"[1] (ZMP) and "Inverted Pendulum"[2].

In model-free approach, which is also called "Dynamics Based", it is common to make use of the sensory information and associate it with motions. No physical model

J. Baltes et al. (Eds.): RoboCup 2009, LNAI 5949, pp. 344–354, 2010.

is used in this method that eases the implementation of the skills. There are three important studies done in this field; Passive Dynamic Walking (PDW) [3], Central Pattern Generator (CPG) [4] and Ballistic Walking [5]. In PDW approach, the robot does not have any actuators model and walks just by utilizing the gravity force. The Ballistic walking is originated from the simple human walking model based on the observation of human walking in which the muscles of the swing leg are activated only at the beginning and the end of the swing phase. In CPG approach, special neural circuits take the role of rhythmic walking controller using the non-linear equations to model the neural activities. Researchers usually focus on complex mathematical models like Hopf [6] or Matsuoka [7] to model these neural activities and generate rhythmic walk patterns (Gaits).

In 2006, Truncated Fourier Series (TFS) formulation is used for gait generation in bipedal locomotion [8]. TFS together with a ZMP stability indicator are used to prove that TFS can generate suitable angular trajectories for controlling bipedal locomotion. It does not require inverse kinematics and stable gaits with different step lengths and stride frequencies can be readily generated by changing the value of only one parameter in the TFS.

Taking the advantages of TFS as a model-free approach, we implemented a TFS in a simulated humanoid robot to generate gait trajectories in three dimensions. In this novel approach, the Particle Swarm Optimization (PSO) technique with constraint handling on angles and time is used to find optimum parameters of TFS and train the robot to achieve fast bipedal walking for the first time.

To overcome inherent noise of the simulator, Resampling algorithm is implied which could lead to robustness in nondeterministic environments. The Genetic Algorithm (GA) is also implemented in the same manner. Learning results of GA and PSO are compared with each other which indicate PSO as a better learning method for this complex problem in non-deterministic environment.

2   Simulator and Biped Model

In this paper, a new approach for walking behavior in a simulated humanoid robot is discussed. However simulation is not always efficient, due to difficulty of the modeling collision between feet and the ground, we still believe that numerical simulation is sufficient to explore and test bipedal locomotion methods.

The simulation is performed by Rcssserver3d simulator which is a generic three-dimensional simulator based on Spark and Open Dynamics Engine (ODE). Spark is capable of carrying out scientific distributed multi agent calculations as well as various physical simulations ranging from articulated bodies to complex robot environments [9]. The robot in this study is a simulated model of NAO that is a real humanoid Robot with two arms, two legs and a head. This robot weighs 4.5kg, stands 57cm high and has 22 degrees of freedom (DOF). There are six DOFs in each leg; two in the hip, two in the ankle and one at the knee. An additional DOF that exists at each leg's hip for yaw causes the legs to rotate outward and inward.

As an appropriate test-bed, in our soccer simulation team MRL we have implemented and tested our new bipedal locomotion approach on simulated NAO robot how the generated software based on this simulator is developed by MRL team from scratch. According to our studies, we found 6 DOFs (three for each leg) more effective than other DOFs to make the robot capable of fast walking. The DOFs of hip, knee and ankle which move on the same plane of forward-backward are the major ones. Although other DOFs are effective in walking behavior, but in fact, their role smoothes the robots walking motion. So here, it's preferred to ignore them to decrease learning search space. Like in [10], Foot was kept parallel to the ground by using ankle joint in order to avoid collision. Therefore ankle trajectory can be calculated by hip and, knee trajectories and its DOF parameters are eliminated.

3   TFS gait Generator

Bipedal walking as a complex motion, involves most of humanoid robot's joints. Researchers attempt to imitate the human walking style as well as its speed. Therefore analyzing human walk pattern has been used for acquiring beneficial information about this motion. Human walk has been investigated from many angles; walking trajectory is one of them. The walking trajectory is divided into several types. Positional trajectory and angular trajectory are two of them. In angular trajectory, the angle of each joint is plotted at a certain time slice. Therefore the angular trajectory is obtained by angular variation of each joint. Biped angular trajectory of two joints; hip and knee captured from human walking are shown in Fig 1.a [11].

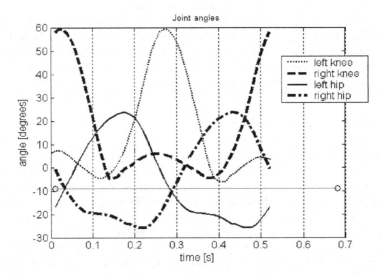

Fig. 1. a Human walking angular trajectory [11]

The angle of each joint in one period of walking signal from t_0 to t_6 is represented in fig 1.b [11] by capturing the main features of fig 1.a and gives a general form to make it applicable to robots. In time range $[t_0, t_2]$ and $[t_5, t_6]$ the left leg is support leg and the right one is swing leg but in range of time $[t_2, t_5]$ the left and

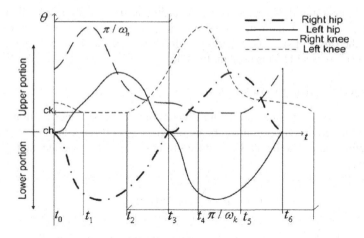

Fig. 1. b gaits elaborated from human gaits features [11]

right legs play the role of support and swing legs respectively. In another word, in two times of t_2 and t_5 the roles of two legs are switched with each other. At time t_3 where two hip trajectories intersect, two thighs cross each other.

3.1 Angular Trajectory Generation

Regarding the fact that all joint trajectories of human walking are periodic and similar to sine or cosine signals [12], the generation of these angular signals can be done by Fourier series.

3.2 Basic Fourier Series

The original definition of Fourier series is described by following formula:

$$F(t) = \frac{a_0}{2} + \sum_{i=1} \left(a_i \cos \frac{i\pi t}{L} + b_i \sin \frac{i\pi t}{L} \right) \tag{1}$$

The first term $(\frac{a_0}{2})$ of equation 1 represents the DC bias a_0 of the signal and the L represents half of the largest period that exists in the signal. By $\omega = \frac{\pi}{L}$ then the frequency form of Fourier series is achieved as follows:

$$F(t) = \frac{a_0}{2} + \sum_{i=1} (a_i \cos(i\omega t) + b_i \sin(i\omega t)) \tag{2}$$

Where ω is frequency of periodic signal, any complicated signal can be produced by this formula when i is considered infinite. But when the value of i is limited to a definite number, precision of generating signal is reduced and this type of Fourier series is called partial sum of the Fourier series. According to fig. 1.a, Human Walking angular trajectories are too complicated to be produced by a definite Fourier series band limited to the second harmonic. Therefore a modified definite Fourier series as a Truncated Fourier series (TFS) is used in this study.

3.3 Trajectory Generation by Using TFS

According to Fig. 1.b., the signals are divided in two parts; upper portion and lower portion. Whereby each portion can be assumed as an odd function, the cosine part of TFS is eliminated. So the TFS is reduced to equation 3 to generate each portion of trajectory.

$$F(t) = a + \sum_{i=1}^{n} b_i \sin(i\omega t) \tag{3}$$

Where ω is fundamental frequency of signal and a is signal offset. Separate production for each portion, caused to generate complex signals with different upper and lower portions. The number of parameters for generating these complex signals is also less than the parameters used in Fourier series. As shown in Fig. 1.b., each signal has an offset. C_h and C_k are hip trajectory and knee trajectory offsets respectively. From t_0 to t_2 the left leg is considered as supporting leg and the variation of its knee angle is so minute that can be assumed fixed. This duration of walking is named lock phase. In addition, the amount of shift phase of the two leg trajectories signal is half of the period of each signal. The trajectories for both legs are identical in shape but are shifted half of the walking period in time. Therefore by figuring out walking angular trajectory of one leg the other leg trajectory is obtained. Using (3) and considering curves of Fig. 1.b., the TFS for generating each portion of hip and knee trajectories are formulated as follow (4):

$$\theta_k^+ = \sum_{i=1}^{n} C_i. \sin(i\omega_k t_2) + C_k \ , \omega_k = \frac{2\pi}{T_k}$$

$$\theta_k^- = C_k \geq 0$$

$$\theta_h^+ = \sum_{i=1}^{n} A_i. \sin(i\omega_h t_3) + C_h \ , \omega_h = \omega_k \tag{4}$$

$$\theta_h^- = \sum_{i=1}^{n} B_i. \sin(i\omega_h t_6) + C_h \ , \omega_h = \omega_k$$

In these equations, the plus (+) sign represents the upper portion of walking trajectory and the minus (-) shows the lower portion. A_i, B_i and C_i are constant coefficients for generating signals. The h and k index stands for hip and knee respectively. C_h and C_k are signal offsets and T_k is assumed as period of knee trajectory. Considering the fact that all joints in walking motion have equal movement frequency [12], the equation $\omega_k = \omega_h = \frac{2\pi}{T_k}$ can be concluded. Parameter t_3 shows the end time of hip trajectory in upper portion and starts its down portion, t_6 shows the end time in down portion. These parameters are not significant since they can be obtained when the hip trajectory intersects the C_h line. But parameter t_2 represents the end time of knee lock phase and must be considered to produce knee trajectory. Therefore Truncated Fourier series parameters to produce trajectories are; C_h, C_k, A_i, B_i, C_i, t_2, and W_k. In this essay there are some constraints to be dealt with as shown in the following equation:

$$0 < t_2 < T_k \, , \omega_k = \frac{2\pi}{T_k} \Rightarrow 0 < t_2 < \frac{2\pi}{\omega_k} \tag{5}$$

$$C_k \geq 0$$

Finally an optimization algorithm is needed to optimize a 7_dimension Problem for finding the best gait generator.

4 PSO Algorithm

The PSO algorithm consists of three steps; generating primitive particle's positions and velocities, velocity update and position update [13]. These parts will be described in sections 4.1, 4.2 and 4.3 respectively.

4.1 Initializing Particles' Positions and Velocities

Equations (6) and (7) are used to initialize particles which Δt are the constant time increment. Using upper and lower bounds on the design variables values, X_{min} and X_{max}, the positions, X_k^i and velocities, V_k^i of the initial swarm of particles can be first generated randomly. The swarm size will be denoted by N. The positions and velocities are given in a vector format where the superscript and subscript denote the ith particle at time k.

$$X_0^i = X_{min} + Rand(X_{max} - X_{min}) \tag{6}$$

$$V_0^i = \frac{X_{min} + Rand(X_{max} - X_{min})}{\Delta t} = \frac{Position}{time} \tag{7}$$

4.2 Updating Velocities

The fitness function value of a particle is used to determine the particle which has the best global value in the current swarm (P_k^g), and to determine the best position of each particle over time (P^i).

The three values that affect the new search direction, namely, current motion, particle own memory, and swarm influence, are incorporated via a summation approach as shown in Equation below with three weight factors, namely, inertia factor, w, self confidence factor, C_1, and swarm confidence factor, C_2, respectively.

$$\underbrace{V_{k+1}^{i}}_{\substack{Velocity\ of\ Particle \\ i\ at\ time\ k+1}} = w\ \underbrace{V_{k}^{i}}_{\substack{Current \\ Motion}} + C_1\ R\ and\ \underbrace{\frac{(P^{i} - X_{k}^{i})}{\Delta t}}_{Particle\ Memory\ Influence} + C_2\ R\ and\ \underbrace{\frac{(P_{k}^{g} - X_{k}^{i})}{\Delta t}}_{Swarm\ Influence} \quad (8)$$

The inertia weight w controls how much of the previous velocity should be retained from the previous step. A larger inertia weight facilitates a global search, while a smaller inertia weight facilitates a local search [14]. A balance can be achieved between global and local exploration to speed up search results using a dynamically adjustable inertia weight formulation. There have been different strategies for determining the value dynamic inertia weight. Introducing a nonlinear decreasing inertia weight as a dynamic inertia weight into the original PSO significantly improves its performance through the parameter study of inertia weight [14]. This nonlinear distribution of inertia weight is expressed as follow:

$$w = w_{init} + U^{-k} \quad (9)$$

Where w_{init} is the initial inertia weight value selected in the range [0, 1] and U is a constant value in the range [1.0001, 1.005], and k is the iteration number.

4.3 Updating the Position

Position update is the final step of each iteration and it is done by using the current particle position and its own updated velocity vector shown in the Equation below.

$$X_{k+1}^{i} = X_{k}^{i} + V_{k+1}^{i}\Delta t \quad (10)$$

In summary, the PSO technique will be:

Let initialization iterative number k = 0, initialization population size.(6),(7) Calculate each particle's fitness value of initialization population, and let first generation Pi be initialization particles, and choose the particle with the best fitness value of all particles as the P_1^{g}.

Repeat
For each particle
  Calculate inertia Weight according to equation (9).
  Update the velocities according to equation (8).
  Update the positions according to equation (10).
  Evaluate its fitness value according to the objective function.
  Update P_k^{g} and P^{i} if necessary.
End for

Until a sufficient good criterion is met, either good fitness or a maximum number of iterations (As in genetic algorithm).

5   Implementation

Bipedal walking is known as a complicated motion since many factors affect Walking style and stability such as robot's Kinematics and dynamics, and collision between feet and the ground. In such a complex motion, relation between Gait trajectory and walking characteristic is nonlinear. In this approach the best parameters to generate angular trajectories for bipedal locomotion must be found. This kind of optimization problem is usually difficult; therefore particle swarm optimization (PSO) seems to be appropriate solution.

In PSO, the parameters of the problems are coded into a finite length of string as a particle. According to section 2, TFS has 7 parameters to generate all joints angular trajectories; there is a 7-dimension search space for the PSO to find the optimum solution.

Fitness function has a critical rule in PSO that is used to judge whether a solution represented by a particle is good or bad. Angular trajectory produced by each particle is used for walking by simulated robot. To use angular trajectory for walking, all individual robot's joints attempt to drive towards their target angles using proportional derivative (PD) controllers. To equip the robot with a fast walking skill a fitness function based on robot's straight movement with limited action time is considered. First the robot is initialized in $x=y=0$ (0, 0) to walk for 15 seconds then fitness function is calculated whenever robot falls or time duration for walking is over. Fitness function formulation is expressed as follow; The *Current Time* in the formula determines time passed since robot has started walking:

```
If ((Current Time >= time duration for walking) or (ro-
bot is fallen))
    Fitness := 10*x ;
End if
```

Due to the fact that there is noise in simulated robot's actuators and sensors, training walking task in this approach can be viewed as an optimization problem in a noisy and nondeterministic environment. Resampling is one of the techniques to improve the performance of evolutionary algorithms (EAs) in noisy environment [15]. In Resampling, the individual set of parameters (particle)y_i, the fitness $F(y_i)$ is measured m times and averaged yielding fitness. According to (11) the strength of noise $\overline{F}$ is reduced by the factor$\sqrt{m}$.

$$\overline{F(y_i)} = \frac{1}{m}\sum_{i=1}^{m} F(y_i), \quad y(i) = const \quad \Rightarrow \quad \overline{\sigma_e} = \sqrt{Var\left[\overline{F(y_i)}\right]} = \frac{\sigma_e}{\sqrt{m}} \quad (11)$$

In this study, for comparing GA and PSO performance as an optimizer, we implemented them by the same mentioned model, fitness function and Resampling factor of m as 3.

5.1 PSO and GA Implementation

Since particles may not be satisfied in constraints during updating position procedure constraint handling is a vital step in PSO algorithm. There are many constraints on parameters in this study (i.e time parameters in TFS must be positive). Therefore Pareto [16] with multi-objective modeling is used for handling constraints.

In Pareto, a solution, $x(2)$, is dominated by solution, $x(1)$, if $x(1)$ is not worse than $x(2)$ in all objectives, and for at least one of the objectives, $x(1)$ is strictly better than $x(2)$. Without loss of generality, these conditions can be expressed as follows for the case where all of the objective functions are going to be minimized:

$$fm\left(x\left(1\right)\right) \leq fm\left(x\left(2\right)\right) for \forall m = 1, 2, ..., M \quad \text{and}$$

$$fm\left(x\left(1\right)\right) \prec fm\left(x\left(2\right)\right) \quad \text{for some } m.$$

Each constraint is assumed as an object in which parameters must be satisfied .So according to Pareto method, a particle can be considered to find P_i, P_k^g when it satisfies objects and constraints. So calculating fitness for particles that cannot satisfy constraint is not necessary.

Salman et.al [17] used the values of 0.9, 2 and 2 for w, C_1 and C_2 respectively. But it is possible that much combination of values lead to much slower convergence or even non-convergence. The tuning of the PSO algorithm values is an issue that warrants proper investigation but is outside the scope of this work. We considered various values for each parameter of the algorithm and tried all possible combinations. Finally we chose the best combination of the parameters regarding the dynamic inertia weight and test results that C_1 and C_2 are assumed as 1, 1.5, w_{init} as 0.8, U as 1.0002 and Δt as 1, respectively. We have also implied a swarm consisted of 100 particles ($N = 100$) and maximum iteration of 10.

In GA implementation, the crossover rate and mutation rate are set to 0.8 and 0.06 respectively and roulette wheel is assumed as selection method. Population for each generation is 100, termination condition is to have a generation counter greater than 10 and Resampling m factor is 3. In another world 3000 trials are needed to find appropriate TFS parameters.

6 Results

4 hours after starting GA on a Pentium IV 3 GHz Core 2 Duo machine with 2 GB of physical memory, 3000 trials were performed. The robot could walk 6.7m in 15s with average body speed of 0.45m/s and the period of 0.41s for each step. Fig. 3 shows the average and best fitness values during these 10 generations.

Running the PSO on the same system with the same parameters of iteration and population, more satisfactory results are achieved. Implying constraint handling, some of the particles that did not satisfy constraint were not tested. So through PSO after 1782 training tests instead of 3000 by GA, the robot could walk 8.7 m in 15 s with average body speed of 5.8 m/s, that's significantly better than GA result. This outcome also proves that PSO has bypassed a local minimum that GA was caught in and it can optimize faster. Fig. 4 illustrates PSO algorithm convergence results.

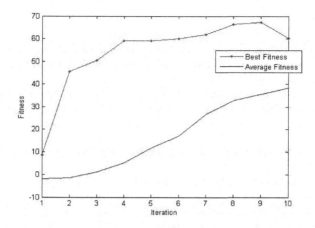

Fig. 2. GA Convergence

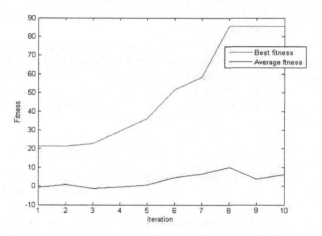

Fig. 3. PSO Convergence

7 Conclusion

In this study for the first time TFS with PSO is implemented in a simulated robot that can walk fast and stable. The technique has some advantages. First, it can be implemented on many humanoid robots as simulated NAO robot to walk based on its walking performance without considering any mathematical modeling. Second, the modified PSO converges sooner than GA to find the best TFS parameters. Since each individual or particle needs a long time to be tested, the higher speed of PSO convergence becomes more significant. On the other hand by using PSO the robot has achieved a faster walk that means PSO performs better than GA in such problems. Resampling technique is also used to overcome uncertainty and noise of the environment.

References

1. Vukobratovic, M., Borovac, B.: Surdilovic. D.:Zero-moment point proper interpretation and new applications. In: Proceedings of the 2nd IEEE-RAS International Conference on Humanoid Robots, pp. 237–244 (2001)
2. Kajita, S., Kanehiro, F., Kaneko, K., Yokoi, K., Hirukawa, H.: The 3D linear inverted pendulum mode A simple modeling for a biped walking pattern generation. In: Proceedings of the 2001 IEEE/RSJ International Conference on Intelligent Robots and Systems, pp. 239–246 (2001)
3. McGeer, T.: Passive dynamic walking. International Journal of Robotics Research 9(2), 62–82 (1990)
4. Pinto, C.M.A., Golubitsky, M.: Central Pattern Generator for Bipedal locomption. J. Math. Biol. 53, 474–489 (2006)
5. Mochon, S., McMahon, T.A.: Ballistic walking. J. Biomech. 13, 49–57 (1980)
6. Buchli, J., Iida, F., Ijspeert, A.J.: Finding Resonance: Adaptive Frequency scillators for Dynamic Legged Locomotion. In: Proceedings of the 2006 IEEE/RSJ International Conference on Intelligent Robots and Systems, pp. 3903–3910 (2006)
7. Matsuoka, K.: Sustained oscillations generated by mutually inhibiting neurons with adaptation. Biol. Cybern. 52, 367–376 (1985)
8. Yang, L., Chew, C.M., Poo, A.N.: Adjustable Bipedal Gait Generation using Genetic Algourithm Optimized fourier Series Furmulation. In: Proceedings of the 2006 IEEE/RSJ International Conference on Intelligent Robots and Systems, pp. 4435–4440 (2006)
9. Boedecker, J.: Humanoid Robot Simulation and Walking Behaviour Development in the Spark Simulator Framework. Technical report, Artificial Intelligence Research University of Koblenz (2005)
10. Kagami, S., Mochimaru, M., Ehara, Y., Miyata, N., Nishiwaki, K., Kanade, T., Inoue, H.: Measurement and comparison of humanoid H7 walking with human being. Robotics and Autonomous Sys. 48, 177–187 (2003)
11. Yang, L., Chew, C.M., Zielinska, T., Poo, A.N.: A Uniform Bipedal Gait Generator with Offline Optimization and Online Adjustable Parameters. Robotica 25, 549–565 (2007)
12. Schot, P.K., Decker, M.J.: The force driven harmonic oscillator model accurately predicts the preferred stride frequency for backward walking. Human movement science 17, 67–76 (1998)
13. Shi, Y., Eberhart, R.: Parameter selection in particle swarm optimization. In: Evolutionary programming VII proceedings of the seventh annual conference on evolutionary programming, New York, pp. 591–600 (1998)
14. Jiao, B., Lian, Z., Gu, X.: A dynamic inertia weight particle swarm optimization algorithm. J. Chaos 37, 698–705 (2008)
15. Beyer, H.G.: Evolutionary algorithms in noisy environments: theoretical issues and guidelines for practice. Comput. Methods Appl. Mech. Engrg. 186, 239–267 (2000)
16. Coello, C.A., Pulido, G.T., Lechuga, M.S.: Handling Multiple Objectives With Particle Swarm Optimization. IEEE Transaction on Evolutionary Computation 8(3), 256–279 (2004)
17. Salman, A., Ahmad, I., Al-Madani, S.: Particle swarm optimization for task assignment problem. Microprocessors and Microsystems 26, 363–371 (2002)

Efficient Behavior Learning by Utilizing Estimated State Value of Self and Teammates

Kouki Shimada[1], Yasutake Takahashi[1], and Minoru Asada[1,2]

[1] Dept. of Adaptive Machine Systems, Graduate School of Engineering
Osaka University
[2] JST ERATO Asada Synergistic Intelligence Project
Yamadaoka 2-1, Suita, Osaka, 565-0871, Japan
{kouki.shimada,yasutake,asada}@ams.eng.osaka-u.ac.jp
http://www.er.ams.eng.osaka-u.ac.jp

Abstract. Reinforcement learning applications to real robots in multi-agent dynamic environments are limited because of huge exploration space and enormously long learning time. One of the typical examples is a case of RoboCup competitions since other agents and their behavior easily cause state and action space explosion.

This paper presents a method that utilizes state value functions of macro actions to explore appropriate behavior efficiently in a multi-agent environment by which the learning agent can acquire cooperative behavior with its teammates and competitive ones against its opponents.

The key ideas are as follows. First, the agent learns a few macro actions and the state value functions based on reinforcement learning beforehand. Second, an appropriate initial controller for learning cooperative behavior is generated based on the state value functions. The initial controller utilizes the state values of the macro actions so that the learner tends to select a good macro action and not select useless ones. By combination of the ideas and a two-layer hierarchical system, the proposed method shows better performance during the learning than conventional methods.

This paper shows a case study of 4 (defense team) on 5 (offense team) game task, and the learning agent (a passer of the offense team) successfully acquired the teamwork plays (pass and shoot) within shorter learning time.

1 Introduction

There have been studies on cooperative/competitive behavior acquisition in a multiagent environment by using reinforcement learning methods, especially in the RoboCup domain. In such a dynamic multi-agent environment, the state and action spaces for the learning can be easily exploded since not only objects but also other agents should be involved in the state and action spaces, and therefore the sensor and actuator level descriptions may cause information explosion that disables the learning methods to be applied within practical learning time. Kalyanakrishnan et al. [4] showed that the learning can be accelerated by sharing

J. Baltes et al. (Eds.): RoboCup 2009, LNAI 5949, pp. 355–365, 2010.

the learned information in the 5 on 4 game task. However, they need still long learning time since they directly use the sensory information as state variables to decide the situation. Noma et al. [6] achieved the cooperative behavior acquisition in the same 5 on 4 game domain within much shorter time by introducing the macro actions and abstracted state variables based on the macro actions and reducing the size of the state-action space. However, the learning time is still too long to realize real robot learning.

Noma et al. [6] presented a method of hierarchical modular learning in a multiagent environment in order to reduce the exploration space, that is, the state space. Learning modules at the lower layer acquire basic skills for soccer play, for example, dribbling and shooting, passing, and receiving behavior, based on reinforcement learning. The module of the top layer takes the state values of the action modules of the lower layer as state variables to construct the state space for learning the cooperative/competitive behavior. The key idea of their work is to utilize the state values of action modules as abstracted state variables instead of using sensory information directly in order to reduce the size of state space. However, the state value can be utilized in more efficient way to reduce the time learning behavior for more complicated cooperative task.

On the other hand, there are case studies in which evaluation of the player situation is designed by hand and the players behave cooperatively based on the evaluation. Isik et al. [3] proposed a multi-robot control system by sharing utility of certain behavior among the players. Mcmillen et al.[5] shows cooperative behavior with AIBOs by sharing the information of the ball on the field among the teammates. Fujii et al. [2] proposed to share the utility for role assignment. Those methods are useful for realizing cooperative behavior among a number of robots, however, there is no room to improve their performance through trial and error as machine learning, especially reinforcement learning, does.

This paper presents more advanced method to learn cooperative behavior in multi-robot environment efficiently. An appropriate initial controller for learning cooperative behavior is generated based on the state value functions of the action modules at the lower layer. The initial controller utilizes the state values of the macro actions so that the learner tends to select a good macro action and not select useless ones. By combination of the ideas and a two-layer hierarchical system, the proposed method shows better learning performance than conventional methods. This paper shows a case study of 4 (defense team) on 5 (offense team) game task, and the learning agent (a passer of the offense team) successfully acquired the teamwork plays (pass and shoot) within shorter learning time.

2   Task and Assumptions

The game consists of the offence team (five players and one of them can be the passer) and the defence team (four players attempt to intercept the ball). The offence player nearest to the ball becomes a passer who passes the ball to one of its teammates (receivers) or shoot the ball to the goal if possible while the opposing team tries to intercept it (see Fig. 1).

 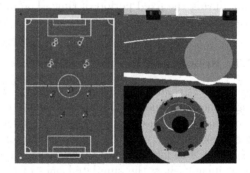

Fig. 1. A passer and the defence formation

Fig. 2. Viewer of simulator

Only the passer learns its behavior while the receivers and the defence team members take the fixed control policies. The receiver becomes the passer after receiving the ball and the passer becomes the receiver after passing the ball. After one episode, the learned information is circulated among team members through communication channel but no communication during one episode. The action and estimation modules are given a priori.

The offence (defence) team color is magenta (cyan), and the goal color is blue (yellow) in the following figures. The game restarts again if the offense team successfully scores a goal, kicks the ball outside of the field, or the defense team intercepts the ball from the opponent.

2.1 Offence Team

The passer who is the nearest to the ball learns the team player behavior by passing the ball to one of four receivers or dribbling and shooting the ball to the goal by itself. After its passing, the passer shows a pass-and-go behavior that is a motion to the goal during the fixed period of time automatically. The receivers face to the ball and move to the positions so that they can form a rectangle by taking the distance to the nearest teammates (the passer or other receivers) (see Fig. 1). The initial positions of the team members are randomly arranged inside their territory.

2.2 Defence Team

The defence team member who is nearest to the passer attempts to intercept the ball, and each of other members attempts to "block" the nearest receiver. "Block" means to move to the position near the offence team member and between the offence and its own goal (see Fig. 1). The offence team member attempts to catch the ball if it is approaching. In order to avoid the disadvantage of the offence team, the defence team members are not allowed inside the penalty area during the fixed period of time. The initial positions of the team members are randomly arranged inside their territory but outside the center circle.

2.3 Robots and the Environment

Robots participating in RoboCup Middle Size League are supposed in this paper. Fig. 2 shows the viewer of the simulator for our robots and the environment. The robot has an omni-directional camera system. A simple color image processing is applied to detect the ball, the interceptor, and the receivers on the image in real-time. The left of Fig. 2 shows a situation the robot can encounter while the right images show the simulated ones of the normal and omni vision systems. The mobile platform is an omni-directional vehicle (any translation and rotation on the plane.)

We suppose that the omni directional vision system provides the robot with 3-D construction of the scene. This assumption is needed for the estimation of the state value of the teammates since it is needed to estimate the sensory information observed by other robots.

3 Multi Module Learning System with Other's State Value Estimation Modules

In this section, we briefly review the work of Noma et al. [6]. Fig.3 shows a basic architecture of the two-layered multi-module reinforcement learning system. The bottom layer (left side of this figure) consists of two kinds of modules: action modules and estimation ones that infer the state value of the teammates. The top layer (right side of the figure) consists of a single gate module that learns which action module should be selected according to the current state that consists of state values sent from the modules at the bottom layer. The selected module then sends action commands based on its policy.

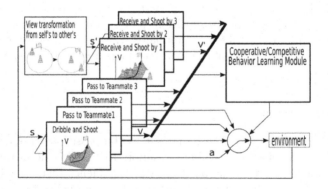

Fig. 3. A multi-module learning system

An action module of the lower layer has a reinforcement learning module which estimates state values for the action. An agent can discriminate a set S of distinct world states. The world is modeled as a Markov process, making stochastic transitions based on its current state and the action taken by the agent based on a policy π. The agent receives reward r_t at each step t. State value V^π, the discounted sum of the reward received over time under execution of policy π, will be calculated as follows:

$$V^\pi = \sum_{t=0}^{\infty} \gamma^t r_t \ .$$

(1)

In case that the agent receives a positive reward if it reaches a specified goal and zero else, then, the state value increases if the agent follows a good policy π. The agent updates its policy through the interaction with the environment in order to receive higher positive rewards in future. For further details, please refer to the textbook of Sutton and Barto[7] or a survey of robot learning[1]. Here, we suppose that the state values in each action module have been already acquired before the learning of the gate module.

As shown in Figure 3, the gate module receives state values of lower modules, that is, the action modules and the other's state value estimation ones, and constructs a state space with them. The state space of the gate module is constructed as direct product of the variables of the state values. In order to adopt a discrete state transition model described above, the state space is quantized appropriately. The action set of the gate module is constructed with all action modules of the lower layer as macro actions. For further details, please refer to [6]. Here, three kinds of action modules are prepared as follows.

- Dribble and Shoot
- Pass to a teammate
- Receive and Shoot

There are 4 "Pass to a teammate" and "Receive and Shoot" modules because there are 4 teammates (receivers) besides the passer. "Pass to teammate 1" module returns the state value when the passer tries to pass the teammate 1. "Receive and Shoot 1" module returns estimated state value of "dribble and shoot" behavior of the teammate 1 if the ball is pass to the teammate 1. Details of those modules are described later.

4   Evaluation of Team Situation

The objective of the team playing soccer is scoring a goal. It is hard to evaluate the situation of the team to score the goal only from positions of teammates, opponents, and a ball. On the other hand, one player situation, how close the player score a goal, can be evaluated based on the state value of the "dribble and shoot" behavior. This behavior can be learned beforehand. If the player can score the goal, then, it means that the team situation is good. Even if the player with the ball cannot score a goal directly because it is far from the opponent goal or the opponent players are close to it, it can pass the ball to one of the teammates that is close to scoring a goal. If the receiver finds that scoring a goal after it receives the ball, then, it is going to find another teammate that is near to scoring a goal. This idea can be applied recursively. The evaluation of team situation from the view point of player possessing a ball can be approximated as follows:

$$
\begin{aligned}
E_i^{team} = \max\{ \ & V_i^{dribble\&shoot}, \\
& \max_j [V_{ij}^{pass} + \beta V_j^{receive\&shoot}], \\
& \max_{jk} [V_{ij}^{pass} + +\beta V_{jk}^{pass} + +\beta^2 V_k^{receive\&shoot})], \\
& \cdots \}
\end{aligned}
\tag{2}
$$

where $V_i^{dribble\&shoot}$, V_{ij}^{pass}, and $V_i^{receive\&shoot}$ indicate state values of player i's behavior "dribble and shoot", "pass" to player j, and "receive and shoot", respectively.

5   Initial Controller Design Based on Team Situation Evaluation

An appropriate initial controller for learning cooperative behavior is generated based on the team situation evaluation. The initial controller utilizes the state values of the macro actions so that the learner tends to select a good macro action and not select useless ones. Based on the approximated team situation evaluation, the initial controller selects one of the macro actions ma as below:

$$
ma = \begin{cases}
ma^{dribble\&shoot} & \text{if } E_i^{team} = V_i^{dribble\&shoot} \\
ma_j^{pass} & \text{if } E_i^{team} = V_{ij}^{pass} + \beta V_j^{receive\&shoot} \\
ma_j^{pass} & \text{if } E_i^{team} = V_{ij}^{pass} + \beta V_{jk}^{pass} + \beta^2 V_k^{receive\&shoot} \\
\cdots
\end{cases}
\tag{3}
$$

It is not possible to calculate all possible options within a limited time. Therefore, the set of the options is limited as only the case of just "dribble&shoot" macro action and a combination of "pass" and "dribble&shoot" ones, in this paper. A concrete pseud algorithm is given at Algorithm 1.

6   Structure of the State and Action Spaces

6.1   Gate Module

The passer is only one learner, and the state and action spaces for the lower modules and the gate one are constructed as follows. The action modules are four passing ones for four individual receivers, and one dribble-shoot module. The other's state value estimation modules are the ones to estimate the degree

Algorithm 1. Initial Controller for Passer

1: MaxEvaluation = $Value_{Dribble\&Shoot}$
2: MaxRobotID = 0
3: N = Number of Receiver
4: **for** $j = 1$ in N **do**
5:     Evaluation(j) = $Value_{Pass(j)} + \beta Value_{Receive\&Shoot(j)}$
6:     **if** Evaluation(j) $\geq$ MaxEvaluation **then**
7:         MaxEvaluation = Evaluation(j)
8:         MaxRobotID = j
9:     **end if**
10: **end for**
11: **if** MaxRobotID = 0 **then**
12:     return **DribbleShoot**
13: **else if** MaxRobotID = 1 **then**
14:     return **Pass(1)**
15: **else if** MaxRobotID = 2 **then**
16:     return **Pass(2)**

17:         $\vdots$
18: **else if** MaxRobotID = N **then**
19:     return **Pass(N)**
20: **end if**

of achievement of ball receiving for four individual receivers, that is how easily the receiver can receive the ball from the passer. These modules are given in advance before the learning of the gate module.

The action spaces of the lower modules adopt the macro actions that the designer specifies in advance to reduce the size of the exploration space without searching at the physical motor level.

The state space S for the gate module consists of the following state values from the lower modules:

- one state value of dribble-shoot action module,
- four state values of passing action modules corresponding to four receivers, and
- four state values of receiver's state value estimation modules corresponding to four receivers.

The reward 1 is given only when the ball is shot into the goal and reward 0 else. When the ball is out of the field or the pre-specified time period elapsed, the game is called "draw" and one episode is over.

6.2 "Dribble&Shoot" Module

In order to reduce learning time for macro actions, one macro action is decomposed into 2 simple behavior. For example, the "Dribble&Shoot" macro action consists of "single Dribble&Shoot" module and "success estimation" module.

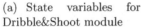

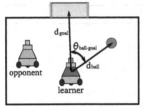

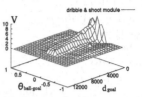

(a) State variables for	(b) State value function of
Dribble&Shoot module	Dribble&Shoot module

Fig. 4. State variables (a) and learned state value function (b) for the dribble and shoot module

The "single Dribble&Shoot" module learns the state value function of the "dribble and shoot" behavior under the environment where a single player shoot a ball without any teammate or opponent. The "success estimation" module estimates success rate of the "single Dribble&Shoot" behavior in case of existence of an opponent. The "Dribble&Shoot" macro action module combines the two basic modules and estimates state value of the behavior accordingly.

The state space of the "single Dribble&Shoot" module S is defined as follows:

- the angle between the opponent goal and the ball
- the distance to the opponent goal, and
- the distance to the ball

Each of these state values is quantized into 31. A CMAC system is adapted with 8 tilings for the approximation of state value.

The state space of the "success estimation" module consists of only the angle between the goal and the opponent. The module learns the state value while the player taking the behavior of the "single Dribble&Shoot" module. Negative reward -1 is given when the opponent takes the ball and zero else. Finally, the "Dribble&Shoot" module estimates state value of the behavior by adding the estimated values of two simple modules.

6.3 "Pass" Module

The state space of the passing module consists of the angle between the receiver and the opponent. The state variable is quantized into 31 levels. A CMAC system is adapted again with 8 tilings for the approximation of state value. The state value map is shown in Fig. 5(c) that indicates the smaller the angle between the receiver and the opponent player is, the lower the state value is.

6.4 "Receive&Shoot" Module

The passer infers each receiver's state that indicates how easily the receiver can shoot the passed ball to the goal by reconstructing its TV camera view of the

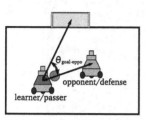 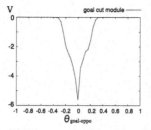 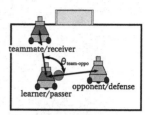

(a) State variables for success estimation module

(b) State value function of "success estimation" module

(c) State variables for pass module

Fig. 5. State variables and learned state value function of "success estimation" and "pass" modules

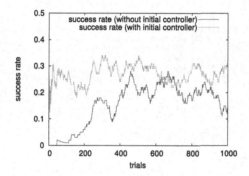

Fig. 6. Curves of success rate

scene from the passer's omnidirectional view. Since we suppose that the passer has already learned the shooting behavior, the passer can estimate the receiver's state value by assigning its own experienced state of the shooting behavior. The "Dribble&Shoot" macro action module is reused for estimation of state value of the "Receive&Shoot" module. This means, the passer estimates the state value of "Dribble&Shoot" behavior on an assumption that the passer successfully pass the ball to the receiver and the receiver controls the ball.

7   Experimental Results

The success rates of case studies with/without the initial controller based on the state value functions of macro action are shown in Fig. 6 where the action selection is 80% greedy and 20% random to cope with new situations. The success rate is moving average during the last 100 trials. The condition of case study without the initial controller is same with the one of Noma et al. [6]. The figure shows the initial controller shows much better performance from the early stage of the learning than the system without the initial controller.

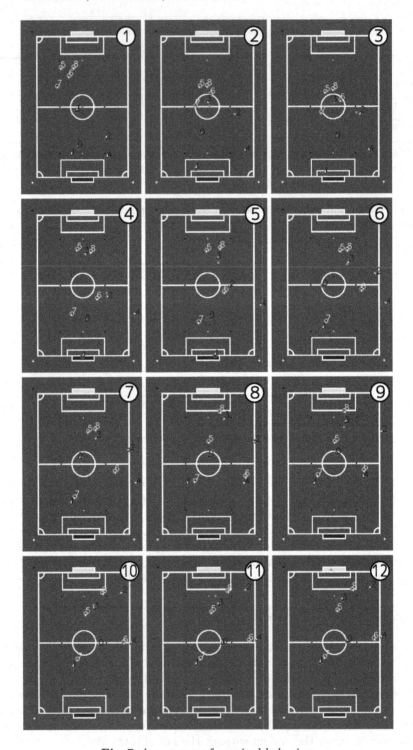

Fig. 7. A sequence of acquired behaviors

An example of acquired behavior is shown in Fig. 7 where a sequence of twelve top views indicates a successful pass and shoot scene.

8 Conclusion

We have utilized the state value functions of macro actions to build a good initial controller for cooperative behavior acquisition instead of learning the behavior from scratch. As a result, we have much improved the performance during the learning compared to the result of the previous method [6].

The initial controller seems to be too good, therefore, the performance of the cooperative behavior during the learning shows little improvement. Further investigation is undergoing for performance improvement of cooperative behavior based on the reinforcement learning.

Real robot experiments are planned in near future because the proposed method reduces actual learning time and it is practical to apply to real robots.

References

1. Connell, J.H., Mahadevan, S.: Robot Learning. Kluwer Academic Publishers, Dordrecht (1993)
2. Fujii, H., Kato, M., Yoshida, K.: Cooperative action control based on evaluating objective achievements. In: Bredenfeld, A., Jacoff, A., Noda, I., Takahashi, Y. (eds.) RoboCup 2005. LNCS (LNAI), vol. 4020, pp. 208–218. Springer, Heidelberg (2006)
3. Isik, M., Stulp, F., Mayer, G., Utz, H.: Coordination without negotiation in teams of heterogeneous robots. In: Lakemeyer, G., Sklar, E., Sorrenti, D.G., Takahashi, T. (eds.) RoboCup 2006. LNCS (LNAI), vol. 4434, pp. 355–362. Springer, Heidelberg (2007)
4. Kalyanakrishnan, S., Liu, Y., Stone, P.: Half field offense in robocup soccer: A multiagent reinforcement learning case study. In: Lakemeyer, G., Sklar, E., Sorrenti, D., Takahashi, T. (eds.) RoboCup 2006 Symposium papers and team description papers, CD–ROM, Bremen, Germany (June 2006)
5. Mcmillen, C., Veloso, M.: Distributed, play-based coordination for robot teams in dynamic environments. In: Lakemeyer, G., Sklar, E., Sorrenti, D.G., Takahashi, T. (eds.) RoboCup 2006. LNCS (LNAI), vol. 4434, pp. 483–490. Springer, Heidelberg (2007)
6. Noma, K., Takahashi, Y., Asada, M.: Cooperative/competitive behavior acquisition based on state value estimation of others. In: Visser, U., Ribeiro, F., Ohashi, T., Dellaert, F. (eds.) RoboCup 2007. LNCS (LNAI), vol. 5001, pp. 101–112. Springer, Heidelberg (2008)
7. Sutton, R.S., Barto, A.G.: Reinforcement Learning: An Introduction. MIT Press, Cambridge (1998)

Sensor and Information Fusion Applied to a Robotic Soccer Team

João Silva, Nuno Lau, João Rodrigues, José Luís Azevedo, and António J.R. Neves

IEETA / Department of Electronics, Telecommunications and Informatics
University of Aveiro, Portugal
{joao.m.silva,nunolau,jmr,jla,an}@ua.pt

Abstract. This paper is focused on the sensor and information fusion techniques used by a robotic soccer team. Due to the fact that the sensor information is affected by noise, and taking into account the multi-agent environment, these techniques can significantly improve the accuracy of the robot world model. One of the most important elements of the world model is the robot self-localisation. Here, the team localisation algorithm is presented focusing on the integration of visual and compass information. To improve the ball position and velocity reliability, two different techniques have been developed. A study of the visual sensor noise is presented and, according to this analysis, the resulting noise variation depending on the distance is used to define a Kalman filter for ball position. Moreover, linear regression is used for velocity estimation purposes, both for the ball and the robot. This implementation of linear regression has an adaptive buffer size so that, on hard deviations from the path (detected using the Kalman filter), the regression converges more quickly. A team cooperation method based on sharing of the ball position is presented. Besides the ball, obstacle detection and identification is also an important challenge for cooperation purposes. Detecting the obstacles is ceasing to be enough and identifying which obstacles are team mates and opponents is becoming a need. An approach for this identification is presented, considering the visual information, the known characteristics of the team robots and shared localisation among team members. The same idea of distance dependent noise, studied before, is used to improve this identification. Some of the described work, already implemented before RoboCup2008, improved the team performance, allowing it to achieve the 1st place in the Portuguese robotics open Robótica2008 and in the RoboCup2008 world championship.

1 Introduction

Robotic soccer is nowadays a popular research domain in the area of multi robot systems. RoboCup[1] is an international joint project to promote artificial intelligence, robotics and related fields that includes several leagues, each one with a different approach, some only at software level, others at hardware, with single or multiple agents, cooperative or competitive [1].

In the context of RoboCup, the Middle Size League (MSL) is one of the most challenging. In this league, each team is composed of up to 6 robots with maximum size

---

[1] http://www.robocup.org/

J. Baltes et al. (Eds.): RoboCup 2009, LNAI 5949, pp. 366–377, 2010.

Fig. 1. Picture of the team robots

of 50x50cm base, 80cm height and a maximum weight of 40Kg, playing in a field of 18x12m. The rules of the game are similar to the official FIFA rules, with required changes to adapt for the playing robots [2]. Each robot is autonomous and has its own sensorial means. They can communicate among them, and with an external computer acting as a coach, through a wireless network. This coach computer cannot have any sensor, it only knows what is reported by the playing robots. The agents should be able to evaluate the state of the world and make decisions suitable to fulfil the cooperative team objective.

CAMBADA, *Cooperative Autonomous Mobile roBots with Advanced Distributed Architecture*, is the Middle Size League Robotic Soccer team from Aveiro University. The project started in 2003, coordinated by the IEETA[2] ATRI[3] group and involves people working on several areas for building the mechanical structure of the robot, its hardware architecture and controllers and the software development in areas such as image analysis and processing, sensor and information fusion, reasoning and control.

To be able to accomplish the objective of playing soccer, it is important that the agent is able to build a good representation of its environment. In the CAMBADA team, this process is called integration. It is a step executed after image analysis and is responsible to take raw information from the vision and other robot sensors and make a sensor and information fusion of all the sources, estimating reliable information of the elements on the field (e.g.: self-localisation, ball position and velocity, obstacles).

For that task it may use the values stored in the previous representation, the current sensor measures (eventually after pre-processing) that has just arrived, the current actuator commands and also information that is available from other robots sensors or world state. This is essentially an information fusion problem. The most common methods to tackle information fusion are based on probabilistic approaches, including Bayes rule, Kalman filter and Monte Carlo methods [3].

All the information available from the sensors in the current cycle is kept in specific data structures (Fig. 2), for posterior fusion and integration, based on both the current information and the previous state of the world.

---

[2] Instituto de Engenharia Electrónica e Telemática de Aveiro - Aveiro's Institute of Electronic and Telematic Engineering.

[3] Actividade Transversal em Robótica Inteligente - Transverse Activity on Intelligent Robotics.

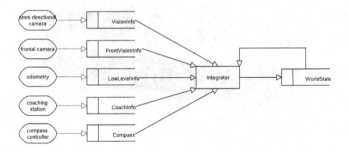

Fig. 2. Integrator functionality diagram

This paper focuses on the description of some sensor and information techniques used in the CAMBADA team. Section 2 describes the fusion of sensorial data for self-localisation. The several aspects of ball integration are described in Section 3. Section 4 presents solutions for identification of visually detected obstacles. Finally, Section 5 presents the conclusion and team achievements.

2   Localisation

Self-localisation of the agent is an important issue for a soccer team, as strategic moves and positioning must be defined by positions on the field. In the MSL, the environment is completely known, as every agent knows exactly the layout of the game field. Given the known mapping, the agent has then to locate itself on it.

The CAMBADA team localisation algorithm is based on the detected field lines, with fusion information from the odometry sensors and an electronic compass. It is based on the approach described in [4], with some adaptations. It can be seen as an error minimisation task, with a derived measure of reliability of the calculated position so that a stochastic sensor fusion process can be applied to increase the estimate accuracy [4].

From the centre of the image (the centre of the robot), radial sensors are created around the robot, each one represented by a line with a given angle. These are called *scanlines*. The image processing, in each cycle, returns a list of positions relative to the robot where the *scanlines* intercept the field line markings [5]. The idea is to analyse the detected line points, estimating a position, and through an error function describe the fitness of the estimate. This is done by reducing the error of the matching between the detected lines and the known field lines (Fig. 3). The error function must be defined considering the substantial amount of noise that affect the detected line points which would distort the representation estimate [4].

Although the odometry measurement quality is much affected with time, within the reduced cycle times achieved in the application, consecutive readings produce acceptable results and thus, having the visual estimation, it is fused with the odometry values to refine the estimate. This fusion is done based on a Kalman filter for the robot position estimated by odometry and the robot position estimated by visual information. This approach allows the agent to estimate its position even if no visual information is available. However, it is not reliable to use only odometry values to estimate the

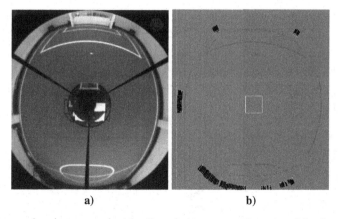

a)   b)

Fig. 3. Captures of an image acquired by the robot camera and processed by the vision algorithms. Left **a**): the image acquired by the camera; Right **b**): the same image after processing with magenta dots over the detected field lines.

position for more than a very few cycles, as slidings and frictions on the wheels produce large errors on the estimations in short time.

The visually estimated orientation can be ambiguous, i.e. each point on the soccer field has a symmetric position, relatively to the field centre, and the robot detects exactly the same field lines. To disambiguate, an electronic compass is used. The orientation estimated by the robot is compared to the orientation given by the compass and if the error between them is larger than a predefined threshold, actions are taken. If the error is really large, the robot assumes a mirror position. If it is larger than the acceptance threshold, a counter is incremented. This counter forces relocation if it reaches a given threshold. Fig. 4 shows situations where the threshold was reached and relocalisation was forced after some cycles.

3 Ball Integration

Within RoboCup several teams have used Kalman filters for the ball position estimation [6,7,8,9]. In [9] and [8] several information fusion methods are compared for the integration of the ball position using several observers. In [9] the authors conclude that the Kalman reset filter shows the best performance.

The information of the ball state (position and velocity) is, perhaps, the most important, as it is the main object of the game and is the base over which most decisions are taken. Thus, its integration has to be as reliable as possible. To accomplish this, a Kalman filter implementation was created to filter the estimated ball position given by the visual information, and a linear regression was applied over filtered positions to estimate its velocity.

3.1 Ball Position

It is assumed that the ball velocity is constant between cycles. Although that is not true, due to the short time variations between cycles, around 40 milliseconds, and given

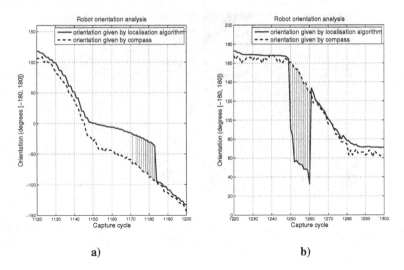

Fig. 4. Illustration of two situations where relocalisation was forced. Left **a**): the camera was covered while the robot moved. The estimated orientation error degrades progressively and after getting higher than the threshold, the cycle count starts and forces relocation; Right **b**): the robot tilted. The estimated orientation error is immediately affected by more than threshold and the cycle count starts and forces relocation.

the noisy environment and measurement errors, it is a rather acceptable model for the ball movement. Thus, no friction is considered to affect the ball, and the model doesn't include any kind of control over the ball. Therefore, given the Kalman filter formulation (described in [10]), the assumed state transition model is given by

$$X_k = \begin{bmatrix} 1 & \Delta T \\ 0 & 1 \end{bmatrix} X_{k-1}$$

where X_k is the state vector containing the position and velocity of the ball. Technically, there are two vectors of this kind, one for each cartesian dimension (x,y). This velocity is only internally estimated by the filter, as the robot sensors can only take measurements on the ball position. After defining the state transition model based on the ball movement assumptions described above and the observation model, the description of the measurements and process noises are important issues to attend. The measurements noise can be statistically estimated by taking measurements of a static ball position at known distances (Fig. 5).

The standard deviation of those measurements is used to calculate the variance and thus define the measurements noise parameter. In practice, the measurements of the static ball were taken while the robot was rotating over itself, to simulate movement and the trepidation it causes, so that the measurements were as close to real game conditions as possible. Some of the results are illustrated in Fig. 5.

A relation between the distance of the ball to the robot and the measurements standard deviation is modeled by the 2nd degree polynomial best fitting the data set in a least-squares sense (Fig. 6). A 1st degree polynomial does not fit the data properly, and

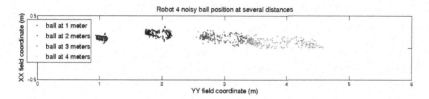

Fig. 5. Noisy position of a static ball taken from a rotating robot

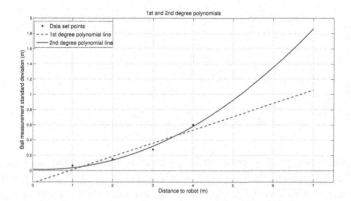

Fig. 6. Representation of the standard deviation value for variable distance to the robot. Data set points as blue dots. 1st degree polynomial as dashed line, 2nd degree polynomial as solid line.

assumes negative values for positive distance, which is not acceptable. Given the few known points, a 3rd degree polynomial would perfectly fit all 4 of them. However, these known points are also estimated and thus cannot be taken as exact. For that reason, a curve that would exactly fit them is not desirable.

As for the process noise, this is not trivial to estimate, since there is no way to take independent measurements of the process to estimate its standard deviation. The process noise is represented by a matrix containing the covariances correspondent to the state variable vector.

Empirically, one could verify that forcing a near null process noise causes the filter to practically ignore the read measures, leading the filter to emphasise the model prediction. This makes it too smooth and therefore inappropriate. On the other hand, if it is too high, the read measures are taken into too much account and the filter returns the measures themselves.

To face this situation, one had to find a compromise between stability and reaction. Given the nature of the two components of the filter state, position and speed, one may consider that their errors do not correlate.

Because we assume a uniform movement model that we know is not the true nature of the system, we know that the speed calculation of the model is not very accurate. A process noise covariance matrix was empirically estimated, based on several tests, so that a good smoothness/reactivity relationship was kept.

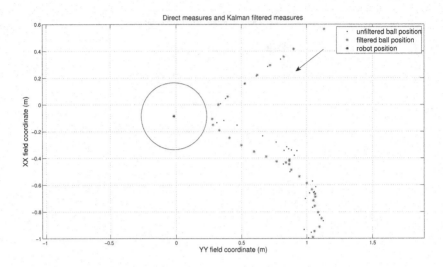

Fig. 7. Plot of a ball movement situation

In practice, this approach proved to improve the estimation of the ball position. Fig. 7 represents a capture of a ball movement, where the black dots are the ball positions estimated by the robot visual sensors and thus are unfiltered. Red stars represent the position estimations after applying the Kalman filter. The ball was thrown against the robot and deviated accordingly and the robot position is represented by the black star in its centre and its respective radius. It is easily perceptible that the unfiltered positions are affected by much noise and the path of the ball after the collision is deviated from the real path. The filtered positions however, seem to give a much better approximation to the real path taken by the ball.

Using the filter *a-priori* estimation, a system to detect great differences between the expected and read positions was implemented, allowing to detect hard deviations on the ball path.

3.2   Ball Velocity

The calculation of the ball velocity is a feature becoming more and more important over the time. It allows that better decisions can be implemented based on the ball speed value and direction. Assuming the same ball movement model described before, constant ball velocity between cycles and no friction considered, one could theoretically calculate the ball velocity by simple instantaneous velocity of the ball with the first order derivative of each component $\frac{\Delta D}{\Delta T}$, being ΔD the displacement on consecutive measures and ΔT the time interval between consecutive measures. However, given the noisy environment it is also predictable that this approach would be greatly affected by that noise and thus its results would not be satisfactory (as it is easily visible in Fig. 8.a).

To keep a calculation of the object velocity consistent with its displacement, an implementation of a linear regression algorithm was chosen. This approach based on linear

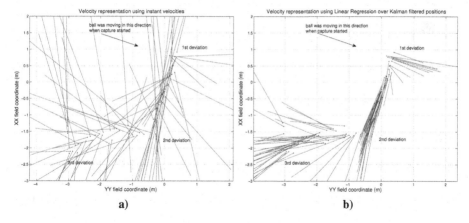

Fig. 8. Velocity representation using: Left, **a**): consecutive measures displacement; Right, **b**): linear regression over Kalman filtered positions

regression [11] is similar to the velocity estimation described in [6]. By keeping a buffer of the last m measures of the object position and sampling instant (in this case buffers of 9 samples were used), one can calculate a regression line to fit the positions of the object. Since the object position is composed by two coordinates (x,y), we actually have two linear regression calculations, one for each dimension, although it is made in a transparent way, so the description in this section is presented generally, as if only one dimension was considered.

When applied over the positions estimated by the Kalman filter, the linear regression velocity estimations are much more accurate than the instant velocities calculated by $\frac{\Delta D}{\Delta T}$, as visible in Fig. 8.b.

In order to try to make the regression converge more quickly on deviations of the ball path, a reset feature was implemented, which allows deletion of the older values, keeping only the n most recent ones, allowing a control of the used buffer size. This reset results from the interaction with the Kalman filter described above, which triggers the velocity reset when it detects a hard deviation on the ball path.

Although in this case the Kalman filter internal functioning estimates a velocity, the obtained values were tested to confirm if the linear regression of the ball positions was still needed. Tests showed that the velocity estimated by the Kalman filter has a slower response than the linear regression estimation when deviations occur. Given this, the linear regression was used to estimate the velocity because quickness of convergence was preferred over the slightly smoother approximation of the Kalman filter in the steady state. That is because in the game environment, the ball is very dynamic, it constantly changes its direction and thus a convergence in less than half the cycles is much preferred.

3.3 Team Ball Position Sharing

Due to the highly important role that the ball has in a soccer game, when a robot cannot detect it by its own visual sensors (omni or frontal camera), it may still know the position of the ball, through sharing of that knowledge by the other team mates.

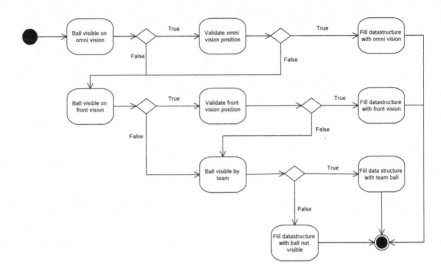

Fig. 9. Ball integration activity diagram

The ball data structure include a field with the number of cycles it was not visible by the robot, meaning that the ball position given by the vision sensors can be the "last seen" position. When the ball is not visible for more than a given number of cycles, the robot assumes that it cannot detect the ball on its own. When that is the case, it uses the information of the ball communicated by the other running team mates to know where the ball is. This can be done through a function to get the statistics on a set of positions, mean and standard deviation, to get the mean value of the position of the ball seen by the team mates and assume it as its own.

Another approach is to simply use the ball position of the team mate that is closer to the ball, being the one that theoretically have more confidence in the detection. Whatever the case, the robot assumes that ball position as its own. When detecting the ball on its own, there is also the need to validate that information. Currently the seen ball is only considered if it is within a given margin inside the field of play as there would be no point in trying to play with a ball outside the field. Also, a maximum detection distance is considered, because of the large image distortion at long distances. Fig. 9 illustrates the general ball integration activity diagram.

4   Obstacle Detection and Sharing

An increasing necessity felt by the team, to improve its performance, is the need for a better obstacle detection and sharing of obstacle information among team mates. This need is important to ensure a global idea of the field occupancy, since the team formation usually keeps the robots spread across the field. With a good cover of field obstacles, passlines and dribbling corridors can be estimated more easily allowing improvements on team strategy and coordination. According to RoboCup rules, the robots are mainly black. Since in game robots play autonomously, every obstacles in the field

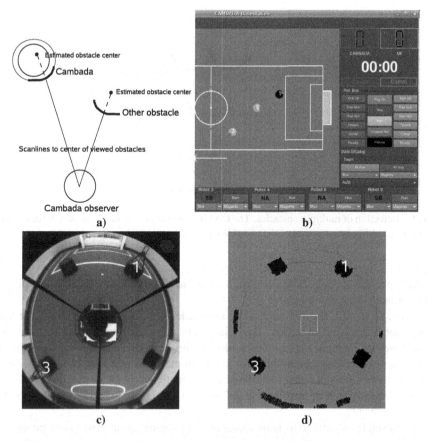

Fig. 10. Identification of single obstacles. Top Left **a**): When a CAMBADA robot is on, the detected obstacles estimated centres are compared with the known position and tested if they are within the robot radius; the left obstacle is within the CAMBADA radius, the right one is not; Top Right **b**): A screenshot of the CAMBADA base station, with 3 robots localised; Bottom Left **c**): an image acquired from the middle robot, with robots 1 and 3 visible and other 2 single obstacles (opponents); Bottom Right **d**): the same image processed where all the single obstacles are detected. 1 and 3 are the correctly detected CAMBADA robots, while the other 2 are marked as opponents.

are the robots themselves (occasionally the referee, which is recommended to have black/dark pants). The vision algorithm take advantage of this fact and detects the obstacles by evaluating blobs of black colour inside the field of play [12]. Through the mapping of image positions to real metric positions, obstacles are identified by their centre and left and right limits. The integration is then responsible for the identification of the obstacles.

In a first step, and since the maximum size of the robots is known, visual obstacles are separated by size. An obstacle can be a candidate to be a robot if it has acceptable dimensions, always considering an error margin, depending on the distance to it. With the known team mates positions (shared via wireless), a matching is tried by testing the

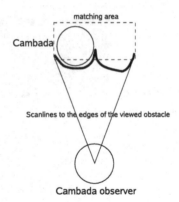

matching area

Cambada

Scanlines to the edges of the viewed obstacle

Cambada observer

Fig. 11. Detection of multiple obstacles. The CAMBADA robot is matched as part of the detected obstacle, resulting in a division of the obstacle in 2 (team mate and opponent).

obstacle estimated centre with the team mate position, considering the robot radius plus an error margin as matching area (Fig. 10.a)).

In a second step, the remaining large obstacles are also compared with the team mates not previously identified. These large obstacles are usually due to the robots being together, forming a unique black blob. In this case, the idea is somewhat opposed to singular obstacles, since in this case, the team mate position is to be tested with the obstacle area. A positive identification of a team mate within the detected obstacle area results in the division of the obstacle in 2 parts, a team mate obstacle and an opponent obstacle (Fig. 11).

The obstacles identified as team mates and opponents can afterwards be treated differently for team cooperation purposes.

5   Conclusion

The work already accomplished concerning sensor and information fusion, especially ball information treatment, helped to maintain a more reliable description of the state of the world.

The techniques chosen for information and sensor fusion proved to be effective in accomplishing their objectives. The Kalman filter allows to filter the noise on the ball position and provides an important prediction feature which allows fast detection of deviations of the ball path. The linear regression used to estimate the velocity is also effective, and combined with the deviation detection based on the Kalman filter prediction error, provides a faster way to recalculate the velocity in the new trajectory.

The increasing reliability of the ball position and velocity lead to a better ball trajectory evaluation. This allowed the development of a more effective goalie action, as well as other behaviours, such as ball interception behaviours and pass reception.

The obtained preliminary results regarding obstacle identification, provide tools for an improvement of the overall team coordination and strategic play.

The accomplished work improved the team performance, allowing it to distinctively achieve the 1st place in the Portuguese robotics open Robótica2008 and the 1st place in the RoboCup2008.

Acknowledgments

This work was partially supported by project ACORD Adaptive Coordination of Robotic Teams, FCT/PTDC/EIA/70695/2006.

References

1. Kitano, H., Asada, M., Kuniyoshi, Y., Noda, I., Osawa, E.: RoboCup: The Robot World Cup Initiative. In: Proceedings of the first international conference on Autonomous agents, pp. 340–347 (1997)
2. MSL Technical Committee 1997-2008: Middle Size Robot League Rules and Regulations for 2008 (2007)
3. Durrant-Whyte, H., Henderson, T.: Multisensor Data Fusion. In: Springer Handbook of Robotics. Springer, Heidelberg (2008)
4. Lauer, M., Lange, S., Riedmiller, M.: Calculating the perfect match: an efficient and accurate approach for robot self-localization. In: Bredenfeld, A., Jacoff, A., Noda, I., Takahashi, Y. (eds.) RoboCup 2005. LNCS (LNAI), vol. 4020, pp. 142–153. Springer, Heidelberg (2006)
5. Neves, A., Martins, D., Pinho, A.: A hybrid vision system for soccer robots using radial search lines. In: Proc. of the 8th Conference on Autonomous Robot Systems and Competitions, Portuguese Robotics Open - ROBOTICA 2008, Aveiro, Portugal, pp. 51–55 (2008)
6. Lauer, M., Lange, S., Riedmiller, M.: Modeling Moving Objects in a Dynamically Changing Robot Application. In: Furbach, U. (ed.) KI 2005. LNCS (LNAI), vol. 3698, pp. 291–303. Springer, Heidelberg (2005)
7. Xu, Y., Jiang, C., Tan, Y.: SEU-3D 2006 Soccer Simulation Team Description. In: CD Proc. of RoboCup Symposium 2006 (2006)
8. Marcelino, P., Nunes, P., Lima, P., Ribeiro, M.I.: Improving object localization through sensor fusion applied to soccer robots. In: Proc. Scientific Meeting of the Portuguese Robotics Open - Robótica 2003 (2003)
9. Ferrein, A., Hermanns, L., Lakemeyer, G.: Comparing Sensor Fusion Techniques for Ball Position Estimation. In: Bredenfeld, A., Jacoff, A., Noda, I., Takahashi, Y. (eds.) RoboCup 2005. LNCS (LNAI), vol. 4020, pp. 154–165. Springer, Heidelberg (2006)
10. Bishop, G., Welch, G.: An Introduction to the Kalman Filter. In: Proc of SIGGRAPH, Course 8. Number NC 27599-3175, Chapel Hill, NC, USA (2001)
11. Motulsky, H., Christopoulos, A.: Fitting models to biological data using linear and nonlinear regression. GraphPad Software Inc. (2003)
12. Neves, A., Corrente, G., Pinho, A.: An omnidirectional vision system for soccer robots. In: Neves, J., Santos, M.F., Machado, J.M. (eds.) EPIA 2007. LNCS (LNAI), vol. 4874, pp. 499–507. Springer, Heidelberg (2007)

Omnidirectional Walking Using ZMP and Preview Control for the NAO Humanoid Robot

Johannes Strom, George Slavov, and Eric Chown

Bowdoin College

Abstract. Fast-paced dynamic environments like robot soccer require highly responsive and dynamic locomotion. We present an implementation of an omnidirectional ZMP-based walk engine for the Nao robot. Using a simple inverted pendulum model, a preview controller generates dynamically balanced center of mass trajectories. To enable path planning, we introduce a system of global and egocentric coordinate frames to define step placement. These coordinate frames allow translation of the CoM trajectory, given by the preview controller, into leg actions. Walk direction can be changed quickly to suit a dynamic environment by adjusting the future step pattern.

1 Introduction

Robust locomotion is crucial to effective soccer play. Successful soccer players must be able to move to the ball quickly, change direction smoothly, and withstand physical interference from opponents. While concepts like omnidirectional walking, Zero Moment Point (ZMP) and constructs like preview control have been explored extensively in the biped walking literature [1,5,4], these discussions often gloss over the realities of implementation. Particularly, these results are often based on simulated experiments, or do not provide the detailed workings of the walk engine. In addition, a system for omnidirectional walking using ZMP and preview control has yet to be presented. This article presents a successful implementation of an omnidirectional walk engine on the Nao robot used in the Standard Platform league.

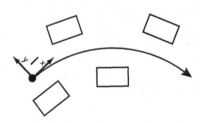

Fig. 1. A sample omnidirectional footstep pattern generated from a constant motion vector (x, y, θ) that in this case, has both a forward and rotational component

J. Baltes et al. (Eds.): RoboCup 2009, LNAI 5949, pp. 378–389, 2010.

1.1   Overview

Omnidirectional walking is crucial in soccer, since a soccer player is constantly changing her objective in a quickly changing environment. Other methods, such as the capability to walk in preset directions which ships with the Nao robot, are not adequate for playing soccer because they do not allow fine-grained control over the direction of motion. The preset trajectories can walk straight, to the side, or turn but cannot combine the three.

Our walk is omnidirectional because we have the capability to place footsteps in any position and orientation: given a desired direction of motion, each successive step is placed along this direction (Figure 1). Given a pattern of steps, a preview controller can use the ZMP balance criterion (discussed in sections 3 and 4) to generate a motion of the center of mass which maintains dynamic balance during the execution of the footsteps [5]. Finally, using the locations of the footsteps and the center of mass trajectory, the motions of the legs can be calculated using inverse kinematics.

ZMP-based approaches to walking that consider the full dynamics of the robot traditionally rely on pre-calculated trajectories and are thus ill-suited for dynamic environments such as robot soccer. Alternatively, dynamically balanced walking patterns can be created at runtime using a simplified model and a preview controller. The preview controller generates valid Center of Mass (CoM) trajectories by examining future foot steps, so motion velocity cannot be changed instantaneously. A certain degree of previewing is absolutely necessary for walking, since it is impossible to change walking vectors instantaneously without falling over. The duration of the preview controller's look ahead determines explicitly which future steps can be safely replaced or updated when the motion vector changes. This allows a quick response to changes in the environment without compromising the robustness of the walk.

What follows is a discussion of each of the components of the walk engine, starting with an overview of step placement, followed by a description of the implementation of a preview controller using the ZMP balance metric, finishing with a discussion of our inverse kinematics system. A schematic overview of the system is shown in Figure 2.

2   Omnidirectional Step Placement

The implementation of an omnidirectional walking system is non-trivial. Translating a series of steps to joint angles requires many layers of abstraction in order to build a well designed system. The central parts of this abstraction are the four homogeneous coordinate frames which we define to allow each part of the system to be expressed in the simplest possible terms (See Table 1, Figure 3 and the following sections for details). The coordinate frames allow step planning, step execution and leg control to be expressed in their natural frame of reference. This ensures that the system stays manageable because each component only acts on a limited amount of information anchored to its appropriate coordinate

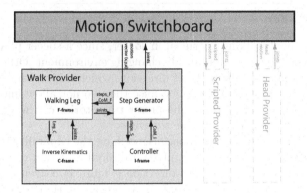

Fig. 2. An overview of the motion architecture. A switchboard manages many modules seeking to provide motion functionality. The walk provider provides the robot with walking capability, while the scripted provider enables execution of scripted motions. The four main components of the walk provider correspond with the four coordinate frames discussed in section 2.

frame. Since each coordinate frame corresponds directly to one of the components of the implementation (see Figure 2), we list the corresponding part of the architecture in brackets in the section headings for the relevant coordinate frame below.

To translate between each coordinate frame, we maintain some transformation matrices which can be applied to move motion trajectories from one coordinate frame to the next (See Appendix A for details). Although matrix multiplication can incur a heavy computational load, they reduce human error and improve maintainability by reducing the complexity of the system. In addition, the matrices are small (3x3), and many are updated only once every walking step – only one matrix must be updated each time step. One alternative to our approach would be to specify the entire walking motion of each leg as a locus relative to the body's CoM. This removes the need for many matrix translations, but the process of integrating the controller is no longer well defined. Additionally, under such a model, omnidirectional walking is very complex. The small overhead potentially incurred by the coordinate transformations is worth avoiding the complexity needed to design the system another way.

2.1   Steps in the S Frame [StepGenerator]

During the walking process, irrelevant steps are discarded and new steps must be planned in the future (as required by the preview controller). Each successive step is generated from the currently desired walk vector in the S frame as shown in Figure 4. The S frame is always offset by H_O towards the CoM from the F frame (See Table 1). Defining steps in this manner allows step planning without needing to consider any history of steps. After each step, the S frame moves to the inside of the next support step, so it is easy to chain multiple steps together.

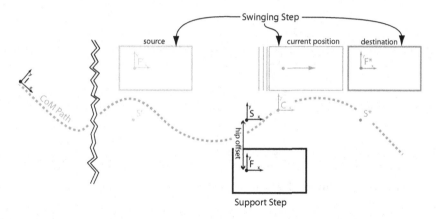

Fig. 3. The various coordinate frames are shown in single support mode while the right leg is supporting and the left leg is swinging from its source to its destination. The support foot is always anchored at F. The previous F coordinate frame is F', and the next one will be F* once the swinging leg arrives at its destination and becomes the next supporting foot. The I coordinate frame is located at the initial starting position of the robot, and does not depend on the footsteps shown above.

Table 1. The four coordinate frames necessary for specifying step placement and leg movements of the robot. H_O is the 50 mm horizontal offset between the CoM and the hip joint. Some of the frames move with the robot and must be updated at various intervals.

Coordinate Frame	Anchor	Updated
C (Center of Mass)	CoM	Every motion time step
F (Foot)	Support Foot	When switching from single to double support
S (Step)	F $\pm H_O$	When switching from single to double support
I (Initial)	World	Never

2.2 CoM Trajectories in the I Frame [Controller]

Using the steps defined in the S frame, the preview controller can calculate the optimal CoM posture which will keep the robot balanced. Since the controller operates in the I coordinate frame, we maintain a transformation matrix from the current S frame to the I frame that gets updated each time a new future step is created.[1] A more detailed discussion of the controller is in section 4.1.

---

[1] The controller runs in a static coordinate frame because if the coordinate frame of the controller were to move with the robot (like the F frame, for example), each of the previewed ZMP values would need to get translated as well, which is expensive. Instead the cost is only that of updating a matrix once per step. The only danger is overflowing the float type, but this will only happen after 500m of walking in a single direction, which is not possible on a soccer field.

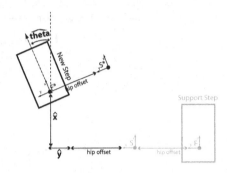

Fig. 4. An additional step being added using a motion vector $(\hat{x}, \hat{y}, \hat{\theta})$

2.3 Leg Trajectories in the F Frame [WalkingLeg]

Trajectories for each leg can be expressed simply in the F frame. Since the F coordinate frame is anchored to the support foot, it provides a consistent frame of reference to define the motion of both legs regardless of how many steps have already been taken. In the F frame the support foot's position always remains at the origin by definition. For the other (swinging) leg, the motion is interpolated between the swinging source and the swinging destination (See Figure 3.) We use a cycloid function to generate a smooth stepping motion which has zero velocity when the foot begins to lift and when it arrives at the destination (inspired by the walk Aldebaran Robotics ships with the robots [7]). In order to eventually specify the motion of the legs in the C frame, we maintain a second transformation matrix from I to F, which is updated at the beginning of each new walking cycle.

2.4 Leg Trajectories in the C Frame [Inverse Kinematics]

Once the leg trajectories are known in the F coordinate frame, they are translated into the C coordinate frame with another transformation matrix. Since the CoM is always moving, this matrix is recalculated each time step from the I to F transformation matrix and the current position and rotation of the CoM in the F frame. The position is easily obtainable from the controller - the current rotation is stored as the robot rotates the support foot relative to the CoM. Targets for the legs can be translated from the C frame into joints using inverse kinematics and the body height z_h (see sections 3 and 4.3).

2.5 Turning

Planning the turning motion on the Nao robot is more complex than on a standard humanoid because each hip does not have 3 linearly independent actuators [7]. Each hip has a pitch (Y-axis) and roll (X-axis) actuator, but both hips share in common a transverse yaw-pitch (ZY-axis) actuator. Without this extra joint, turning is impossible, since the legs are unable to rotate around the Z axis with respect to the body.

To achieve turning, we manually control the hip-yaw actuator to rotate the swinging leg open with respect to the support foot during one step, and then close it again on the next step. By alternating these types of steps, we can induce a good turning motion during a sequence of steps (see Figure 4). Using inverse kinematics, we are able to compensate for the Y-axis hip rotation introduced by employing the yaw-pitch actuator (see Section 4.3).

3   Modeling the Dynamics of the Robot

Given a series of steps, our goal is to specify a trajectory for the CoM which will allow the robot to remain upright and balanced. A good criterion for determining whether a robot will fall is the Zero Moment Point, which is widely used in biped walking [10,5,4]. To simplify the calculation of the ZMP, we follow [5] to simplify the dynamics of the robot by modeling it as an inverted pendulum with the entire mass of the robot concentrated at the CoM.

3.1   The Zero Moment Point

The Zero Moment Point (ZMP), is the point on the support polygon of the robot where the moments acting on the robot are balanced by an opposing moment from the ground [10]. When this point exists (i.e. it is inside the support polygon), then the robot will not rotate about the edges of the foot and will remain upright. Given a pattern of steps, we are thus interested in defining the motion of the robot such that the ZMP always remains near the center of the robot's supporting foot during single support. In double support, when both feet are in contact with the ground, our aim is to quickly pass the ZMP to the other foot.

3.2   Cart Table/Inverted Pendulum Model

Calculating the ZMP of the robot using its full dynamics is computationally intensive and not suitable for online computation. Instead, we simplify the model of the robot as an inverted pendulum [5].

This model allows the ZMP in one dimension to be calculated easily from the position and acceleration of the center of mass of the robot:

$$p = x - \frac{z_h}{g}\ddot{x} \tag{1}$$

Where x is the position of the CoM, $\ddot{x}$ is its acceleration, z_h is the constant height of the CoM from the ground, and g is 9.81, the magnitude of gravity.

This simplification has obvious drawbacks, since it does not account for the complete dynamics of the robot. However, these simplifications can be dealt with in the controller by incorporating sensor feedback, as in [4] – see Section 5.

4   Controlling the Dynamics of the Robot

In the previous section, we discussed how to model the robot in order to use the
ZMP as a balance metric. Though this model allows us to discover if a certain
movement will maintain the robot's dynamic balance, it will not allow us to
calculate motions which are inherently balanced. In fact, what we need is an
inverse to the ZMP equation above, which is not symbolically obtainable [5]. To
calculate the inverse numerically, we can use a preview controller which is able
to generate motions which result in a specified ZMP trajectory.

4.1   Preview Control

Kajita proposes to solve the inverse by casting the problem as a servo control
problem using preview control [5]. A preview controller acts on a ZMP reference
function $p_{ref}(k)$, which defines the location of the desired ZMP at time $k\Delta t$,
where Δt is the duration of a motion time step. $p_{ref}(k)$ is determined by ensuring
the ZMP remains over the support foot during a series of steps (see section 4.2).
The state of the robot is modeled in one dimension using its position, velocity
and ZMP as $[x, \dot{x}, p]^t$. The controller works to converge p, from the state vector,
with p_{ref} given by the reference function. Given proper preview values, two
controllers can work in parallel to generate the states needed to follow the 2-
dimensional reference ZMP necessary for walking. This is effectively an inverse
to the ZMP equation, (1). One controller in the lateral direction is shown in
Figure 5.

The preview control state update is given by

$$\mathbf{x}(k+1) = \mathbf{A}\mathbf{x}(k) + \mathbf{b}u(k) \tag{2}$$

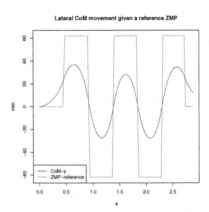

Fig. 5. Lateral CoM movement given a reference ZMP plotted during a sequence of
five steps

Where the optimal system control is given as:

$$u(k) = \sum_{j=1}^{N} G_d(j) p_{ref}(k+j) \tag{3}$$

Where N is the number of previewed time steps. $\mathbf{A}$ is the state matrix, $G_d(j)$ is the preview gain function, and $\mathbf{b}$ is a constant vector. The optimal values for these are defined in [5,4,6,9]. The process for numerically obtaining them relies on an off-line numerical computing environment such as Matlab or Scilab. We believe the following additional references would be useful in this pursuit [9,2].

4.2   Computing ZMP Reference from Steps

A crucial part of the preview controller is previewing the reference ZMP in the future. In order to generate the reference ZMP, we turn each desired step into a sequence of reference ZMP values. As mentioned in section 3.1, when the robot is in single support, we desire the ZMP to rest in the center of the foot - when the robot is in double support, we want to quickly pass the ZMP to the next foot, before beginning to lift the new swinging leg. Though [4] uses a Bézier curve to have a smooth reference ZMP passing between support feet during double support, we have found that a simple linear interpolation from one foot to the next is sufficient, since the controller naturally smoothes the state transitions. The preview values are initially expressed in the S frame since they are generated from steps, but are then translated into the I frame for use in the controller. To facilitate this, we maintain another transformation matrix which is updated each time a future step is generated.

4.3   Kinematics

The final component of controlling the robot is translating leg trajectories from the C frame into joint angles which can be sent to the actuators. This process is called inverse kinematics, but is often used implicitly in the literature with little or no explanation. The method we use is iterative. That is, an initial solution is improved by perturbing the joint angles until the error between the goal (x, y, z) of the end-effector and its current position, which is calculated through forward kinematics, is minimized. The initial solution is the current joint angles of the robot. This is convenient because the net change between subsequent goals of the end-effector during walking is very small.

Let J be the joint space and $\mathbb{R}^3$ be the 3D space. Then we can define forward kinematics as a function $f : J \to \mathbb{R}^3$ which takes a set of joint angles $\boldsymbol{\theta} = (a_1, a_2, \ldots, a_6)$ to the position of the end-effector in space (x, y, z). This function would normally be defined using a set of linear transformations defined by the modified Denavit Hartenberg convention [8]. We used *Mathematica* to symbolically perform the matrix algebra for forward kinematics for a generic set of joint angles. Evaluating the resultant expression at a $\boldsymbol{\theta}$ is significantly faster than performing the matrix algebra that would otherwise be necessary.

In order to minimize the number of iterations in the algorithm, we follow the path of largest decrease in the error. First we define a vector $e = t - s$ where t is the target and s is the current position of the end-effector. e is then the desired direction in $\mathbb{R}^3$. What we need is the desired direction in J. A Jacobian matrix, denoted by M, helps us accomplish this change of variables. Each partial derivative in M is evaluated at the current iteration's θ. x, y and z are the components of the output of f and can be viewed as functions of θ as well. These partial derivatives were computed symbolically using *Mathematica*. Only their evaluation is performed online. We then have an expression ready for the calculation of $\Delta\theta$ which is the desired amount of perturbation in each joint value.

$$\Delta\theta = (M^t M + \lambda^2 I)^{-1} J^t e \qquad (4)$$

λ is a dampening factor which increases the number of iterations required for the algorithm to converge but increases numeric stability. $\Delta\theta$ is clipped because it only provides a direction, not a magnitude. Thus, the final part of the algorithm is deciding how long it should follow this direction in this iteration. Our implementation of this algorithm uses $\lambda = 0.4$ and $max\Delta\theta = 0.5$ radians. These were chosen through trial and error be selecting for quick convergence. The latter is used as a maximum for each component of $\Delta\theta$. We can perform tens of thousands of calls to inverse kinematics in a second, so it does not present a significant efficiency bottleneck. For balanced walking, we also impose a second condition that the foot remain parallel to the ground. We accomplish this by splitting the leg chain into two end-effectors: the ankle and the heel, each with its separate goal. The algorithm achieves high levels of both accuracy and precision. A more detailed presentation of this approach is presented here [3].

As discussed in Section 2.5 dealing with the peculiar kinematics of the Nao robot is necessary to achieve turning motion. Since the addition of the yaw-pitch actuator as a variable to the inverse kinematics algorithm greatly increases the number of possible joint combinations for any given (x, y, z) end effector target, we hold the yaw-pitch joint constant during the iterations. This allows explicitly setting that actuator as required by the turning algorithm, but also results in more regular movements of the legs, since the algorithm keeps the thigh and shin generally aligned to the ground.

4.4   From Theory to Practice

A crucial part of our implementation was bridging the gap from theory to practice. Imperfections such as asymmetry in the robot's joints can cause us to fall. To compensate for this we introduced some adjustments in addition to the core parameters of the walking engine. The most important adjustment we made was inspired by the walk Aldebaran ships with the robots. The actuators struggle to give enough power to the hip joints in order to lift the swinging leg from the ground. To compensate for this, we gradually add a sinusoidal offset to each individual hip-roll (hip lateral swing) joint during the swinging phase. Since the offset is distinct for each hip, this provides considerable help in offsetting asymmetries in the robots, which may have slightly stronger left or right legs.

In addition, extreme values of this adjustment can be used to lift the feet from the ground. By lifting from the hips, the robot does not build up downward momentum with its feet as it steps down, and thus experiences a much smoother gait.

Another offset we introduced helped balance the robot by moving the reference ZMP laterally away from the inside of the foot, inducing a greater hip swing. This helps to compensate for the simplicity of the inverted pendulum model.

The final breakthrough we had was to dramatically reduce the duration of the steps, as well as to reduce the portion of the step spent in double support. By doing this, the robot was able to balance better and move faster because the magnitude of the hip swing was reduced.

5   Results

Using our system of coordinate frames coupled with preview control, we are currently able to achieve maximum forward walking speeds of 10.5 cm/s, which is comparable to the maximum walk speed of the Aldebaran walk engine. However, at such speeds, the robot is not very stable. In practice, we prefer a gait which has a maximum speed of 7 cm/s, with a step frequency of 1Hz, which is much less prone to falling, even during large changes in the motion vector.

We have also been able to extend the preview controller with an observer as described in [4]. However, estimating the sensor ZMP from the accelerometers while minimizing the lag time is non-trivial. In practice, using an observer informed by lagging, noisy sensor values adds instability to the walk even while visibly controlling larger disturbances. This trade-off makes the closed loop controller perform mostly on par with the open loop one. Further refinement of sensor based state estimation is being actively researched.

Videos of our implementation can be found on our team's blog[2].

The code implementation of our system, written in C++ using Boost, is publicly accessible under the LGPL using using `git`[3]. However, as of this writing, no stable release candidate has yet been designated.

6   Conclusion

Since humanoid robots are best suited to coexist with humans, there is an increasing emphasis on humanoid robots. In RoboCup, this reflects the desire to compete on even terms with humans. A critical part of that competition will rely on developing motion engines which are at least as quick and agile as humans are. Among the necessary advances are developing motion systems capable of executing omnidirectional motion in real time. This paper provides an implementation of omnidirectional walking which will serve to help those who are arriving in this field for the first time. Furthermore, it attempts to fill in some

---

[2] `http://robocup.bowdoin.edu/blog`
[3] `http://github.com/northern-bites/nao-man.git`

of the gaps which have been left open by other papers in the field (particularly in the implementation, and testing on real robots). The elegant nature of the preview control comes with some draw backs due to its computational simplicity, however, they can theoretically be overcome using the observer.

References

1. Behnke, S.: Online trajectory generation for omnidirectional biped walking. In: 2006 IEEE International Conference on Robotics and Automation (2006)
2. Benner, P., Sima, V.: Solving algebraic riccati equations with slicot. In: Proceedings of the 11th Mediterranean Conference on Control & Automation MED 2003 (June 2003), http://www-user.tu-chemnitz.de/~benner/pub/med03.pdf
3. Buss, S.R.: Introduction to inverse kinematics with jacobian transpose, pseudoinverse and damped least squares methods (April 2004), http://math.ucsd.edu/~sbuss/ResearchWeb/ikmethods/iksurvey.pdf
4. Czarnetzki, S., Kerner, S., Urbann, O.: Observer-based dynamic walking control for biped robots. Elsevier, Amsterdam (2008) (to appear)
5. Kajita, S., Kanehiro, F., Kaneko, K., Fujiwara, K., Harada, K., Yokoi, K., Hirukawa, H.: Biped wlaking pattern generation using preview control of zero-moment point. In: Proceedings of the 2003 IEEE International Conference on Robotics and Automation (2003)
6. Katajama, T., Ohki, T., Inoue, T., Kato, T.: Design of an optimal controller for a discrete-time system subject to previewable demand. International Journal of Control 41(3), 677–699 (1985)
7. Aldebaran Robotics, http://www.aldebaran-robotics.com
8. Siciliano, B., Khatib, O.: Springer Handbook of Robotics (Spring 2008), http://books.google.com/books?id=Xpgi5gSuBxsC
9. Slavov, G.: The math of dynamically balanced humanoid locomotion, Senior Thesis, Bowdoin College (2009)
10. Vukobratović, M., Borovac, B.: Zero moment point – thirty five years of its life. International Journal of Humanoid Robotics 1(1), 157–173 (2004)

A   Translating between Coordinate Frames

The four key coordinate frames used to generate walking motions are discussed
in section 2. The matrices to do the transformations are given below

$$\mathbf{T_{if}}(n) = \mathbf{F_n} \times \mathbf{F_{n-1}} \times \cdots \times \mathbf{F_2} \times \mathbf{F_1}$$

$$\mathbf{F_i} = \begin{bmatrix} 0 & 0 & 0 \\ 0 & 0 & \pm H_O \\ 0 & 0 & 1 \end{bmatrix} \begin{bmatrix} \cos(-\theta) & -\sin(-\theta) & 0 \\ \sin(-\theta) & \cos(-\theta) & 0 \\ 0 & 0 & 1 \end{bmatrix} \begin{bmatrix} 0 & 0 & -s_x \\ 0 & 0 & -s_y \\ 0 & 0 & 1 \end{bmatrix}$$

$$\begin{bmatrix} x_f(n\varDelta S + t) \\ x_f(n\varDelta S + t) \\ 1 \end{bmatrix} = \mathbf{T_{if}}(n) \begin{bmatrix} x_i(n\varDelta S + t) \\ y_i(n\varDelta S + t) \\ 1 \end{bmatrix}$$

$$\mathbf{T_{fc}}(n\varDelta S + t) = \begin{bmatrix} \cos(\phi(n\varDelta S + t)) & -\sin(\phi(n\varDelta S + t)) & 0 \\ \sin(\phi(n\varDelta S + t)) & \cos(\phi(n\varDelta S + t)) & 0 \\ 0 & 0 & 1 \end{bmatrix} \begin{bmatrix} 0 & 0 & -x_f(n\varDelta S + t) \\ 0 & 0 & -y_f(n\varDelta S + t) \\ 0 & 0 & 1 \end{bmatrix}$$

$$\begin{bmatrix} destx_c \\ desty_c \\ 1 \end{bmatrix} = \mathbf{T_{fc}} \begin{bmatrix} destx_f \\ desty_f \\ 1 \end{bmatrix}$$

$$(A\text{-}1)$$

Where $\mathbf{T_{if}}(n)$ is the transformation matrix between coordinate frames I and F
after n steps. H_O is the horizontal offset between the CoM and the hip joint,
and $\mathbf{F_i}$ is the matrix to translate from the F(i-1) coordinate frame to the next
F(i) coordinate frame given the ith step (s_x, s_y, θ). $(x_i(n\varDelta S + t), y_i(n\varDelta S + t))$
is the position of the CoM in the I coordinate frame at time t after the nth step
was started ($\varDelta S$ is the duration of a step). $\phi(n\varDelta S + t)$ is the rotation of the
center of mass at time t between the C frame and the F frame. $\mathbf{T_{fc}}$ is the transformation matrix between the F and C coordinate frames, and $destx_c, desty_c$ is
the destination of a heel in the c coordinate frame.

RoboCup@Home: Results in Benchmarking Domestic Service Robots

Thomas Wisspeintner[1], Tijn van der Zan[2], Luca Iocchi[3], and Stefan Schiffer[4]

[1] Department of Mathematics and Computer Science
Freie Universität Berlin, Berlin, Germany
[2] Department of Artificial Intelligence
University of Groningen, Groningen, The Netherlands
[3] Dipartimento di Informatica e Sistemistica
University "La Sapienza", Rome, Italy
[4] Knowledge-Based Systems Group
RWTH Aachen University, Aachen, Germany

Abstract. Benchmarking robotic technologies is of utmost importance for actual deployment of robotic applications in industrial and every-day environments, therefore many efforts have recently focused on this problem. Among the many different ways of benchmarking robotic systems, scientific competitions are recognized as one of the most effective ways of rapid development of scientific progress in a field. The ROBOCUP@HOME league targets the development and deployment of autonomous service and assistive robot technology, being essential for future personal domestic applications, and offers an important approach to benchmarking domestic and service robots.

In this paper we present the new methodology for benchmarking DSR adopted in RoboCup@Home, that includes the definition of multiple benchmarks (tests) and of performance metrics based on the relationships between key abilities required to the robots and the tests. We also discuss the results of our benchmarking approach over the past years and provide an outlook on short- and mid-term goals of @HOME and of DSR in general.

1 Introduction

Creating a personal Domestic Service Robot (DSR) is a very complex task that requires cooperation between many scientific disciplines. DSRs have to operate in realistic and unconstrained environments which include humans. They must acquire on-line knowledge about both the animate and inanimate world and need to be very robust to unpredictable and changing environments and safe in the interactions with humans and the environment. This requires the integration of many abilities and technologies including: HRI, reasoning, planning, behavior control, object recognition, object manipulation or tracking of objects. Regarding artificial intelligence, the systems should contain adaptive but robust behaviors and planning methods, social intelligence, and on-line learning capabilities.

The recent increase in availability, accessibility, and compatibility of essential robot hardware and software components allows research groups not only to address a small

J. Baltes et al. (Eds.): RoboCup 2009, LNAI 5949, pp. 390–401, 2010.
© Springer-Verlag Berlin Heidelberg 2010

subset of the mentioned above challenges in DSR, but also to address the problem as a whole, without having to sacrifice a focus in a specific research topic.

This progress is also confirmed by the presence of some rather specialized service robotic applications on the market. Such applications include floor cleaning (e.g. Roomba and Scooba), lawn mowing (e.g. Robomow), and surveillance (e.g. Robowatch). Still, these applications are missing some essential properties of a multi-purpose autonomous and intelligent domestic service robot. Prominent examples of domestic and personal assistant robot research projects are ReadyBot[1] and PR2[2], while Wakamaru[3] and Pa-PeRo[4] focus more on social interaction studies. Many of these projects address relevant aspects of DSR. Still, what appears to be missing is a joint continuous international and multi-disciplinary research and development effort which also includes the aspect of application-oriented benchmarking of systems in DSR.

The ROBOCUP@HOME league [1] targets development and deployment of autonomous service and assistive robot technology being essential for future personal domestic applications. It is part of the international RoboCup initiative and it is the largest annual service and home robotic competition world-wide. ROBOCUP@HOME aims to be a combination of inter-disciplinary community building, scientific exchange, and competition, that iteratively defines benchmarks and performance metrics on which service robots can be evaluated and compared in a realistic domestic environment.

Since the real world is not standardized, measuring the performance of non standardized robots acting in it is a difficult task. The experimental paradigm to evaluate complex robotic systems has to use consequent scientific analysis to improve on itself. Measuring the performance of the robots requires continuous reconsideration of the methodologies used, since both the robots (their capabilities) and their operation environment (and the robot's tasks) will definitely change over time. In our case, the tools are specific benchmarks testing certain robot abilities and the measurement of the robots' performance in the tests. We firmly believe that creating and applying this experimental paradigm can greatly improve and accelerate the development in DSR as it already is the case in other robotic areas like Rescue Robotics or the Robot Soccer leagues.

This paper presents a new methodology for benchmarking DSR. The proposed approach defines multiple benchmarks (tests) related to DSR and performance metrics based on the relationships between key abilities required to the robots and the tests. We also discuss the results of the ROBOCUP@HOME benchmarking in the past years and provide an outlook on short- and mid-term goals of @HOME and of DSR in general.

2   Benchmarking Domestic Service Robotics

Benchmarking has been recognized as a fundamental activity to advance robotic technology [2,3] and many activities are in progress, such as the EURON Benchmarking

---

[1] ReadyBot (http://www.readybot.com/)

[2] PR2 (http://www.willowgarage.com/)

[3] Wakamaru (http://www.mhi.co.jp/kobe/wakamaru/english/)

[4] PaPeRo (http://www.nec.co.jp/robot/english/robotcenter_e.html)

Initiative[5], the international workshops on Benchmarks in Robotics Research and on Performance Evaluation and Benchmarking for Intelligent Robots and Systems, held since 2006[6], the Rawseeds project[7], which aims at creating standard benchmarks specially for localization and mapping, the RoSta project[8], which focuses on standardization and reference architectures, etc.

Benchmarking can be distinguished in two classes: *system benchmarking*, where the robotic system is evaluated as a whole, and *component benchmarking*, where single functionality is evaluated. Component benchmarking is very important to compare different solutions to a specific problem and to identify the best algorithms and approaches. However, it is not sufficient to assess the general performance of a robot with respect to a class of applications. Indeed, the best solution for a specific problem may be unfeasible or inconvenient when integrated with other components that compose a robotic application.

On the contrary, system benchmarking offers an effective way to measure the performance of an entire robotic system in the accomplishment of complex tasks, as such tasks require the interplay of various sub-systems or approaches. Thus, standard reference environment, reference tasks and related performance metrics are to be defined. However, when defining standard benchmarks two common problems arise: Firstly, the difficulty of defining a benchmark that is commonly accepted by the community (this is due to different view points of different research groups about a problem) and secondly, the risk of causing specialized solutions for a certain benchmark or problem that can not be applied in real world applications. To avoid these problems, scientific competitions have proven to be a very adequate method, because benchmarks are usually discussed and then accepted by all the participants. Especially annual competitions provide a constant feedback on a yearly basis about the increase in performance and allow for setting up medium-term projects.

Among the many robotic competitions, AAAI Mobile Robot Competitions was one of the first, being established in 1992 [4]. It has provided significant scientific and technological contributions but its focus and benchmarks change domain every year.

RoboCup (founded in 1997) [5] has currently the largest number of participants (e.g., 440 teams with over 2600 participants from 35 countries in 2006). The RoboCup soccer competitions offer evaluation through competition in the robotic soccer domain and have contributed to significant scientific achievements in the last ten years. However, the special focus on soccer tends to steer the solutions towards specialized robotic architectures.

Robot rescue games started in 2000 within the AAAI Mobile Robot Competition [7] and then since 2001 within the RoboCup Rescue initiative [8]. There is a large focus on benchmarking robots in an abstract and standardized environment. Common metrics for HRI have been defined [9] and effective evaluation of HRI techniques have been carried out [10,11], but the type of HRI via an operator station is different to the more direct kind of interaction desired in DSR.

---

[5] http://www.euron.org/activities/benchmarks/index

[6] All these workshops are summarized in http://www.robot.uji.es/EURON/en/index.htm.

[7] http://www.rawseeds.org/

[8] http://www.robot-standards.eu/

The DARPA Grand Challenge[9] is probably the most recognized competition in terms of public and media attention, and the one that is most directly application oriented. However, there is little relation to DSR, the benchmarking setting is very difficult to reproduce (participation and organization were very costly), and the continuation of this competition is uncertain.

Finally, educational contests, such as EUROBOT[10] or RoboCup Junior[11], are organized with the main goal of presenting robotics to young students and thus they deal with simpler tasks and robotic platforms.

Competitions that have a more direct relation to DSR mainly focus on a single task. The AHRC Vacuum Contest[12] and the 2002 IROS Cleaning Contest[13] [12] are focused only on floor cleaning, while ROBOEXOTICA[14] focusses on robots preparing and serving cocktails. The ICRA HRI Challenge[15] has a broader scope, namely the effectiveness of HRI, but lacks evaluation criteria for the performance.

Following the successful lines of RoboCup competitions and the experiences offered by other competitions related to DSR, the ROBOCUP@HOME annual competition has been set up as a *system benchmarking* activity for domestic service robotics.

3 The @HOME Initiative

ROBOCUP@HOME is a combination of scientific exchange and competition that provides standard benchmarks (called "Tests") and performance metrics on which personal domestic service robots can be evaluated and compared in a realistic domestic environment. This section briefly summarize the main concepts behind the @HOME competitions, that are useful for the following analysis. Details on the rules and on the organization can be found in the ROBOCUP@HOME web site[16]. In particular, in this section we will discuss the key features that we identified to be relevant for DSR and the @HOME competitions, and the score system of @HOME.

3.1 Key Features

An initial set of robot key features (abilities and properties) was derived from an analysis of the state of the art in DSR and from experiences and observations of other robotic competitions. These features help to design the benchmarks and the score system for the competition. Furthermore, these features allow for a later analysis of the teams' performances and help to develop and later enhance the competition in a structured way. As the competition with its benchmarks is expected to evolve over time, also the key features and their weights in the competition are expected to be adapted. The key features are divided in two groups: *functional abilities* and *system properties*.

---

[9] http://www.darpa.mil/grandchallenge/index.asp

[10] http://www.eurobot.org/

[11] http://rcj.sci.brooklyn.cuny.edu/

[12] http://www.botlanta.org/

[13] http://robotika.cz/competitions/cleaning2002/en

[14] http://www.roboexotica.org/en/mainentry.htm

[15] http://lasa.epfl.ch/icra08/hric.php

[16] http://www.robocupathome.org/

Functional abilities. *Functional abilities* include specific functionalities that must be implemented on the robot in order to perform decently in the tests. Each test requires a certain subset of these abilities as they are also directly represented in the score system. Teams must thus decide which of these abilities to implement and up to which degree of performance, depending on their background and the kind of tests they intend to participate in. *Functional abilities* currently are:

- *Navigation*, the task of path-planning and safely navigating to a specific target position in the environment, avoiding (dynamic) obstacles
- *Mapping*, the task of autonomously building a representation of a partially known or unknown environment on-line
- *Person Recognition*, the task of detecting and recognizing a person
- *Person Tracking*, the task of tracking the position of a person over time
- *Object Recognition*, the task detecting and recognizing (known or unknown) objects in the environment
- *Object Manipulation*, the task of grasping or moving an object
- *Speech Recognition*, the task of recognizing and interpreting spoken user commands (speaker dependent and speaker independent)
- *Gesture Recognition*, the task of recognizing and interpreting human gestures

System properties. *System properties* include demands on the entire robotic system that we consider of general importance for any domestic service robot. They can be described as "Soft Skills" which need to be implemented for an effective system integration and a successful participation in the @HOME competition. *System abilities* currently are:

- *Ease of Use* - Laymen should be able to operate the system intuitively and within little amount of time
- *Fast Calibration and Setup* - Simple and efficient setup and calibration procedures for the system
- *Natural and multi-modal interaction* - Using natural modes of communication and interaction like, e.g. using natural language, gestures or intuitive input devices like touch screens.
- *Appeal and Ergonomics* - General appearance, quality of movement, speech, articulation or HRI
- *Adaptivity / General Intelligence* - Dealing with uncertainty, problem solving, on-line learning, planning, reasoning
- *Robustness* - System stability and fault tolerance
- *General Applicability* - Solving a multitude of different realistic tasks

The *system properties* can not be benchmarked as directly as the *functional abilities*, because it is not possible to relate actual portions of the score to them. However, they are considered as integral and implicit part of the competition and the tests, because teams are required to provide their robot with these properties. We thus believe that ROBO-CUP@HOME tests allow to measure improvements in the *system properties* through improvements in the *functional abilities*.

3.2 Implementation

The competition is organized in a multi-stage system where teams perform from 5 to 10 tests. The tests are oriented along realistic and useful tasks for a domestic service robot. Each test evaluates certain key features, as shown in the next section.

Two types of tests exist: pre-defined tests, which are specified in terms of the task to solve and the scoring; open tests, in which teams can either freely choose what to show (the Open Challenge and the Finals) or a topic is given according to which teams can do a demonstration (Demo Challenge). The following pre-defined tests were implemented in the 2008 competitions:

- Fast Follow: A person guides the robot through the scenario using voice and gesture commands.
- Lost&Found: Find certain objects in the scenario.
- Fetch&Carry: Find and bring and object to the user.
- Who's Who: Find,remember and distinguish unknown persons.
- Partybot: Find persons, receive orders and serve a drink.
- Supermarket: An unknown user has make the robot to retrieve certain objects from a shelf.
- Walk&Talk: Teach in locations in an unknown environment by showing the robot around.
- Cleaning up: Detect and arrange unknown objects on the floor.

The scoring in the pre-defined tests is oriented along the key features mentioned earlier, while the scoring in the open tests is based on an evaluation by a jury and on a list of criteria along which the jury should evaluate.

Scoring for the pre-defined tests uses a *partial score system*, in which a team receives a part of the total score for showing a part of the task's specification. Each of the partial scores is connected to one or more of the functional abilities and/or system properties. This does not only allow for assessing the fulfillment of these features individually, but it is also an incentive for teams to participate in a test even if they know that they cannot solve it completely.

4 Evaluation of Results and Discussion

One important objective for an annual scientific competition is to provide a common benchmark to many teams that allows for measuring the advances of performance over time and to develop relevant scientific solutions and results. In this section we describe and discuss the results obtained by the ROBOCUP@HOME teams both in terms of performance in the tests and in terms of scientific achievements.

4.1 Representation of Key Features in the Benchmarks

The score system of ROBOCUP@HOME allows for directly relating the desired abilities of the robots with scores that are gained during the competition and adapting future benchmarks accordingly.

Table 1 relates the *functional abilities* defined in Section 3.1 with the pre-defined tests described above. It quantifies the maximum score distribution per test with respect to the contained functional abilities. For ease of notation, we use abbreviations

Table 1. Distribution of test scores related to functional abilities

Test	Nav	Map	PRec	PTrk	ORec	OMan	SRec	GRec	Total
FF	550	0	0	450	0	0	0	0	1000
FC	375	0	0	0	150	400	75	0	1000
WW	350	0	550	0	0	0	100	0	1000
LF	550	0	0	0	450	0	0	0	1000
PB	1000	0	700	0	0	300	0	0	2000
SM	0	0	0	0	400	1000	200	400	2000
WT	918	416	0	250	0	0	416	0	2000
CL	1000	0	0	0	550	450	0	0	2000
Tot	4743	416	1250	700	1550	2150	791	400	16000

as follows. For the tests we have Fast Follow (FF), Fetch & Carry (FC), Who is Who (WW), Lost & Found (LF), PartyBot (PB), Supermarket (SM), Walk & Talk (WT), and Cleaning Up (CL). The abilities are Navigation (Nav), Mapping (Map), Person Recognition (PRec), Person Tracking (PTrk), Object Recognition (ORec), Object Manipulation (OMan), Speech Recognition (SRec), and Gesture Recognition (GRec).

Since the competition involves mobile robots, navigation is currently the most dominant ability represented in the score. Object manipulation and recognition also play an important role since service robots are useful if they can effectively manipulate objects in the environment. Person recognition, tracking, and speech/gesture recognition are needed to implement effective human-robot interaction behaviors. As gesture recognition was introduced as a new (and optional) ability in 2008, its weight in the total score still is comparably low. Finally, mapping plays a more limited role: such an ability is used in the Walk & Talk test, where the environment is completely remodeled during the test, so the robot enters in an unknown environment, while for other tests only minor modifications of the environment are done right before the tests and thus pre-computed maps (either built off-line by the robot or manually drawn) can be used.

It is important to observe that these values have been chosen after discussion among the members of the Technical Committee, taking into account the feedback from the teams. Consequently, the values implicitly contains a compromise between pushing towards new capabilities and rewarding more difficult functionalities (Technical Committee) and measuring current performance of the robots (feedback from teams). It is even more important to notice that our benchmarking approach is not to look for an optimal set of values, but to make them evolve over time in order to gradually improve the performance of DSR.

This table is important in order to define the weight of each ability in a test and in order to distribute the abilities among the tests. Furthermore, one can actually measure and analyze the performance of the teams and the difficulty of the tests after a competition, allowing for an iterative and constant development of the benchmarks.

Similar relations between system properties and the tests exist. However, this relation can not be quantified in scores as easily, as the system properties are of more implicit meaning for the tests.

System properties are instead represented in the general rules, in overall require-ments, and special properties in certain tests. By using laymen to operate the robots in the Supermarket test, the Who is Who test, and the PartyBot test, *Ease of Use* (EUse) is fostered. The restrictions on setup time and procedures demands for *Fast Calibration and Setup* (FCal). *Natural Interaction* (NInt) and *Multi-modal input* is currently re-warded in the Supermarket test and by the common use of speech and gestures. *Appeal and Ergonomics* (App) are part of the evaluation criteria in the Introduce test, the Open Challenge, and the Finals. *Adaptivity* (Adap) is especially requested in the Cleaning Up test. The limited amount of specifications in the tests and the environment and the fact that persons who interact with the robot are chosen randomly in many tests demands for *Robustness* (Rob). Finally, a team can only reach the *Finals* if their robot performs well in many tests with different tasks to solve. This stimulates the claim for *General Applicability* (GAppl).

4.2 Analysis of Team Performance

In the following, we analyze the performance of the teams in these abilities during ROBOCUP@HOME 2008 competition.

Table 2 presents the scores actually gained by the teams during the competition and the percentage with respect to the total score available, related to each of the desired abilities. The third column shows the best result obtained by some team, while the fourth one is the average of the results of the five finalist teams. This table allows for many considerations, such as: 1) which abilities have been mostly successfully imple-mented by the teams; 2) how difficult are the tests with respect to such abilities; 3) which tests and abilities need to be changed in order to guide future development into desired directions.

From the table it is evident that teams obtained good results in navigation, speech recognition, mapping, and person tracking. Notice that the reason for a low percentage score in navigation is not related to inabilities of the teams, but to the fact that part of the navigation score was available only after some other task was achieved. Speech recognition worked quite well, especially considering that the competition environment is much more challenging than a typical service or domestic application due to a large

Table 2. Available points for the desired abilities

Ability	Available points	Achieved score [max]	Achieved score [avg]
Navigation	4743 (40%)	1892 (40 %)	1178 (25%)
Object Manipulation	2150 (18%)	75 (3%)	15 (1%)
Object Recognition	1550 (13%)	450 (29%)	125 (8%)
Person Recognition	1250 (10%)	400 (32%)	190 (15%)
Speech recognition	791 (7%)	692 (87%)	293 (37%)
Person Tracking	700 (6%)	700 (100%)	570 (81%)
Mapping	416 (3%)	416 (100%)	183 (44%)
Gesture recognition	400 (3%)	0 (0%)	0 (0%)
Total	**12000 (100%)**	**4909 (41%)**	**2554 (21%)**

amount of people and a lot of background noise. The good achievements in mapping and person tracking may instead be explained by a limited difficulty of the corresponding tasks in the tests.

On the other hand, in some abilities, teams were not very successful. Object manipulation is a hard task, specially when an object is not known in advance and calibration time is limited or null. Because of the large proportion of score available, many teams have attempted manipulation but only a few were successful. A similar analysis holds for object and person recognition, that reported slightly better results with the same difficulties arising from operating under natural environment conditions (i.e., lighting) with small or null calibration time. Finally, gesture recognition has not been implemented by teams, probably for the small amount of points available.

An evaluation of system properties is more complicated since they are difficult to quantify precisely. Our current approach is to test for system properties through general requirements and by enforcing the combination of functional abilities.

An analysis of these results is very helpful for the future development of the @HOME competition. It gives direct, quantitative feedback on the performance of the teams with respect to the key abilities and tasks. This allows us to identify abilities and respective tests which need to be modified, to adjust the weights of certain abilities with respect to the total score. Possible modifications involve:

– Increasing the difficulty if the average performance is already very high
– Merging of abilities into high-level skills, more realistic tasks
– Keeping or even decreasing difficulty if the observed performance is not satisfying
– Introducing new abilities and tests

As the integration of abilities will play an increasingly important role for future general purpose home robots, this aspect should be especially considered in the future competition.

In addition to the analysis of the last competition we have conducted an analysis of presence and performance of teams over the years. Since 2006, a total of 25 teams (12 from Asia, 8 from Europe, 4 from America, 1 from Australia), have participated to the three editions of the annual ROBOCUP@HOME world championship. The percentage of @HOME teams in the RoboCup increased from 2.7% in 2006 to 3.7% in 2008. For RoboCup 2009 we expect 23 teams from 14 countries.

Moreover, it is interesting to notice that some teams adapted their robots designed and built for other RoboCup Leagues to compete in @HOME and that one team in 2006 and 2007 used the same robot in both the soccer Four-Legged and @HOME leagues and one team in 2008 used the same robot in both the Rescue and @HOME leagues.

Another important parameter to assess the results of an annual competition is the increase of performance of the teams over the years. Obviously, it is difficult to determine such measure in a quantitative way: the constant evolution of the competition with its iterative modification of the rules and of the partial scores do not allow a direct comparison.

However, it is possible to define some metrics of general increase of performance. In Table 3, the first row holds the percentage of unsuccessful tests, i.e., tests where no score was achieved at all, dropping from 83% in 2006 to 41% in 2008. The second row shows the increase in the total number of tests per competition. The third row holds the average number of tests that teams participated in successfully (i.e., with a

Table 3. Measures indicating general increase of performance

Measure	2006	2007	2008
Percentage of 0-score performance	83%	64%	41%
Total amount of tests	66	76	86
Avg. number of succ. tests p. team	1.0	2.5	4.9

non-zero score). The enormous increase from 1.0 tests in 2006 to 4.9 in 2008 is a strong indication for an average increase in robot abilities and in overall system integration.

4.3 Scientific Achievements

Besides numerical analysis about performance in the tests, relevant scientific achievements have been obtained by teams participating in the competition. Regarding the evolution of robot hardware and software architectures, we found special focus on Human-Robot-Interaction (e.g. [13]), on personal assistive robots (e.g. [14]) and on high level programming for domestic service robots (e.g. [15]). Regarding specific functionalities, speech recognition evolved from artificial and unnatural interaction with headset and portable laptop (2006-2007) to speaker independent speech recognition with effective noise cancellation using on-board microphones (2008) [16][17]. Face recognition has been made robust in presence of spectators standing at the border of the scenario [17,18] and tuned for real-time use [19]. Object recognition under natural and dynamic light condition has improved significantly: Techniques using different feature extractors and matching procedures have been tested (e.g. [20]), reaching a level in which the robot can reliably memorize an object shown by a user (by holding it in front of the robot) and then recognize it among several others (2008). Gesture detection and recognition has also been studied in order to communicate with the robot, using an effective approach based on active learning [21]. Finally, object manipulation has evolved from gathering a newspaper from the floor (2006), to grasping cups from a table (2007), and grasping different objects on various heights (2008).

5 Conclusion and Outlook

This paper presented the ROBOCUP@HOME initiative as a community effort to iteratively develop and benchmark domestic service robots through scientific competitions, by evaluating robot performance in a realistic, complex and dynamic environment. Starting with the first competition in 2006, the overall development of the initiative with respect to the increase of performance, the growing community, knowledge exchange and public awareness is very promising. @HOME has become the largest international competition for domestic service robots with currently 5 national competitions in China, Japan, Germany, Iran and Mexico besides the annual world championships.

The future development of the @HOME competition is highly iterative, as it involves constant feedback from the community, adjustments on the focus of desired abilities

---

[17] Best Student Paper Award at RoboCup International Symposium 2008.

and changes of the rules. Tests, functional abilities and desired system properties will evolve over the years and will be combined to form more realistic high-level tasks. At the moment a focus lies on physical and sensory capabilities, such as manipulation, human recognition and navigation. In the future, more focus will be put on Artificial Intelligence, high-level autonomy and mental capabilities in the context of HRI. This includes situation awareness, online learning, understanding and modeling of the surrounding world, recognizing human emotions and having appropriate responses. Moreover, as one of the main issues for domestic robots is their safety, we will consider in the future also evaluation methods for robot safe operation in domestic environments.

Still, concrete goals are necessary as they help to identify and to approach specific problems in the large real-world problem space in a structured way.

Rule changes in 2009 will involve an increased focus on HRI, e.g combined use of speech and gestures, robot operation by laymen or following previously unknown persons. Application scenarios will become more realistic, e.g. the demo challenge will involve robots serving drinks and food at a real party setting involving many people unfamiliar with the robots.

Mid-term goals include the search, identification, design and use of a common robot software architecture or framework to better exchange and reuse of software components already developed in the community and beyond. The same holds true for hardware, where companies or groups with relevant hardware components like sensors, actuators, or even standard robot platforms will be identified and asked to join and to support the community. Gradually testing the robots in the real world like e.g. going shopping in a real supermarket or taking the public transport is another mid-term goal.

The future @HOME scenario will contain more ambient intelligence, which the robots can interact with. The use of the Internet as a general knowledge base, communication with household devices, TVs, or external video cameras are some examples. Moreover, usability, safety and appearance of the robots will be of higher importance if one wants to increase their public acceptance.

References

1. van der Zant, T., Wisspeintner, T.: RoboCup@Home: Creating and Benchmarking Tomorrows Service Robot Applications. In: Robotic Soccer, pp. 521–528. I-Tech Education and Publishing (2007)
2. del Pobil, A.: Why do We Need Benchmarks in Robotics Research? In: Proc. of the Workshop on Benchmarks in Robotics Research, IEEE/RSJ International Conference on Intelligent Robots and Systems (2006)
3. Sabanovic, S., Michalowski, M., Simmons, R.: Robots in the wild: observing human-robot social interaction outside the lab. In: 9th IEEE International Workshop on Advanced Motion Control, pp. 596–601 (2006)
4. Balch, T., Yanco, H.A.: Ten years of the aaai mobile robot competition and exhibition: looking back and to the future. AI Magazine 23(1), 13–22 (2002)
5. Kitano, H., Asada, M., Kuniyoshi, Y., Noda, I., Osawa, E., Matsubara, H.: RoboCup: A Challenge Problem for AI. AI Magazine 18(1), 73–85 (1997)
6. The RoboCup Federation: RoboCup (2008), http://www.robocup.org/ (Retrieved 09/2008)

7. Meeden, L., Schultz, A.C., Balch, T.R., Bhargava, R., Haigh, K.Z., Bohlen, M., Stein, C., Miller, D.P.: The aaai 1999 mobile robot competitions and exhibitions. AI Magazine 21(3), 69–78 (2000)
8. Kitano, H., Tadokoro, S.: RoboCup Rescue: A Grand Challenge for Multiagent and Intelligent Systems. AI Magazine 22(1), 39–52 (2001)
9. Steinfeld, A., Fong, T., Kaber, D.B., Lewis, M., Scholtz, J., Schultz, A.C., Goodrich, M.A.: Common metrics for human-robot interaction. In: Proc. of HRI, pp. 33–40 (2006)
10. Yanco, H.A., Drury, J.L., Scholtz, J.: Beyond Usability Evaluation: Analysis of Human-Robot Interaction at a Major Robotics Competition. Human-Computer Interaction 19, 117–149 (2004)
11. Drury, J.L., Yanco, H.A., Scholtz, J.: Using competitions to study human-robot interaction in urban search and rescue. Interactions 12(2), 39–41 (2005)
12. Prassler, E., Hägele, M., Siegwart, R.: International Contest for Cleaning Robots: Fun Event or a First Step towards Benchmarking Service Robots (2006)
13. Savage, J., Ayala, F., Cuellar, S., Weitzenfeld, A.: The use of scripts based on conceptual dependency primitives for the operation of service mobile robots. In: Proceedings of the International RoboCup Symposium 2008 (CD-ROM Proceedings) (2008)
14. Ruiz-del-Solar, J.: Personal robots as ubiquitous-multimedial-mobile web interfaces. In: Proc. of 5th Latin American Web Congress (LA-WEB), pp. 120–127 (2007)
15. Schiffer, S., Ferrein, A., Lakemeyer, G.: Football is coming Home. In: Proc. of International Symposium on Practical Cognitive Agents and Robots (PCAR 2006). University of Western Australia Press (2006)
16. Doostdar, M., Schiffer, S., Lakemeyer, G.: A Robust Speech Recognition System for Service-Robotics Applications. In: Iocchi, L., Matsubara, H., Weitzenfeld, A., Zhou, C. (eds.) RoboCup 2008. LNCS (LNAI), vol. 5399, pp. 1–12. Springer, Heidelberg (2009)
17. Correa, M., Ruiz-del-Solar, J., Bernuy, F.: Face recognition for human-robot interaction applications: A comparative study. In: Proceedings of the International RoboCup Symposium 2008 (CD-ROM Proceedings) (2008)
18. Knox, W.B., Lee, J., Stone, P.: Domestic interaction on a segway base. In: Proceedings of the International RoboCup Symposium 2008 (CD-ROM Proceedings) (2008)
19. Belle, V., Deselaers, T., Schiffer, S.: Randomized trees for real-time one-step face detection and recognition. In: Proceedings of the 19th International Conference on Pattern Recognition (ICPR 2008), December 8-11. IEEE Computer Society, Los Alamitos (2008)
20. Loncomilla, P., Ruiz-del-Solar, J.: Robust object recognition using wide baseline matching for robocup applications. In: Visser, U., Ribeiro, F., Ohashi, T., Dellaert, F. (eds.) RoboCup 2007. LNCS (LNAI), vol. 5001, pp. 441–448. Springer, Heidelberg (2008)
21. Francke, H., Ruiz-del-Solar, J., Verschae, R.: Real-time hand gesture detection and recognition using boosted classifiers and active learning. In: Mery, D., Rueda, L. (eds.) PSIVT 2007. LNCS, vol. 4872, pp. 533–547. Springer, Heidelberg (2007)

Connecting the Real World with the Virtual World - Controlling AIBO through Second Life

Evan Wong[1], Wei Liu[1,*], and Xiaoping Chen[2]

[1] School of Computer Science and Software Engineering
University of Western Australia, Crawley, WA 6009
[2] Department of Computer Science
University of Science and Technology China, Anhui, China
wei@csse.uwa.edu.au

Abstract. The main aim of this project is to develop middleware so that the Second Life online virtual space (virtual world) can be used to simulate and control the movements of a Sony AIBO robot (real world) in a wireless environment. This paper details the design of an immersive teleoperation system, and the rationale behind the design. The prototype proves that the concept of teleoperation with greater sense of immersion is achievable and can lead to future work in application domains such as smart home and immersive remote operating machinery in the industry such as mining.

1 Introduction

The Internet, as a communication backbone of modern society is further exploited, in the past decade, in the scenarios of *teleoperation*[1], to ensure the safety of personnel in high-risk industries such as mining, aerospace and defence. Broadly speaking, teleoperation means controlling and operating a device remotely by an operator from afar.

Benali et al. [1] identified that teleoperation over the Internet faces major challenges due to the fact that there is no guaranteed upper bound to the potential large time delay and the consequent data loss. Therefore, in the past, researchers are mainly focusing on addressing the network delay issues. Benali et al. [1] proposed a system with a network analyser. It measures the packet round-trip time for evaluating the quality of service, and employs a control mode manager to make decisions on whether to continue a given task or not. Xue et al. [2] attempts to guarantee a stable data stream by using a handshaking protocol between the server and the client. Both Xue et al. and Benali et al. concluded that the time delay can be managed and resolved through either human intervention [1] or using a fuzzy controller triggered only by a sensor event [2].

Therefore, this project takes the manual control approach to counter the effect of the network delay. This allows us to centrate on the idea of providing

---

* Corresponding Author.

[1] Telelabs Project: http://telerobot.mech.uwa.edu.au/information.html

J. Baltes et al. (Eds.): RoboCup 2009, LNAI 5949, pp. 402–413, 2010.

a higher level immersion for what we termed as *immersive teleoperation*. It is widely agreed that, comparing to the traditional human computer interface with buttons and drop-down menus, being able to operate machinery in an immersive 3D virtual environment will enhance job satisfaction, which is especially attractive to trainees.

The overall aim of this project is to use a popular and accessible 3D virtual online environment - Second Life as a medium to teleoperate the Sony AIBO (short for **A**rtificial **I**ntelligence ro**BO**t), more specifically the Sony AIBO ERS-7 robot. Second Life (SL), created by Linden Lab, is a 3D online virtual environment that attempts to model the surface of an Earth-like world in a reasonably life-like way [3]. Users create models or avatars, which are essentially the "people" of the virtual world. The Sony AIBO is an artificial intelligence robotic pet dog designed and manufactured by Sony. The AIBO is an autonomous agent able to gain information about the environment and make decisions without human intervention [4]. The AIBO also has a built-in wireless adapter receiving and transmitting data wirelessly [5]. The joints as well as the vision, acoustic and motion sensors can be accessed directly via the OPEN-R platform provided by Sony [6] through programming in C++. This allows the AIBO to be programmed to perform specific behaviours. The design objective of this project is that users can control the movements of the AIBO in a physical environment through moving an avatar in Second Life.

This paper reports the design and development of such a system with technical details on how to intercept Second Life packets and how the different modules in the system communicate. We hope as a preliminary system, this paper can offer some insights and starting points for similar projects focusing on remote machinery operations via immersive virtual environment. Section 2 provides a general overview of existing architectures adopted by the state of the art teleoperation systems. In Section 3, we propose the design and development of the system with detailed discussions on the roles of each module and how they are implemented and communicate with each other. Section 4 presents experiments results. The paper concludes with an outlook to future work in Section 5.

2 Related Work

2.1 General System Architecture

The system architectures widely adopted are variations on the Internet-based client-server structure [2,7,8]. The control architectures typically comprise of a *central main controller* sitting on the server that talks to *embedded controllers* within the robot. For example, the architecture by Bensoussan et al. [9] uses two controllers; a central controller on the server computer and a real-time controller which sits on the robot (in the context of their paper, the robot is a vehicle). The central controller manages the sensors, camera, ultra-sound, radio communication; the real-time controller in the robot, on the other hand, controls of locomotion of the vehicle. Benali et al. [1] and Xue et al. [2] designed the controllers as classical fuzzy controllers using fuzzy logic. Hohl et al. [10] take this one step further by introducing the notion of cross-compilation of the controller so that the robot can run independently of the host computer once the main controller is compiled.

To enable teleoperation by human operators, a common practice is to use a graphical user interface (GUI) that sits on the client computer as an desktop or a Web application. This GUI is usually coded as a Java applet with VRML (Virtual Reality Modelling Language). VRML is a language that can be used to visualise and build virtual worlds which include 3D objects, light sources, animations, and user representation via avatars [7]. This allows for manual control of both the simulated and the real robots. The idea is that the real robot will mimic whatever that is shown on the screen by having the GUI communicating with the controllers on the server and the robot. This mode of manual control is what we adopted for this project. However, it can be easily extended to various levels of control, including, semi-automatic, fully automatic [2], and a hybrid of all three [1]. It is also noted that it is possible to use multiple hosts simulating and controlling the robot concurrently [8].

2.2   Second Life Networking Architecture

Second Life implements a client-server architecture. Separate servers are employed to handle different tasks, including Login Server, Data Server, Map Server and etc. Among these, the servers of interest to us are the *simulators*.

The world of Second Life is made up of many *simulators*. Simulators are basically servers that simulate a 256x256 metre region each. When the avatar in Second Life moves through the Second Life world, it moves from one simulator to another [11]. The simulator keeps track of where everything is and sends the location of objects in Second Life to the client (a.k.a. viewer). The simulator is in charge of running the physics engine in the Second Life world and it does collision detection as well.

The *viewer*, on the other hand, does not handle any collision detection. It sends velocities and simple physics information to the simulator, keeping track of avatar movements. When collisions occur, updates are sent from the simulator to the viewer, which is then updated accordingly on the viewer [11].

All these communication amongst the simulators as well as the communication between the simulators and viewers are done via UDP through *circuits*. A circuit is basically a two-way UDP connection between two nodes.

Packets transmitted on Second Life between the client and the server have a consistent layout. A packet is divided into three parts: the header, body and the appended acks. The *header* contains information regarding the packet itself [12]. The *body* of a Second Life packet contains the message number, which is a numeric encoding of the message types, followed by the actual message data [12]. For example, the message storing information about the movement of the avatar is an `AgentUpdate` message. Finally, the packet might have acknowledgements appended. Acknowledgments from previous reliable messages "piggyback" and fill up as much of the packet as it can fit [12].

3   System Design and Development

3.1   System Overview

As show in Figure 1, our design follows the widely adopted client-server
architecture discussed in Section 2.
The GUI is, in this case, the Second
Life client installed on a client computer. The only difference is that the
central main controller (the VRInterface in Figure 1) resides in the client
computer, rather than a separate
server machine. This is because the
Linden Lab's Second Life simulator
servers are dedicated and close source,
unlike the systems discussed in the
literature, which have servers located
close the robot to minimise extra net-

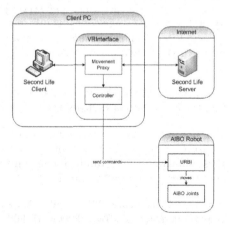

Fig. 1. System Overview

work delay. The embedded real-time controller we choose to use is the URBI
platform, which is loaded to the programmable memory stick on AIBO. Detailed discussions of each module are available in the following subsections.

3.2   Virtual Real Interface - VRInterface

The core module of our system that enables intercepting Second Life packets
and sending commends to the robot is the VRInterface. It consists of two sub-
modules, the MovementProxy and the Controller.

Movement Proxy is an application we developed based on a third party
software library: the SLProxy of the libsecondlife libraries. It sits in between
the Second Life client and the Second Life servers, analysing the packets that
are being transmitted along the stream.

SLProxy is a library that allows applications to act as a proxy between the
Second Life client and the servers. SLProxy tracks the circuits, modifying the
sequence numbers and acknowledgments when changes are made to the packet
stream from a third-party application that makes use of the SLProxy library [13].
Therefore, applications using the SLProxy library can read, inspect and modify
any packet that is transmitted and received between the Second Life client and
the servers. It can also remove packets or inject new packets into the packet
stream [13].

The Movement Proxy shown in Figure 1 only reads the packets but theoretically, based on the capabilities of the SLProxy library, it can also be used to inject
packets back into the Second Life client. Injecting packets can be used to realise
scenarios such that the Second Life avatar moves to resemble the autonomous
movement of the AIBO in the real environment.

Movement in Second Life is handled by both the client and the servers. The
client sends packets which contains information of the movements of the avatar

to the servers up to 20 times per second [14]. The packet that we are interested in is the `AgentUpdate` message [2]. The `AgentUpdate` message contains information from the camera in the Second Life viewer which is sent to the simulator [15] at a rate of up to 20 times per second. The `ControlFlags` variable inside the `AgentUpdate` message contains information about the movements of an avatar in Second Life. Each movement is given a constant integer which is used to shift 0x1 by the given constant. As an example, for a movement in the forward (positive) direction,

```
const U32 CONTROL_AT_POS_INDEX  = 0;
const U32 AGENT_CONTROL_AT_POS  = 0x1 << CONTROL_AT_POS_INDEX;
```

The `ControlFlags` variable will contain 0001, which is the decimal 1.

However, a "nudge" is sent by the client first, followed by the normal key press shown in the example above. This "nudge" is to impart velocity when brief keypresses are made by the client [14]. For a "nudge" in the forward direction,

```
const U32 CONTROL_NUDGE_AT_POS_INDEX  = 19;
const U32 AGENT_CONTROL_NUDGE_AT_POS  = 0x1 << CONTROL_NUDGE_AT_POS_INDEX;
```

The `ControlFlags` variable will contain 1000 0000 0000 0000 0000, which is the decimal 524288.

If the forward key continues to be held down after the "nudge", Second Life will store a combination of the forward movement with a "fast" movement (the avatar is walking faster) into the `ControlFlags` variable.

```
const U32 CONTROL_AT_POS_INDEX  = 0;
const U32 CONTROL_FAST_AT_INDEX = 10;
const U32 AGENT_CONTROL_AT_POS  = 0x1 << CONTROL_AT_POS_INDEX;
const U32 AGENT_CONTROL_FAST_AT = 0x1 << CONTROL_FAST_AT_INDEX;
```

The `ControlFlags` variable will now contain a combination of the forward movement and the "fast" movement 0100 0000 0001, which is the decimal 1025.

What this allows for is that a combination of keyboard presses stored and transmitted within one packet. If, for example, the avatar makes a diagonal movement (forward and left), the two movements can be combined, which means less packets are required to be transmitted, thus reducing any unnecessary delays from repeatedly sending a packet to move one step forward, followed by a packet to turn left, and then yet another packet to move one step forward, yet another to turn left, and so on.

When a packet containing the movement information is sent to the corresponding server, the server computes the current position and transmits the position back to the client. As mentioned in Section 2.2, the client does not perform collision detection and it is the job of the simulator or server to do that task. The position of the avatar seen on the viewer is therefore velocity and acceleration interpolated [14]. This means that if there is a long delay in the network, the avatar in the Second Life viewer may appear to unrealistically walk through an obstacle such as a wall. The position of the avatar is then again

---

[2] For a template of the `AgentUpdate` message please see [15].

unrealistically "corrected" at the time when the viewer eventually receives the packets containing correct position calculated by the server, after long network delays. One will see that the avatar is pushed back (after walking through a wall) to reflect that there is an obstacle in front of it.

The **Controller**'s job is to analyse the packets from the Movement Proxy and translate the movements of the avatar in Second Life stored in the `ControlFlags` variable into commands that is understood by the AIBO robot.

To translate the movements of the avatar in the `ControlFlags` variable into commands understood by the AIBO robot, a lookup table is used to map the `ControlFlags` variable into URBI commands for the AIBO robot. URBI (Universal Real-Time Behavior Interface) is a software platform by Gostai supporting development of robotics and AI applications [16]. It is chosen in preference to other platforms such as Tekkotsu because it is a universal platform that works with not only the Sony AIBO robots, but also a variety of other robots, independent of operating systems and programming languages [17].

We implement this lookup table as a hash map data structure, in which the keys would be the possible values (unsigned 32-bit integers) that the `ControlFlags` variable might hold, while the values of the hash map would be the URBI commands the robot understands.

Table 1. Hash Map

Key	Value
1025	walk.go(1)
1026	walk.go(-1)
256	walk.turn(30)
512	walk.turn(-30)

Commands sent to the AIBO robot will be repeated and continuous if the three latest packets received in a row contain the same `ControlFlags` variable. If a new movement is made, the `ControlFlags` variable of the new packet will not be the same as the `ControlFlags` variable of the last two packets, forming the stopping rule for the robot. A stop command is then sent to the robot followed by commands for the new movement.

In order to send commands to the AIBO through URBI, a new socket connection is established at start-up between the Controller and the AIBO robot on port 54000, which is the port for URBI commands on the robot. TCP is used as the protocol for the connection between the Controller and the AIBO robot due to its more reliable nature as compared to UDP.

3.3   Creating a Dog-Like Avatar in Second Life

The default avatar in Second Life is a basic human avatar. In order to create a closer representation of the AIBO robot in the Second Life virtual environment, a dog-like avatar is created. Objects in Second Life are created from basic primitive objects such as a cube, cylinder, prism, pyramid etc. These objects can be transformed by stretching, shrinking and then put together to form a larger object. Using the primitive objects provided in Second Life, objects such as arms, legs, chest, head and so on can be created and then attached onto the avatar to form a dog-like avatar, thereby creating a closer representation of the AIBO robot in Second Life.

3.4 Communications between the Modules

Figure 2 illustrates the sequence of events and interactions that occur between the different services when the Movement Proxy is started up and connections are established with the Second Life server and the Sony AIBO robot.

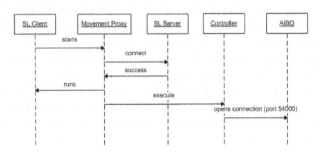

Fig. 2. Sequence of Events at Startup

The sequence of events and interactions between the services during runtime is quite similar to those during execution with slight differences being that the services in the system perform different actions.

4 Experiments and Results

4.1 Experiment Environment and Settings

The experimental environment was kept stable by ensuring the following parameter stayed as constant as possible throughout the experiments:

The speed of the Internet and networks cannot remain constant and fluctuates based on the amount of traffic on the network. Long network delay can result in lag and potential loss of packets. Therefore, in cases when the Internet speed drops to a level deemed unsuitable, the experiments are abandoned until speeds are improved and deemed suitable again.

Table 2. Experimental Environment

Robot:	Sony AIBO ERS-7
Robot Platform:	URBI
Workstation:	Intel Core 2 Duo 2.20Ghz processor
	2GB of memory
	NVIDIA GeForce 8400M GS video card
Workstation Platform:	Windows XP Professional
Second Life Version:	Release 1.19.0(5)
Wireless Network Protocol:	802.11b
Internet Speed:	> 1.5Mbps

4.2   Setting Up The Movement Proxy

To allow the Movement Proxy to capture and analyse data being transmitted between the client and the server, the client is set to connect to the server through the Movement Proxy, which is on the same workstation as the Second Life client through port 8080. An extra flag is added to the Second Life client to connect to the Second Life server through the Movement Proxy. This can be done by appending an `loginuri` when running a Second Life client:

```
C:\Program Files\SecondLife\SecondLife.exe -loginuri http://localhost:8080
```

This ensures that before running the Second Life client executable, the Movement Proxy is loaded up first. When the Movement Proxy is loaded up, it establishes a connection to the Second Life server. At the same time the Controller, which is compiled and executed together with Movement Proxy as part of the same executable file, opens a connection to the AIBO robot through URBI on port 54000. When everything is loaded up and the connections are established successfully, the AIBO robot will make a sound and the Movement Proxy will display that it has been loaded up successfully as shown in Figure 3.

Fig. 3. Movement Proxy

4.3   Movement Testing and Observations

To test the success of the system in terms of achieving the main objective of being able to teleoperate the AIBO robot in the real world and mirror the movements of its virtual representation in Second Life, basic movements were performed on an avatar in Second Life and results were observed on the AIBO robot. These basic movements performed are: Moving Forwards (Figure 4) and Backwards (Figure 5) in a straight line; Turning 90° to the Left (Figure 6) and to the Right (Figure 7).

The corresponding figures of each of the four basic movements mentioned above show frame-by-frame shots of the AIBO robot performing the movements with the Second Life avatar moving in the Second Life client visible in the background.

As part of the experiments to test the forward and backward movements, markers were placed on Second Life that were 10.0 Second Life units apart in a straight line (Y-axis). Markers were also placed some distance apart to form a rectangular region in the real world. The purpose of placing the markers in Second Life and the real world was to measure the closeness in the representation of the movements of the AIBO in the real world with that in the Second Life virtual environment. The coordinates of the markers in Second Life were:

Marker 1 X: 140.0 Y: 90.0 Z: 27.157

Marker 2 X: 140.0 Y: 100.0 Z: 27.157

The Z coordinate represents the distance of the marker from sea level in Second Life. Both the X and Z coordinates were unchanged while the difference between the Y coordinates between markers 1 and 2 was 10.0. A distance of 10.0 units in Second Life reflects a distance of approximately 0.6m in the real world.

As one can observe from Figure 4, while the avatar is moving closer to the identified marker in Second Life, the AIBO robot is also moving closer. In Figure 5, while the avatar is moving away from the identified marker in Second Life, the AIBO robot is also moving away from the markers. Based on the observations, the straight line movements (moving forwards and backwards) of the AIBO robot mimicked that of the avatar in Second Life very closely. This also includes faster and slower walking movements in a straight line.

In Figure 6 and Figure 7, using the identified stable object in Second Life, we can say that turning movements performed by the AIBO robot did mimic the turning movements made on the avatar in Second Life. However, it was not mimicked as close as we would have liked them to be. This might be due to the lack of detailed and fine calibrations of the turning angles in Second Life and in the real world. The representation of the AIBO robot as a 2-legged avatar and the fact that the AIBO is a 4-legged robot had some negative impact on accurately calibrating and representing the movements of the avatar in Second Life for the AIBO robot. If the bounding box of the avatar is proportional to the size of the robot in the real world, a much closer representation of the AIBO robot in Second Life can be achieved.

More complex movements such as the movement of individual joints and movement of the head were not performed due to the limitation of Second Life in creating more sophisticated avatar. This is further discussed in Section 5.

As stated in Section 4.1, the Internet speed was monitored to ensure that it is fast enough (> 1.5Mbps) so that the delays would not have a great impact on the experiments. However, given the dynamic nature of wireless networks and the Internet, occasional delays on the network imply that the latency cannot always stay constant. The experiments were aborted if the Internet speed is deemed to

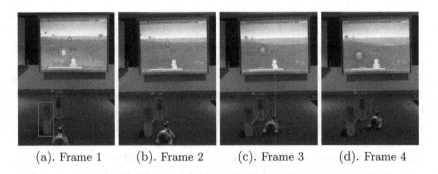

(a). Frame 1   (b). Frame 2   (c). Frame 3   (d). Frame 4

Fig. 4. Forward Movement - Walking Forward in a Straight Line

(a). Frame 1   (b). Frame 2   (c). Frame 3   (d). Frame 4

Fig. 5. Backward Movement - Walking Backward in a Straight Line

(a). Frame 1   (b). Frame 2   (c). Frame 3   (d). Frame 4

Fig. 6. Left Turn Movement - Turning 90° to the Left

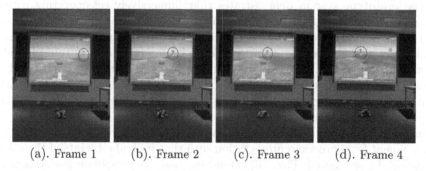

(a). Frame 1   (b). Frame 2   (c). Frame 3   (d). Frame 4

Fig. 7. Right Turn Movement - Turning 90° to the Right

be unacceptable to ensure that the Internet speed does not greatly affect the outcome of the experiments.

We measured that the average time delay from the time the avatar moves in Second Life and when the AIBO robot responds and starts moving is in a range of 500 milliseconds to 3 seconds. This range was obtained by sending ICMP packets ("pinging") to the Second Life server and the AIBO robot from the client workstation, in which the Second Life client as well as the Movement Proxy and Controller resides. The round trip time in which the packets were sent out by the client and when the response is received is averaged and recorded. This procedure was repeated 20 times during different times of the day to take into account the variation of network traffic during the course of a day.

The impact of delays on the system can be minimised by the use of a buffer that temporarily stores a set number of packets at a given time and then filtering the packets out to ensure that re-sent packets that were delayed do not interfere with the current packet stream that is being received, processed and sent to the `Controller` by the client.

5   Conclusions and Future Work

This project is a software engineering exercise to connect the real world (the Sony AIBO robot) with the virtual world (the Second Life virtual environment) in which movements made by the avatar in Second Life were to be mirrored in the real world through the Sony AIBO robot.

Based on the experiments and the results obtained, the main objective of the project has been met. The delay as noted above is within an acceptable range of 500 milliseconds to 3 seconds. This range could be made smaller and improved in the future. There are other 3D virtual environment servers (open source) available which can be installed locally or within an intranet to significantly reduce the delay. A buffer can also be implemented to ensure that delays are taken into consideration.

The prototype system demonstrates that immersively teleoperating robots through the Second Life virtual environment is feasible. Immersion for the operator is the key attractor of this type of 3D virtual environment. For example, a mining site could be replicated within Second Life and machines could be teleoperated through the Second Life environment. There is also growing trend in Second Life where houses are built to replicate the real houses of individuals. Through immersive teleoperation, users can control robots in their homes to carry out various tasks remotely.

In future work, we are planning to import an 3D model of AIBO into a 3D environment for a precise representation such that individual joints of the robot could also be controlled. This will extend the basic movements reported here to complex ones require the coordination of body parts such as sitting and dancing. We are also interested in feeding packets into Second Life so that the robot's autonomous movements in real life are mirrored on Second Life. The preliminary testing has been carried out successfully by injecting short chat messages through the proxy which is then shown in Second Life.

Acknowledgement. This work is supported by Australian Academy of Science and Chinese Academy of Science through the Scientific Visiting Awards, and the University of Western Australia Research Grant Scheme 2008.

References

1. Benali, A., Wasiak, V., Fontaine, J.G.: Remote robot teleoperation via internet. a first approach. In: Wasiak, V. (ed.) Proceedings. 10th IEEE International Workshop on Robot and Human Interactive Communication, 2001, pp. 306–312 (2001)
2. Xue, X., Yang, S.X., Meng, M.Q.H.: Remote sensing and teleoperation of a mobile robot via the internet. In: Yang, S.X. (ed.) 2005 IEEE International Conference on Information Acquisition, p. 6 (2005)
3. Ondrejka, C.R.: A piece of place: Modeling the digital on the real in second life. SSRN eLibrary (2004)
4. Flake, G.: The Computational Beauty of Nature: Computer explorations of fractals, chaos, complex systems, and adaptation. MIT Press, Cambridge (2000)
5. Sony Entertainment Robot Europe: Aibo brochure eng 2004 (2004), http://support.sony-europe.com/AIBO/downloads/en/brochure_04lr_en.pdf
6. Serra, F., Baillie, J.C.: Aibo programming using OPEN-R SDK Tutorial. ENSTA (2003)
7. Hoyer, H., Jochheim, A., Rohrig, C., Bischoff, A.: A multiuser virtual-reality environment for a tele-operated laboratory. IEEE Transactions on Education 47(1), 121–126 (2004)
8. Puente, S.T., Torres, F., Ortiz, F., Candelas, F.A.: Remote robot execution through www simulation. In: Torres, F. (ed.) Proceedings of 15th International Conference on Pattern Recognition, 2000, vol. 4, pp. 503–506 (2000)
9. Bensoussan, S., Parent, M.: Computer-aided teleoperation of an urban vehicle. In: Parent, M. (ed.) ICAR 1997. Proceedings of 8th International Conference on Advanced Robotics, 1997, pp. 787–792 (1997)
10. Hohl, L., Tellez, R., Michel, O., Ijspeert, A.J.: Aibo and webots: Simulation, wireless remote control and controller transfer. Robotics and Autonomous Systems 54(6), 472–485 (2006)
11. Second Life Wiki: Server architecture (2008), http://wiki.secondlife.com/wiki/Server_architecture
12. Second Life Wiki: Packet layout (2008), http://wiki.secondlife.com/wiki/Packet_Layout
13. axial: Slproxy - libsecondlife (2008), http://www.libsecondlife.org/wiki/SLProxy
14. Second Life Wiki: How movement works (2008), http://wiki.secondlife.com/wiki/How_movement_works
15. libsecondlife: Linden lab development message templates (2008), http://www.libsecondlife.org/template/release/1.19.1.4.txt
16. Wikipedia: Urbi (2008), http://en.wikipedia.org/wiki/URBI
17. Gostai: The urbi platform (2008), http://www.gostai.com/urbi.html

A Hybrid Agent Simulation System of Rescue Simulation and USARSim Simulations from Going to Fire-Escape Doors to Evacuation to Shelters

Masaru Okaya, Shigeru Yotsukura, and Tomoichi Takahashi

Meijo University, Tenpaku, Nagoya, 468-8501, Japan
{e0427566,e0527080}@ccalumni.meijo-u.ac.jp, ttaka@ccmfs.meijo-u.ac.jp

Abstract. Disaster & rescue simulations handle complex social issues, the macro level modeling of which is difficult. Agent-based social simulation provides a platform to simulate such social issues. It is ideal that the simulations cover various evacuation patterns and the results are used to make effective plans against disasters. This requires that the behaviors of a numbers of heterogeneous agents are simulated at urban size areas in hostile environments. Representing all buildings of the area by 3D model requires a large amount of computer resources and computing the behaviors of a number of agents takes a lot of computation time. These make it difficult to simulate rescue behaviors at disasters in real scale.

We propose a hybrid agent simulation system that switches systems that is suitable for situations during simulations. A hybrid system of two simulations with different time and space resolution makes it possible to simulate urban size human behaviors and indoor movements with less computational resources than doing by one system. This paper presents protocols that connect two systems that are used in RoboCup Rescue Simulation League, Rescue Agent Simulation and USARSim. The prototype system provides a simulation of people's evacuation from going to fire-escape doors to moving to shelters.

1 Introduction

The Great Hanshin-Awaji earthquake of 1995 led researchers to apply their technologies for decreasing damages from disasters. Subsequent disasters including the 9/11 on the World Trade Center of 2001, 2004 Indian Ocean earthquake, and 2008 Sichuan earthquake China have driven to start disaster & rescue related projects around the world. Several systems that support decision of rescue operations or prompt planning for disaster mitigation have been presented. Their functions are to ensure prompt planning for disaster mitigation, risk management, and support of IT infrastructures at disasters [1][4].

In RoboCup, the rescue agent competition league has started since 2001 using RoboCup Rescue Simulation (RCRS). RCRS was designed to simulate the rescue operations and disasters simultaneously at the Hanshin-Awaji earthquake disaster. In competitions, rescue agents contest their performances at various disasters

J. Baltes et al. (Eds.): RoboCup 2009, LNAI 5949, pp. 414–424, 2010.

situations on various cities. At 2005, virtual robot competition has started [5]. USARSim is a high fidelity simulator based on the Unreal Tournament game engine and has provided environments to develop models of new robotic platforms, sensors and test environments and to develop control algorithms that are seamlessly migrated to systems in fields [6]. Commander training systems or other simulation systems have been presented using the RoboCup system [7][3][8].

The last disasters have made the purposes of disaster & rescue simulations more clear. It includes the simulations are used as emergency management system of local governments and the disaster & rescue simulations need more functions to that end. For example, when disasters occur, an urban size simulation is used to deploy rescue agents at the first stage of rescue operations. After the agents arrive at sites, simulations of inside devastated houses are useful to search victims.

It requires a huge amount of computation power and resources to simulate the behavior of agents at wide area with fine resolutions. We propose a hybrid agent simulation system of RCRS and USARSim. USARSim simulates people's evacuation from going to fire-escape doors with fine resolution and RCRS simulates the behaviors of moving to shelters. Section 2 describes rescue scenarios using multi agent simulation systems (MAS). A framework of hybrid system that executes evacuation systems is described in section 3. Protocols to connect two MAS and to support communication among agents at different MASs are described in section 4. Section 5 shows the simulation results of our prototype system. The summary of our proposal and discussions are described in Section 6.

2   Rescue Scenarios Simulated by MAS

Disaster & rescue simulations handle complex social issues, the macro level modeling of which is difficult. MAS is good to simulate such issues and it is ideal that disaster & rescue simulations can simulate various evacuation patterns and rich human interactions to make effective plans against disasters. It requires followings, (1) simulation of behavior of a numbers of heterogeneous agents, (2) at building inside and urban size areas, (3) under hostile environments caused by disasters, (4) with interactions of others including rescue operations.

Table 1 shows one of rescue scenarios when people evacuate from buildings. The scenario consists of three stages.

A. *the initial stage of disasters:* People in buildings try to evacuate from houses. Rescue teams rush to devastated houses.
B. *rescue operations at devastated houses:* The rescuers execute their actions to save lives, fight fire and do related actions. They use robots to search and rescue victims from the houses.
C. *evacuation to shelters:* People who get out of the houses evacuate to shelters. Rescue headquarters allocate shelters and announce rescue teams.

The rescue scenario contains indoor and outdoor environments. Indoor or open space people behavior are simulated by free space model and the traffic of outside

Table 1. Rescue scenario where agents go in and out of buildings

simulation scenario		outside		inside
stage	behaviors of agents	RCRS		USARSim
A.	When disasters occur,	A, B		x, y, z
1	people evacuate from buildings,	X	⇐	x
2	call for help from inside house to rescue teams.	A	←	y
3	People inside buildings help and communicate each other.			y, z
4	Rescuers rush to the sites according to their headquarters.	A, B		
B. 1	The rescuers arrive at devastated houses.			
2a	Some rescue teams start fire-fighting.	A		
2b	Others start searching by robots,	B	→	r
3b	confirm conditions of rescue operations in the houses,	B	←	y, z
4b	execute search-and-rescue operations.	B	⇒	y, z
5	They enter the houses.	B	⇒	b
6	They communicate each other in houses,	(B	↔	b), y, z
7	or colleagues outside.	A	↔	b
C. 8	All move to outside and evacuate to refuges.	B, Y, Z	⇐	b, y, z

Capital letter and low case letter represent agents in RCRS and USARSim, respectively.
⇐, ⇒ show agents' transfer to the other system,
←, → represents communication among agents.

movements are simulated on a road network. Representing all buildings in three dimensional (3D) models and computing the behaviors of a number of agents require a large amount of computer resources and take a lot of computation time.

Our hybrid system can simulate (1) behaviors of indoor environments by USARSim, (2) the evacuation behaviors after exiting the buildings or going to refuges by RCRS. Figure 1 shows our idea of combining RCRS and USARSim. The two systems are agent based systems (ABS) with different resolutions of space and time. RCRS handled two dimensional (2D) urban size simulations of disaster & rescue operations and USARSim handles rescue robot motions at 3D buildings. Table 2 shows properties of RCRS and USARSim.

3   A Hybrid System to Execute Evacuation Scenarios

3.1   Requirements for a Hybrid Agent System

Followings are the RCRS commands of rescue agents to move into a building or to do rescue operations in it.

AK_MOVE: An agent submits it to enter into a building.
AK_EXTINGUSH: A firefighter submits it to extinguish fires.
AK_RESCUE: An ambulance team submits it to rescue a buried humanoid.
AK_LOAD: An ambulance team submits it to load an injured humanoid.

These commands are executed in one cycle of RCRS, and they don't reflect the facts that rescue operations change according to inside disaster situations. It

Fig. 1. Image of simulation systems that cover from wide rage disasters and local devastated houses. (The left shows the snapshots of RCRS simulation (a), a video picture of a building (b), and USARSim simulation of the building (c). The right shows an image of hybrid system. RCRS simulates 2D world, USARSim simulates 3D world, respectively.)

Table 2. Comparison of two rescue simulations

items	RCRS	USARSim
purposes	planning of disaster prevention, verification of rescue plans	providing platforms to develop rescue robot
agent type	civilian, rescuer humans	rescue robot
agent number	$O(100)$	$O(10)$
area size	$O(km^2)$	O(house size)
disaster simulations	fire, collapse building	–
simulation time	72 hours	real time
map model	2D network*	3D model

*: Dwelling environments vary from country to country. For example, our town has 135,000 and 120,000 inhabitants at daytime and night respectively, 70,000 households, and is $21.6km^2$. The 2D road network has 6,000 nodes and 4,000 edges.

is difficult to simulate the rescue operations inside houses at the resolution of RCRS. We employ USARSim in order to simulate the indoor rescue operations in more detailed way than RCRS. USARSim also simulates the inside evacuation behaviors of agents.

USARSim is a real-time simulator and RCRS was originally designed to simulate situations of 72 hours after earthquakes occur within a specified time. Followings are required to reflect the simulation results of USARSim and RCRS each other.

– *synchronization between simulation systems with different Scales:*
  The progress of simulation is paced by wallclock time [2]. The simulation step is mapped to wallclock time by the following formula.

$$T^s_{present} = T^s_{start} + Scale \times (T^w_{present} - T^w_{start}) \tag{1}$$

where T^s, T^w are simulation time and wallclock time, respectively. *Scale* is a factor that shows how fast or slow the simulation advances the wallclock time, for example, $Scale = 2$ indicates that the simulator runs twice as fast as wallclock time.

- *communications among agents that are in different ABSs:*
  Rescue agents communicate each other to cooperate with properly. When some rescue teams enter to devastated buildings, they report the inside situations to their commanders outside or receive orders from them. Communications among agents are supported whether the agents are in RCRS or in USARSim.

- *management of agents when they move to a different ABS:*
  The kernel and USARSim server manage the status of which agents are connected to themselves. When the agents enter to or exit from a building, they switch connection from RCSR to USARSim or vice verse, respectively. The servers change the data of the connected agents.

- *reflection of a disaster to other ABSs:*
  RCRS simulates disasters such as aftershocks or fire, and rescue actions such as firefighting. These change the situations of houses where USARSim provides to its agents. Reflecting these changes to USARSim makes USARSim environments dynamic ones.

3.2   Protocols Systems for Hybrid System

New protocols are designed to connect RCRS and USARSim and to enable agents to switch servers. Agents consist of parts that connect RCRS and USARSim servers. They use following protocols in addition to the original ones. Table 3 shows the protocols newly added.

AK/KA. Prefix A stands for agent and K for kernel of RCRS. For example, when an agent is in front of building of RCRS, the agent can switch connection from RCRS to USARSim by AK_USASIM_ENTER. And they enter the corresponding 3D building of USARSim.

KU/UK. Prefix U stands for USARSim Controller. USARSim Controller spawns USARSim Client. The clients are connected to a USARSim server that supports the 3D environments. Commands with this prefix have a role to bridge two systems. The kernel submit KU_connect [1] command with the building ID to USARSim Controller that sets up USARSim for the building.

AU/UA. AU_TELL and UA_HEAR commands serve communications among clients in the environments created by one USARSim Server. AU commands are submitted and received at every USARSim time step.

Figure 2 shows architecture of agents and the hybrid system.

- USARSim is connected to RCRS as one of simulators. When agents move from RCRS to USARSim, they submit AK commands and the kernel passes it

---

[1] KU commands are implemented in the operand part of the RCRS commands.

Table 3. Added Protocols to bridge RCRS and USARSim

Commands		specification
Switch systems from RCRS to USARSim, vice versa		
AK_USARSIM_ENTER	A → K	An agent submits it to enter a building (to switch a server to USARSim) at its entrance node.
AK_USARSIM_EXIT	A → K	An agent submits it to get out of the building (to switch a server to RCRS) near its exits.
KA_USARSIM_ENTER_OK KA_USARSIM_ENTER_ERROR	K → A	The kernel notify whether USARSim connection succeeds or not.
KA_USARSIM_EXIT_OK KA_USARSIM_EXIT_ERROR	K → A	The kernel notify whether USARSim disconnection succeeds or not.
UK_ENTER_OK UK_ENTER_ERROR	U → K	USARSim Controller returns IP address and port number that USARSim Client uses when connection succeeds, errors why it fails,
UK_EXIT_OK UK_EXIT_ERROR	U → K	USARSim Controller returns OK when that the agent could disconnect to USARSim, ERROR and the reason that the agent cannot disconnect to USARSim.
Control the corresponding USARSim object		
AU_FORWARD, AU_BACKWARD	A → UC	The agent moves the object forward/backward.
AU_LEFT,AU_RIGHT	A → UC	The agent turns the object left/right.
AU_STOP	A → UC	The agent stops the object in USARSim.
AU_MOVE	A → UC	The agent moves the object to specified position in USARSim.
Communication between agents		
AK_TELL KA_HEAR	A → K K → A	They are basically the same as RCRS. They also support communications to USARSim or within via Communication Center.
AU_TELL	A → UC	The agent sends messages to other agents via Communication Center.
UA_HEAR	UC → A	Communication Center sends received messages to all agents that are in USARSim.
UK_HEAR	UC → A	Communication Center sends a list of received messages at the step of the kernel.

A, K, U, and UC represent agent, kernel, USARSim Controller, and USARSim Client, respectively. UC→ A shows command flow from USARSim Client to agents.

to USARSim Controller. After USARSim Controller receives it, the connection between the kernel and USARSim Controller and the communication are the same way as other sub simulators.

– USARSim Servers supply 3D simulation environments of buildings to the agents. The buildings in RCRS and USARSim are linked with the same ID number. USARSim Servers are set for every building and USARSim Controller supervises these servers.

– While agents connect to USARSim, the agents control the corresponding objects (avatar) in USARSim with AU commands. The commands express their wills how the avatars behave in the USARSim world. The behavior of avatars is simulated by a physical engine of USARSim.

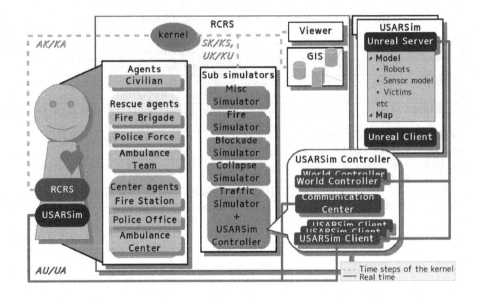

Fig. 2. System architecture of hybrid system connected by protocols. The left human figure shows that agent consists of parts to connect RCRS and USARSim.

4 Management of Changing ABS and Communication among Agents in Different ABSs

4.1 Protocols Systems for Changing ABS

Figure 3 shows a timing chart when an agent enters into a building and a sequence of protocols associated with it. The columns below Agents, kernel US-ARSim Controller show time steps. Time a, c, e, f correspond to RCRS time points and b, d correspond to USARSim time points.

a : When an agent arrives at an entrance node of a building, it submits AK_USARSIM_ENTER to enter the building. Receiving the command from the kernel, USARSim controller
   1. when this is the first entry to the building, spawn USARSim server that maintains the corresponding 3D USARSim world according to entries of its configuration file,
   2. place a corresponding avatar in the USARSim world.
c : The agent switches connection from RCRS to USARSim, when it receives AK_USARSIM_ENTER_OK commands with a port number to communicate with USARSim Client.
d : The agent submits AU commands. The commands control its avatar to move or rescue in the building of USARSim.
e : When the agent get out of the building, it submits AK_USARSIM_EXIT near the entrance of the building.

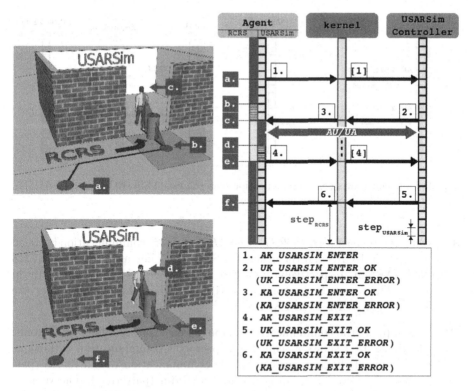

Fig. 3. Time chart for agent management change(The left shows the movements of agent. The middle is time sequence, and the slots show the time steps of RCRS and USARSim. The right one is the sequence of commands. Commands in parenthesis are ones that are embodied in operands of other protocols.)

f : Receiving AK_USARSIM_EXIT, USARSim deletes the avatar and returns AK_USARSIM_EXIT_OK. Receiving the AK_USARSIM_EXIT_OK, the agent returns to RCRS world and it position is the entrance node of the building.

4.2 Protocols for Communication to Agent in ABSs

Changing the connecting server requires to support communications between agents that are in different servers. The scope of commutation expands from within one ABS to between different ABSs. The rescue simulation scenario of Table 1 contains three patterns of communications.

1. *communication within RCRS:* A.4 is the same pattern of communication as the original RCRS. Communications of C.8 are between agents in RCRS. However, Y and Z are USARSim agents initially at USARSim.
2. *communication within USARSim:* Conversely, A.3 is communication between agents, y and z, that are USARSim agents initially. B.6 is communication to b that is a RCRS agent initially.

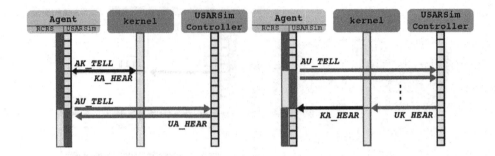

Fig. 4. Time chart for agent communication between two ABSs, RCRS and USARSim. The ratio of *Scale*s is 8 in this figure.

3. *communication between RCRS and USARSim:* A.2 is communication between one agent in RCRS and the other in USARSim. They are initially in RCRS and USARSim, respectively. In a case of B.7, the agent, b, is initially in RCRS. The agent A communicates with B without knowing whether the agent B is in RCRS or in USARSim.

Protocols support the communications among agents that are different systems. The left of Figure 4 shows a sequence of protocols communication within one system. The upper diagram corresponds to communication within RCRS (pattern 1), and the lower diagram is within USARSim (pattern 2). The right of Figure 4 shows a sequence of protocols that agents in RCRS and USARSim communicate. Agents in USARSim tell at USARSim time step, and agents in RCRS hear at RCRS time step.

5 Evacuation Simulation from Indoor to Outdoor Shelters

A subset of the rescue scenario in Table 1 is simulated. Figure 5 are snapshots of simulations by our prototype system. The simulation conditions are followings:

situation: Fires occur at a university campus. Students in school buildings evacuate to an outdoor refuge.

map: The road network and buildings of the university campus are presented by 2D RCRS map. Two of the school buildings are linked to USARSim and represented by its 3D models.

agent: Six and four student agents[2] are at the two buildings respectively, and total 150 agents evacuate to one refuge by following the instructions of teacher agents. There are ten teacher agents instruct the students evacuation routes.

Two PCs are used in our experiment system. One PC is Core2 Duo of 2.2GHz with 2GB Memory and the other one is Pentium 4 of 3GHz with 1GB Memory.

---

[2] Student and teacher agents are the civilian and police agents of RCRS.

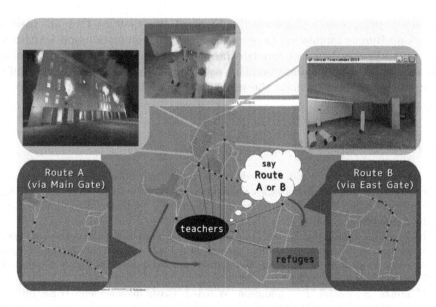

Fig. 5. Execution of evacuation scenario (Two buildings are linked to USARSim, cylinder figure avatars are student agents.)

RCRS and one USARSim server are run at the first PC and the other USARSim server is run on the second PC. The two USARSim servers take charge of the buildings that are linked to RCRS. The ratio of *Scales*, USARSim to RCRS, is set to 60 (one simulation time of RCRS corresponds to one minute).

The upper three figures are snapshots of USARSim. The left one is an exterior view of one school building and right two figures display the behavior of students inside the buildings. The student avatars in USARSim are represented by a cylinder figure robots. The figures of left bottom and right bottom show the results of simulations which the teachers instructed different routes.

The results show that all ten students inside get out of the building and evacuate to the refuge. They show different patterns according to the teachers' instructions. These indicate communications among agents work well.

6 Discussion and Summary

It is ideal that disaster & rescue simulations handle rescue behaviors at disasters in real scale. This requires simulations of a huge number of agents that behave at wide areas with fine resolutions. The requirements lead a huge of computer resources and powers.

We propose an idea of switching systems during simulations to choose one that is suitable for situations. A hybrid system of two different simulations with different time and space resolution makes it possible to simulate urban size human behaviors and indoor movements by reasonable computational resources.

This paper presents protocols that connect two systems that are used in RoboCup Rescue Simulation League, Rescue Agent Simulation and USARSim. The prototype system can simulate people's evacuation from going to inside fire-escape doors to moving to outside shelters. The result shows an possibility that the hybrid system make the simulations feasible ones that takes a lot of computer recourses.

References

1. Van de Walle, B., Turoff, M. (eds.): Emergency response information systems: Emerging trends and technologies. Communications of the ACM 50(3), 28–65 (2007)
2. Fujimoto, R.M.: Parallel and Distributed Simulation Systems. John Wiley & Sons, Inc., Chichester (2000)
3. Kuwata, Y., Takahashi, T., Ito, N., Takeuchi, I.: Design of human-in-the-loop agent simulation for disaster simulation systems. In: Proc. SRMED 2006 (Third International Workshop on Synthetic Simulation and Robotics to Mitigate Earthquake Disaster), pp. 9–14 (2006)
4. Mehrotra, S., Znatri, T., Thompson, W.(eds.): Crisis management. IEEE Internet Computing 12(1), 14–54 (2008)
5. RoboCupRescue, http://www.robocuprescue.org/
6. Wang, J., Balakirsky, S., Carpin, S., Lewis, M., Scrapper, C.: Bridging the gap between simulation and reality in urban search and rescue. In: Lakemeyer, G., Sklar, E., Sorrenti, D.G., Takahashi, T. (eds.) RoboCup 2006: Robot Soccer World Cup X. LNCS (LNAI), vol. 4434, pp. 1–12. Springer, Heidelberg (2007)
7. Schurr, N., Marecki, J., Kasinadhuni, N., Tambe, M., Lewis, J.P., Scerri, P.: The defacto system for human omnipresence to coordinate agent teams: The future of disaster response. In: AAMAS 2005, pp. 1229–1230 (2005)
8. Takeuchi, I., Kakumoto, S., Goto, Y.: Towards an integrated earthquake disaster simulation system. In: First International Workshop on Synthetic Simulation and Robotics to Mitigate Earthquake Disaster (2003),
   http://www.dis.uniroma1.it/~rescue/events/padova03/papers/index.html

SSL-Vision: The Shared Vision System for the RoboCup Small Size League

Stefan Zickler[1], Tim Laue[2], Oliver Birbach[2], Mahisorn Wongphati[3],
and Manuela Veloso[1]

[1] Carnegie Mellon University, Computer Science Department,
5000 Forbes Ave., Pittsburgh, PA, 15213, USA
{szickler,veloso}@cs.cmu.edu
[2] Deutsches Forschungszentrum für Künstliche Intelligenz GmbH,
Safe and Secure Cognitive Systems, Enrique-Schmidt-Str. 5, 28359 Bremen, Germany
{Tim.Laue, Oliver.Birbach}@dfki.de
[3] Chulalongkorn University,
254 Phyathai Road, Patumwan, Bangkok, 10330, Thailand
mahisorn.w@gmail.com

Abstract. The current RoboCup Small Size League rules allow every team to set up their own global vision system as a primary sensor. This option, which is used by all participating teams, bears several organizational limitations and thus impairs the league's progress. Additionally, most teams have converged on very similar solutions, and have produced only few significant research results to this global vision problem over the last years. Hence the responsible committees decided to migrate to a shared vision system (including also sharing the vision hardware) for all teams by 2010. This system – named SSL-Vision – is currently developed by volunteers from participating teams. In this paper, we describe the current state of SSL-Vision, i.e. its software architecture as well as the approaches used for image processing and camera calibration, together with the intended process for its introduction and its use beyond the scope of the Small Size League.

1 Introduction

Given the current rules of the RoboCup Small Size League (SSL) [1], every team is allowed to mount cameras above or next to the field. There has also been an option of using local instead of global vision, but this turned out to be not competitive. For adequately covering the current field, most teams prefer to use two cameras, one above each half. This configuration bears one major problem (implicating a set of sub-problems): The need for long setup times before as well as during the competition. Having five teams playing on a field, ten cameras need to be mounted and calibrated. During these preparations, a field cannot be used for any matches or other preparations. Due to this situation, teams are always bound to their field (during one phase of the tournament) and unable to play any testing matches against teams from other fields. Hence the Small Size

J. Baltes et al. (Eds.): RoboCup 2009, LNAI 5949, pp. 425–436, 2010.
© Springer-Verlag Berlin Heidelberg 2010

League needs as many fields as it has round robin groups. Flexible schedules as in the Humanoid or the Standard Platform League – which have more teams but need less fields – are currently impossible.

In the future, these problems might become even worse since the current camera equipment already reached its limits. Having a larger field – which is a probable change for 2010, given the common field with the Humanoid and the Standard Platform League –, every team will be forced to set up four cameras above the field. This would significantly increase preparation times during a competition and decrease time and flexibility for scheduling matches.

To overcome this situation, the SSL committees decided to migrate to a shared vision system, i. e. to a single set of cameras per field which are connected to an image processing server which broadcasts the vision output to the participating teams. The software for this server needs to be flexible, i. e. scalable for future changes and open to new approaches, as well as competitive, i. e. performant and precise, to not constrain the current performance of the top teams. This system, named SSL-Vision, is now developed by a group of volunteers from the SSL. This paper describes the current state and the future of this project.

The paper is organized as follows: Section 2 describes the overall architecture of the system. The current approaches for image processing and camera calibration are presented in Section 3. The paper concludes with a description of the system's introduction and the resulting implications in Section 4.

2   Framework

SSL-Vision is intended to be used by all Small Size League teams, with a variety of camera configurations and processing hardware. As such, configurability and robustness are key design goals for its framework architecture. Additionally, the project's collaborative openness and its emphasis on research all need to be reflected in its framework architecture through an extendable, manageable, and scalable design.

One major design goal for the framework is to support concurrent image processing of multiple cameras in a single seamless application. Furthermore, the application should integrate all necessary vision functionality, such as configuration, visualization, and actual processing. To achieve better scalability on modern multi-core and hyper-threaded architectures, the application uses a multi-threaded approach. The application's main thread is responsible for the Graphical User Interface (GUI), including all visualizations, and configuration dialogs. Additionally, each individual camera's vision processing is implemented in a separate thread, thus allowing truly parallel multi-camera capture and processing. The application is implemented in C++ and makes heavy use of the Qt toolkit [2], to allow for efficient, platform-independent development.

Fig. 1 shows an overview of the framework architecture. The entire system's desired processing flow is encoded in a *multi-camera stack* which fully defines how

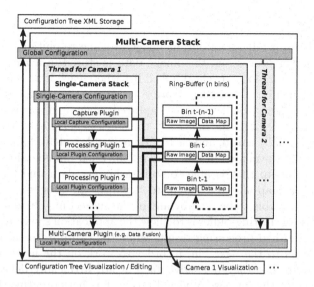

Fig. 1. The extendible, multi-threaded processing architecture of SSL-Vision

many cameras are used for capturing, and what particular processing should be performed. The system has been designed so that developers can create different stacks for different robotics application scenarios. By default, the system will load a particular multi-camera stack, labeled the "RoboCup Small Size Dual Camera Stack" which we will elaborate on in the following section. However, the key point is that the SSL-Vision framework provides support for choosing any arbitrarily complex, user-defined stack at start-up, and as such becomes very extendible and even attractive for applications that go beyond robot soccer.

Internally, a multi-camera stack consists of several threads, each representing the processing flow of a corresponding capture device. Each thread's capturing and processing flow is modeled as a *single-camera stack*, consisting of multiple *plugins* which are executed in order. The first plugin in any single-camera stack implements the image capturing task. All capture plugins implement the same C++ capture interface, thus allowing true interchangeability and extendibility of capture methods. The framework furthermore supports unique, independent configuration of each single-camera stack, therefore enabling capture in heterogeneous multi-camera setups. Currently, the system features a capture plugin supporting IEEE 1394 / DCAM cameras, including higher bandwidth Firewire 800 / 1394B ones. Configuration and visualization of all standard DCAM parameters (such as white balance, exposure, or shutter speed) is provided through the GUI at run-time, thus eliminating the need for third-party DCAM parameter configuration tools. The system furthermore features another capture plugin supporting capturing from still image and video files, allowing development on machines which do not have actual capture hardware. Additional capture plugins for Gigabit Ethernet (GigE) Vision as well as Video4Linux are under construction as well.

2.1   The Capture Loop

A capture plugin produces an output image at some resolution, in some color-space. For further processing, this image data is stored in a *ring-buffer* which is internally organized as a cyclical linked-list where each item represents a *bin*, as is depicted in Fig. 1. On each capture iteration, the single-camera stack is assigned a bin where it will store the captured image and any additional data resulting from processing this image. As the stack is being executed, each of its plugins is sequentially called, and each of them is able to have full read and write access to the data available in the current bin. Each bin contains a *data map*, which is a hash-map that is able to store arbitrary data under a meaningful label. This data map allows a plugin to "publish" its processing results, thus making them available to be read by any of the succeeding plugins in the stack.

The purpose of the ring-buffer is to allow the application's visualization thread to access the finished processing results while the capture thread is allowed to already move on to the next bin, in order to work on the latest video frame. This architecture has the great advantage of not artificially delaying any image processing for the purpose of visualization. Furthermore, this ring-buffered, multi-threaded approach makes it possible to prioritize the execution schedule of the capture threads over the GUI thread, thus minimizing the impact of visualization on processing latency. Of course it is also possible to completely disable all visualizations in the GUI for maximum processing performance.

In some processing scenarios it is necessary to synchronize the processing results of multiple camera threads after all the single-stack plugins have finished executing. This is done through optional *multi-camera plugins*. A typical example would be a plugin which performs the data fusion of all the threads' object detection results and then sends the fused data out to a network.

2.2   Parameter Configuration

Configurability and ease of use are both important goals of the SSL-Vision framework. To achieve this, all configuration parameters of the system are represented in a unified way through a variable management system called *VarTypes* [3]. The VarTypes system allows the organization of parameters of arbitrarily complex types while providing thread-safe read/write access, hierarchical organization, real-time introspection/editing, and XML-based data storage.

Fig. 1 shows the hierarchical nature of the system's configuration. Each plugin in the SSL-Vision framework is able to carry its own set of configuration parameters. Each single-camera stack unifies these local configurations and may additionally contain some stack-wide configuration parameters. Finally, the multi-camera stack unifies all single-camera stack configurations and furthermore contains all global configuration settings. This entire configuration tree can then be seamlessly stored as XML. More importantly, it is displayed as a data-tree during runtime and allows real-time editing of the data. Fig. 2 shows a snapshot of the data-tree's visualization.

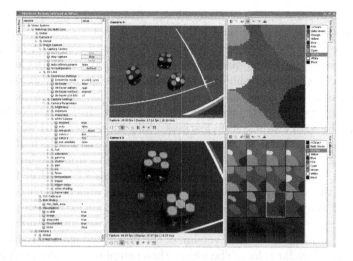

Fig. 2. Screenshot of SSL-Vision, showing the parameter configuration tree (left), live-visualizations of the two cameras (center), and views of their respective color thresholding YUV LUTs (right)

3   RoboCup SSL Image Processing Stack

The system's default multi-camera stack implements a processing flow for solving the vision task encountered in the RoboCup Small Size League. In the Small Size League, teams typically choose a dual-camera overhead vision setup. The robots on the playing field are uniquely identifiable and locatable based on colored markers. Each robot carries a team-identifying marker in the center as well as a unique arrangement of additional colored markers in order to provide the robot's unique ID and orientation. In the past, each team was able to determine their own arrangement and selection of these additional markers. However, with the introduction of the SSL-Vision system, it is planned to unify the marker layout among all teams for simplification purposes.

The processing stack for this Small Size League domain follows a typical multi-stage approach as it has been proven successful by several teams in the past. The particular single-camera stack consists of the following plugins which we explain in detail in the forthcoming sections: image capture, color thresholding, runlength encoding, region extraction and sorting, conversion from pixel coordinates to real-world coordinates, pattern detection and filtering, and delivery of detection results via network.

3.1   CMVision-Based Color Segmentation

The color segmentation plugins of this stack, namely color thresholding, runlength-encoding, region extraction and region sorting, have all been implemented by porting the core algorithms of the existing CMVision library to the new SSL-Vision plugin architecture [4]. To perform the color thresholding, CMVision assumes the existence of a lookup table (LUT) which maps from the input image's 3D color space (by default YUV), to a unique color label which is able to represent any of the marker colors, the ball color, as well as any other desired colors. The color thresholding algorithm then sequentially iterates through all the pixels of the image and uses this LUT to convert each pixel from its original color space to its corresponding color label. To ease the calibration of this LUT, the SSL-Vision system features a fully integrated GUI which is able to not only visualize the 3D LUT through various views, but which also allows to directly pick calibration measurements and histograms from the incoming video stream. Fig. 2 shows two example renderings of this LUT. After thresholding the image, the next plugin performs a line-by-line runlength encoding on the thresholded image which is used to speed up the region extraction process. The region extraction plugin then uses CMVision's tree-based union find algorithm to traverse the runlength-encoded version of the image and efficiently merge neighboring runs of similar colors. The plugin then computes the bounding boxes and centroids of all merged regions and finally sorts them by color and size.

3.2   Camera Calibration

In order to deduce information about the objects on the field from the measurements of the cameras, a calibration defining the relationship between the field geometry and the image plane is needed. Depending on the applied calibration technique, current teams use a variety of different calibration patterns, leading to an additional logistic effort while attending tournaments. Furthermore, many such calibration procedures require the patterns to cover the field partially or as a whole, making the field unusable for other teams during the setup.

For the calibration procedure of SSL-Vision, no additional accessories are required. Instead, the procedure uses solely the image of the field and the dimensions defined in the league's rules. Because SSL-Vision uses two independent vision stacks, we calibrate both cameras independently using the corresponding half field. To model the projection into the image plane, a pin-hole camera model including radial distortion is used. The corresponding measurement function h projects a three-dimensional point M from the field into a two-dimensional point m in the image plane. The model parameters for this function are, intuitively, the orientation q and the position t, transforming points from coordinate system of the field into the coordinate system of the camera, and the intrinsic parameters f, (u_0, v_0) and κ indicating the focal-length, image center and radial distortion, respectively.

In a typical Small Size League camera setup, estimating such a set of calibration parameters by using only the field dimensions is actually ill-posed, due to the parallelism of the image plane and the field plane (which is the reason for the frequent use of calibration patterns). The estimator cannot distinguish whether the depth is caused by the camera's distance from the field (encoded in t_z) or the focal length (encoded in f). To circumvent this problem, a manual measurement of the camera's distance from the field is performed and the parameter is excluded from the estimation algorithm.

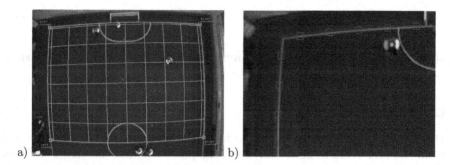

Fig. 3. Camera calibration: a) Result of calibration. The points selected by the user are depicted by labeled boxes, the field lines and their parallels are projected from the field to the image plane. b) Detected edges for the second calibration step.

The actual calibration procedure consists of two steps:

1. The user selects the four corner points of the half-field in the camera image. Based on the fact that the setup is constrained by the rules, rough but adequate initial parameters can be determined in advance. Based on these initial parameters, a least squares optimization is performed to determine a set of parameters corresponding to the marked field points [5] (cf. Fig. 3a). Thus, we want to minimize the squared difference of the image points m_i that were marked by the user in the image plane and corresponding field point M_i, projected into the image plane using the measurement function mentioned above:

$$\sum_{i=1}^{4} |m_i - h(M_i, q, t, f, u_o, v_o, \kappa)|^2 \qquad (1)$$

Since this is a nonlinear least squares problem, the Levenberg-Marquardt algorithm [6] is used to find the optimal set of parameters.

2. After this first estimate, the parameters are refined by integrating segments of field lines into the estimation. Since the field lines contrast with the rest of the field, an edge-detector is applied to find the lines using their predicted position computed from the estimate and the field dimensions as a search window (cf. Fig. 3b). A reasonable number of edges on the lines is then used to extend the least squares estimation. For this, we introduce a new to be estimated parameter α for each measurement and minimize the deviation of the measured point on the field line and the projection of the point $(\alpha\, p_1 + (1 - \alpha)\, p_2$ between the two points p_1, p_2 constraining the line segment. The term to be minimized now reads

$$\sum_{i=1}^{4} |m_i - h(M_i, p)|^2 + \sum_{i=1}^{n} |m_i - h(\alpha_i\, L_{i,1} + (1 - \alpha_i)\, L_{i,2}, p)|^2 \qquad (2)$$

where $L_{i,1}$ and $L_{i,2}$ constrain the line segment and α_i describes the actual position of measurement i on this line. Please note, that multiple measurements may lie on the same line. For better readability, the camera parameters were combined into p.

After this calibration procedure, the inverted measurement function h^{-1} can be used to transform pixel coordinates to real-world coordinates.

3.3   Pattern Detection

After all regions have been extracted from the input image and all their real-world coordinates have been computed, the processing flow continues with the execution of the pattern recognition plugin. The purpose of this plugin is to extract the identities, locations, and orientations of all the robots, as well as the location of the ball. The internal pattern detection algorithm was adopted from the CMDragons vision system and is described in detail in a previous paper [7].

Although this pattern detection algorithm can be configured to detect patterns with arbitrary arrangements of 2D colored markers, the Small Size committees are intending to mandate a standard league-wide pattern layout with the transition to SSL-Vision, for simplification purposes.

3.4   System Integration and Performance

After the pattern detection plugin has finished executing, its results are delivered to participating teams via UDP Multicast. Data packets are encoded using Google Protocol Buffers [8], and contain positions, orientations, and confidences of all detected objects, as well as additional meta-data, such as a timestamp and frame-number. Furthermore, SSL-Vision is able to send geometry data (such as camera pose) to clients, if required. To simplify these data delivery tasks, SSL-Vision provides a minimalistic C++ sample client which teams can use to automatically receive and deserialize all the extracted positions and orientations of the robots and the ball. Currently, SSL-Vision does not perform any "sensor fusion", and instead will deliver the results from both cameras independently, leaving the fusion task to the individual teams. Similarly, SSL-Vision does not perform any motion tracking or smoothing. This is due to the fact that robot tracking typically assumes knowledge about the actual motion commands sent to the robots, and is therefore best left to the teams.

Table 1 shows a break-down of processing times required for a single frame of a progressive YUV422 video stream of 780×580 pixel resolution. These numbers represent rounded averages over 12 consecutive frames taken in a randomly configured RoboCup environment, and were obtained on an Athlon 64 X2 4800+ processor.

3.5   GPU-Accelerated Color Thresholding

The traditional sequential execution of CMVision's color thresholding process is – despite its fast implementation through a LUT – a very computationally intensive process. The performance values in Table 1 clearly show that color

Table 1. Single frame processing times for the plugins of the default RoboCup stack

Plugin	Time
Image capture	1.1 ms
Color thresholding (CPU)	3.6 ms
Runlength encoding	0.7 ms
Region extraction and sorting	0.2 ms
Coordinate conversion	< 0.1 ms
Pattern detection	< 0.1 ms
Other processing overhead	0.4 ms
Total frame processing	**< 6.2 ms**

Table 2. Single frame processing times for the naïve GPU-accelerated color thresholding

Component	Time
Copy data to texture memory	3.0 ms
Color thresholding (GPU)	32 μs
Copy data from frame buffer	11.0 ms
Total thresholding time	**< 15 ms**

thresholding currently constitutes the latency bottleneck of the processing stack by a significant margin. One of the best ways to overcome this latency problem is by exploiting the fact that color thresholding is a massively parallelizable problem because all pixels can be processed independently. However, even with the reasonable price and performance in current Personal Computers, only 2 or 4 physical CPU cores are available for parallel computing which in our case are already occupied by each camera's capture threads, the visualization process, and other OS tasks. Nevertheless, modern commodity video cards which feature a programmable Graphic Processing Unit (GPU) have become widely available in recent years. Because GPUs are inherently designed to perform massively parallel computations, they represent a promising approach for hardware-accelerated image processing. In this section we will provide initial evaluation results of a GPU-accelerated color thresholding algorithm which may be included in a future release of SSL-Vision.

To implement the GPU-accelerated thresholding algorithm, we selected the OpenGL Shading Language (GLSL), due to its wide support of modern graphics hardware and operating systems. GLSL allows the programming of the graphics hardware's *vertex processor* and *fragment processor* through the use of small programs known as *vertex shaders* and *fragment shaders*, respectively [9]. Vertex shaders are able to perform operations on a per-vertex basis, such as transformations, normalizations, and texture coordinate generation. Fragment shaders (also commonly referred to as *pixel shaders*), on the other hand, are able to perform per-pixel operations, such as texture interpolations and modifications.

Because we are interested in performing color thresholding on a 2D image, we implement our algorithm via a fragment shader. Fig. 4 shows an overview of this GPU-accelerated color thresholding approach. First, before any kind of video processing can happen, we need to define a thresholding LUT. This LUT is similar to the traditional CMVision version in that it will map a 3D color input (for example in RGB or YUV) to a singular, discrete color label. The difference is however, that this LUT now resides in video memory and is internally represented as a 3D texture which can be easily accessed by the fragment shader. As modern video hardware normally provides 128MB video memory or more,

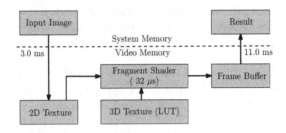

Fig. 4. Block diagram of color thresholding using GLSL

it is easily possible to encode a full resolution LUT (256x256x256, resulting in approximately 17MB). In order to perform the actual color thresholding processing, any incoming video frame first needs to be copied to the video hardware to be represented as a 2D texture that the shader will be able to operate on. The fragment shader's operation then is to simply replace a given pixel's 3D color value with its corresponding color label from the 3D LUT texture. We apply this shader by rendering the entire 2D input image to the frame buffer. After the render process, we now need to transfer the labeled image from the frame buffer back to system memory for further processing by any other traditional plugins.

Table 2 shows the average time used by each step of the color thresholding process using an NVIDIA Geforce 7800 GTX video card under Linux, on the same CPU that was used for the measurements in Table 1. The input video data again had a size of 780×580 pixels. The values clearly indicate that the actual thresholding step is about 100 times faster than on the normal CPU. Interestingly, however, this approach has introduced two new bottlenecks in the upload and download times between system memory and video memory which, in total, makes this approach more than four times slower than the traditional color thresholding routine.

A potential approach for improving this performance would be to convert most or all other image-processing related plugins, which follow color thresholding, to the GPU. This way, there would be no requirement to transfer an entire image back from video memory to system memory. Instead, a major portion of the image processing stack would be computed on the GPU, and only the resulting data structures, such as final centroid locations, could be transfered back to system memory. For this process to work however, the color segmentation algorithms would need to be majorly revised, and as such this approach should be considered future work.

4   Further Steps and Implications

Beyond a proper technical realization, as described in the previous sections, the introduction of a shared vision system for the Small Size League bears several organizational issues as well as implications for future research, even for other RoboCup leagues.

4.1   Schedule of Introduction

The first release of SSL-Vision has been published in spring 2009. Since then, all teams are free to test the application in their labs, to review the code, and to submit improvements. At the upcoming regional competitions as well as at RoboCup 2009, the usage of the system is voluntary, i. e. teams can run it on their own computers or decide to share the vision system with others. However, everybody is free to keep using their own system. After this transition phase, which has been established to provide a rehearsal under real competition conditions, the usage of SSL-Vision will become obligatory, in time for RoboCup 2010.

4.2   Implications for Research

By introducing a shared vision system for all teams, one degree of individuality for solving the global task "Playing Soccer with Small Size Robots" becomes removed. However, during the last years, most experienced teams have converged towards very similar sensing solutions, and have produced only few significant research results regarding computer vision. De facto, having a performant vision system does not provide any major advantage, but should rather be considered a minimum requirement as sophisticated tactics and precise control are dominant factors in the SSL. On the other hand, new teams often experience problems having an insufficient vision application which strongly decreases their entire system's performance. Thus, SSL-Vision will directly benefit all newcomers, allowing them to base their tactics on a robust global vision sensor.

Furthermore, the transition to a shared vision system does not imply a stagnation in vision-related research. In fact, due to its open and modular architecture (cf. Sect. 2), SSL-Vision allows researchers to develop and "plug in" novel image processing approaches without needing to struggle with technical details (e.g. camera interface control or synchronization). Therefore, new approaches can be fairly and directly compared with existing ones, thus ensuring a continuing, community-driven evolution of SSL-Vision's processing capabilities and performance.

Whereas the system's impact for the Small Size League is obvious, it might also become directly useful for teams in other RoboCup leagues. Many researchers in local vision robot leagues require precise reference data – so-called *ground truth* – to evaluate their results during development, e. g. of localization algorithms or for gait optimization. One example for tracking humanoid soccer robots with an SSL vision system is shown in [10]. Due to the standardized field size, SSL-Vision becomes an off-the-shelf solution for the Humanoid as well as the Standard Platform League.

Finally, it needs to be strongly emphasized that SSL-Vision's architecture is not at all limited to only solving the task of robot soccer vision. Instead, the system should really be recognized as a framework which is flexible and versatile enough to be employed for almost any imaginable real-time image processing task. While, by default, the system provides the stacks and plugins aimed at the

RoboCup domain, we are also eagerly anticipating the use and extension of this system for applications which go beyond robot soccer.

5   Conclusion

In this paper, we introduced the shared vision system for the RoboCup Small Size League, called SSL-Vision. We presented the system's open software architecture, described the current approaches for image processing and camera calibration, and touched upon possible future improvements, such as GPU-acceleration. Finally, we discussed SSL-Vision's scheduled introduction and its impact on research within the RoboCup community. We strongly believe that the system will positively affect the Small Size League by reducing organizational problems and by allowing teams to re-focus their research efforts towards elaborate multi-agent systems and control issues. Because SSL-Vision is a community project, everybody is invited to participate. Therefore, SSL-Vision's entire codebase is released as open-source [11].

References

1. RoboCup Small Size League: SSL Web Site (2009),
   http://small-size.informatik.uni-bremen.de
2. Nokia Corporation: The Qt Toolkit, http://www.qtsoftware.com/
3. Zickler, S.: The VarTypes System, http://code.google.com/p/vartypes/
4. Bruce, J., Balch, T., Veloso, M.: Fast and inexpensive color image segmentation for interactive robots. In: Proceedings of the 2000 IEEE/RSJ International Conference on Intelligent Robots and Systems (IROS 2000), vol. 3, pp. 2061–2066 (2000)
5. Zhang, Z.: A flexible new technique for camera calibration. IEEE Transactions on Pattern Analysis and Machine Intelligence 22(11), 1330–1334 (2000)
6. Marquardt, D.: An algorithm for least-squares estimation of nonlinear parameters. SIAM Journal on Applied Mathematics 11(2), 431–441 (1963)
7. Bruce, J., Veloso, M.: Fast and accurate vision-based pattern detection and identification. In: Proceedings of the 2003 IEEE International Conference on Robotics and Automation, ICRA 2003 (2003)
8. Google Inc.: Protocol Buffers, http://code.google.com/p/protobuf/
9. Rost, R.J.: OpenGL Shading Language, 2nd edn. Addison-Wesley Professional, Reading (2006)
10. Laue, T., Röfer, T.: Particle filter-based state estimation in a competitive and uncertain environment. In: Proceedings of the 6th International Workshop on Embedded Systems, Vaasa, Finland (2007)
11. SSL-Vision Developer Team: RoboCup Small Size League Shared Vision System Project Home (2009), http://code.google.com/p/ssl-vision/

Heuristic Formation Control in Multi-robot Systems Using Local Communication and Limited Identification

Michael de Denus, John Anderson, and Jacky Baltes

Autonomous Agents Laboratory
Dept. of Computer Science, University of Manitoba
Winnipeg, Manitoba, Canada R3T2N2
{mdedenu,andersj,jacky}@cs.umanitoba.ca
http://aalab.cs.umanitoba.ca

Abstract. Groups of individuals often use formations as a means of providing orderly movement while distributing members in a manner that is advantageous to the group's activities. A particular formation may offer a defensive advantage over moving individually, for example, exposing only some of the agents to the proximity of enemies, or might increase group abilities by allowing individuals to limit perceptual focus to one small part of the environment. Formations are used throughout the natural world and in many organized human groups, and are equally valuable to multi-robot systems. Most formation control in multi-robot systems is extremely limited compared to the formations we see in nature: formations are precisely defined, and mechanisms for forming and maintaining formations often require unique labels for individuals and broadcast communication. In this paper, we explore a method for creating heuristic formations - where agents create an overall formation, but forgiveness exists for small variations in form - using only local rules for creating formations and allowing only local communication. Our approach defines relative positions in terms of a probability given the position of one's nearest neighbor, and improves on prior work by assuming that all agents do not begin knowing the unique labels of others in the group. The approach also assumes heterogeneity in sensing, in that agents may not be able to perceive the unique labels of others, and thus may require assistance from those who can. This assumptions make formations robust to the failure of individual agents, and allow previously unknown agents to join an existing formation. An evaluation of this approach is illustrated using Player/Stage, a commonly accepted simulation package for multi-robot systems, for controlled experimentation.

1   Introduction

The ability to move into and maintain formations is an important property in many groups. The potential advantages formations bring are many and diverse, and can be see in human organizations and in the natural world. In a military unit, for example, a formation can be used as a defensive structure, exposing only a limited number of agents on the periphery to potential enemies (e.g. a square vs. a straight line). In humans this is seen in military situations as well as many sports, but the same advantage can be seen in the much less geometric formation of a school of fish. Formations may

J. Baltes et al. (Eds.): RoboCup 2009, LNAI 5949, pp. 437–448, 2010.

be used to decrease the amount of work individuals must expend (e.g. aerodynamic increases achieved through the formation of a bird flock [1]) or make better use of limited sensory abilities by allowing individuals to focus only on a given area depending on their position in the formation [2]. They may also allow assumptions for increased ability to navigate and decreased decision-making time (e.g. a flight pattern for a group of jets allows one member to predict the likely positions and motion of others, decreasing the number of factors that must be considered when making a change in movement).

While these are most obviously seen in humans in military situations, they are seen throughout the natural world simply because other creatures that form groups experience significantly more vulnerable situations than do modern humans in their daily lives. For the same reason, formations are very useful for groups of robots: like non-human creatures, they are restricted in their decision-making and perceptual abilities compared to humans, and lack the common-sense knowledge to function in highly complex domains. Moreover, many of the applications that we consider robots amenable to - military environments and search-and-rescue settings, for example - present the same dangers that have led human military units to adopt the use of formations.

There are a number of important active areas of research on formations in multi-robot systems: forming various formations based on global or local interactions between robots; changing from one formation to another (again, based on global or local interactions and motivations), and maintaining formations in the face of hazards to navigation, for example. Most of this work is performed using ideal geometric patterns that are similar to those used in military applications (diamonds, squares, etc.). Comparatively little work takes advantage of the fact that formations in nature are rarely perfectly geometric: while schools of fish for example do form ellipses and other approximations to geometric shapes, for example, these are based on flexible local rules that result in a loose aggregate rather than a perfect geometric formation [3]. Similarly, while a flock of birds may form a V-shape, local rules do not necessarily dictate a precise angle, nor whether one side is symmetric with the other. From the standpoint of creating a formation, those in nature are more flexible and less constrained than that with which we are familiar with militarily. Thus, expecting a formation of robots to achieve something close a stated formation but not rigidly precise should similarly allow the benefits of formation-based movement while making the effort and infrastructure required to achieve and maintain formation reasonable.

In this paper, we describe a method for achieving formations in multi-robot systems where formations are formed heuristically, rather than to a precise, pre-defined pattern. This is done by assigning relative positions in the formation a probability, given the position of a nearest neighbor, rather than demanding an exact placement for particular individuals. The technique we employ to achieve formations requires only a simple set of local rules governing the angle and position between any agent and its nearest neighbor, and unlike other approaches, does not require all agents to know a unique identifier for all others in the group, nor to have the ability to broadcast to all members of the group. In our approach, agents begin knowing nothing about the identities of other agents, and some agents will have the ability to perceive the identities of others. Direct inter-agent communication is the only form of communication required to apply this technique to create heuristic formations in a group of robots.

We begin by reviewing related work on formation control in multi-robot systems, and then describe our approach in relation to this work. We then describe an implementation using Player/Stage [4], a well-accepted simulation system for multiple robots, and examine the performance of our approach, comparing the use of limited identification and communication with a baseline group that is given the ability to identify all individuals uniquely. Current work involves a study of the performance of this approach using Citizen Micro-robots in a mixed reality environment, and this and other future work is then discussed.

2 Related Work

In previous work, Yamaguchi [5] describes a method of formation control requiring only local information. Agents establish a link with one or two neighbors. Each agent then updates a formation vector based on the positions of its neighbors This method is successfully demonstrated in simulation and with real robots. One notable limitation of this method is the fact that it works by maintaining relative distances only. Our work makes use of both distance and angle, allowing for a larger range of potential formations.

Balch and Arkin [2] describe a behavior-based system, in which groups of robots of known size and configuration can move together in formation. Their method relies on knowing the number of other agents and their positions. Our approach calculates the heading and speed of each individual agent as the weighted vector sum of several independent behaviors, as Balch and Arkin's approach does (this is common among behaviour-based agents). The work presented here extends that of Balch and Arkin by not requiring that each agent have a known spot in the formation. Balch and Arkin's work was also important in categorizing formation control approaches by the means with which an agent calculates its appropriate position: *Unit-Center Referenced*, where the center of the formation is determined and positions are taken relative to this; *Leader-Referenced*, where positions are taken relative to a unique leader, and *Neighbor-referenced*, where positions are taken relative to one other predetermined robot. Like many other approaches, our work on formation control is neighbor-referenced, in order to rely more on local information and avoid the bottleneck and failure-recovery problems associated with a unique leader.

Fredslund and Mataric [6] propose a method of formation control similar to that presented here. They assign each agent a unique ID. Each agent passes its ID to a function which determines its desired neighbor and the relative position at which this neighbor should be kept. While this allows for some types of formations that our method cannot currently accomplish, it relies on the ability to locate a unique individual in the group. It also does not have a mechanism for failure recovery. Our proposed method has neither of these limitations.

Howard et al. [7] perform simple formation control with heterogeneous agents. *Follower* agents have sensors to track and follow *Leader* agents. In their approach, sensing differences are strongly tied to specific roles. In our work, all agents share the same role of "formation participant". Agents can perform in this role with different levels of success, however, depending on their sensing capabilities, and thus agents are still heterogeneous.

Hattenberger et al. [8] describe a method of dynamically adapting a formation to changing environmental conditions. The primary limitation of their technique is the dependence on a lead agent to calculate the relative positions of all other agents. Although it ensures that the intended formation is achieved, it creates a bottleneck that does not exist in our decentralized approach.

Other researchers [9,10] discuss methods of using only local sensor information combined with simple rules to create formations. We expand on these techniques by allowing the rules to change as our knowledge of the environment grows. This allows us to create formations where different agents obey different formation conditions. The conditions that they obey can also vary dynamically. This is an advantage that neither of these systems offer.

3 Heuristic Formation Control with Limited Knowledge of Others

As mentioned previously, our approach is neighbor-referenced. This means that each agent takes a position in the formation based on that of a particular neighbor, as opposed to via a secondary frame of reference, which in turn allows rules for positioning to be defined locally. This also allows a team of agents to remain in formation when as little as one other agent is within sensor range.

In our approach to formation control, a formation is defined as a set of one or more *formation conditions*. A formation condition describes a particular relationship between two neighboring agents in an overall formation, in terms of distance and angle. A set of formation conditions must describe all types of relationships between two neighbors to describe the structure of a formation. Since the correct position to occupy may not be the same condition that is being used by a neighbor, there will be relationships between formation conditions as well. For example, a V formation consists of three different formation conditions: one describing the angular relationship on one side of the V, the other the inverse forming the other side of the V, and the third the centermost position where the agent is following no one. Agents joining the formation attempt to query their nearest neighbor (which may or may not be possible, depending on whether they know that neighbor's ID) for advice on a space to occupy in the formation. The neighbor responds with a set of probabilities indicating which formation condition(s) best describe the relationship the querying agent should physically assume if it joins the the formation following the agent being queried. Representationally, a formation condition thus consists of two components: a vector specifying the desired relative angle and distance to the nearest neighbor, and a list of probabilities (one per formation condition) describing the probability that the respective condition correctly defines how an agent should position itself. In turn, the formation condition adopted then defines the answers that new agent will give to queries from future agents joining the formation.

For example, in a V formation, each agent requires the information shown in Fig. 1. Each row represents one formation condition, and the vector information used will depend on the desired size and spread of the formation, and the size of the robots. The probabilities in each formation condition represent the information that will be imparted to a querying agent, advising it as to which formation condition it should likely follow, given the formation condition the encountered neighbor is following. For example, if the encountered neighbor is on the left side of a V formation, the encountering

Condition	Name	Angle (degrees)	Distance (metres)	P1	P2	P3
1	Right	30	2	1	0	0
2	Left	-30	2	0	1	0
3	Center	–	–	0.5	0.5	0
4	Null	–	–	0	0	1

Fig. 1. Formation conditions for a V formation

agent must be also, without exception: the probability of following the same formation condition is one, and the others, zero. In a V formation, the only condition that offers an alternative is that of following the central position, since an agent following this could take either side. While this formation is simple, other formations (e.g. a diamond) offer more choice points, since the diamond will branch back in at a given position as well. In addition, there is the possibility of an implicit extra (*null*) formation condition in any formation, that is followed when no information is available as to the condition that should be followed. In a V formation, this can be defined with probabilities 0, 0, and 1 respectively, allowing a new formation to be formed with that agent occupying the center. Thus, each formation can be given a logical starting point. If at any time, an agent has no visible neighbors, it will revert to this state.

For the current implementation, we consider only fixed values for probabilities and vector components, as opposed to those where these values can be defined as a function of those values in a neighbor. This means that each agent potentially involved in a formation must share the same table of formation conditions.

Unlike some other approaches, ours does not assume that each agent will always be able to uniquely identify and address all others in communication. An agent that cannot uniquely identify others is limited in its ability to participate in the formation, as it cannot direct communications without an ID. Even if it could, it would not be able to precisely determine the physical origin of the response. All messages in the system, however, contain the identity of the sender, so it is possible to reply to others who initiate communication. We make use of this fact by using the *capability* message. Whenever an agent that can determine identifiers encounters a new neighbor, it sends a capability message, asking for the other agent's sensing capabilities. The original sender then uses this knowledge of capabilities to transmit useful information to that agent: in this case, a listing of all other agents that should be visible, and their IDs and positions relative to the receiving agent. The receiving agent can then use this information to its ability to communicate and participate in the formation.

When a new agent enters the vicinity of the formation, it first locates its nearest neighbor. It then attempts to communicate with this neighbor. It can only do so if it knows the ID of that agent, since all communication is directed. If communication is successful, that agent will respond with a description of the likely alternatives for the new agent in the formation, and the agent can select one of these. If an agent cannot communicate with its nearest neighbor, it selects a random formation condition, which may cause a local aberration in the overall formation, but still allows others to build an overall approximation of the intended formation. No prior knowledge of other agents beyond the common knowledge of formations is required, and an agent need not know anything ahead of time about the size of the formation or the other agents involved.

Because an agent can find a neighbor at any given time, new agents can similarly be added in an *ad hoc* manner. Similarly, failure recovery can easily be handled within the agents themselves. When an agent fails to the point where it stops all movement, one of two things will happen: the following agent will collide (or detect a collision with) the agent it was following, or the following agent will stop as well, and no movement on its part will occur. Provided these two conditions are covered in an implementation, failure recovery is assured. An agent must view the lack of movement as not contributing toward its goals, and the stationary former neighbor as an obstacle. This will then allow the agent to begin moving and looking for a new point to join the formation. The ability to deal with failure in this manner means that no member of a formation is ever essential. If any one member fails, it can either be replaced, as described above, or the formation will adapt to its absence. If the center position in a V fails, for example, we will ultimately have two diagonal lines, and the front of each of those will no longer be following any agent, violating their formation conditions. They will both revert to the null formation condition, causing them to recognize themselves as the new center positions. They will then act independently, and will likely merge at some point.

4   Implementation

While the previous section described the overall operation of our approach to formation control, any implementation of this approach requires consideration of agent abilities and the architecture through which agents are designed. This section briefly overviews the decisions we made for the implementation used to examine the performance of the approach (though other forms of implementation are certainly possible). Because Player/Stage was chosen as a platform for evaluating our approach, largely for reasons of control, we employed agents that were easily constructed using available components in Player/Stage [4]. These are simulated Pioneer 2DX robots, using laser scanners with fiducial tracking.

   Given that the core ability to form good formations is the ability to communicate directly with others, we want to examine the ability to form adequate formations using groups of agents with the ability to sense the IDs of other agents (i.e. ultimately have the ability to query them) and those that cannot. To achieve this in Player/Stage, we employ on a subset of the population a laser scanner that can read a fiducial marker attached to an agent. These form the agents that can query an ID and communicate. All agents also have sensors that can determine the distance and angle to another agent or obstacle, independent of determining identity.

   In order to simulate movement towards a goal, the agents are given the ability to self-localize. This information is used to simplify movement towards a goal position. In a real-world setup, this could be replaced by some sort of goal marker, or a distributed path planning system. For the purposes of this research, we will simply assume that some way of agreeing on a goal position exists. This localization information is used only to simplify the selection of a common goal position.

4.1   Agent Behaviors

The agents used in our implementation are behaviour-based, which involve a set of interacting weighted behaviours that ultimately determine the control values for the agent. This is also similar to the agents employed in [2]. Our agents employ three behaviors:

> Keep Formation
> Go to Goal
> Avoid Obstacles

The *Keep Formation* behavior is the heart of the approach. It computes the agent's desired relative position (in terms of a distance and angle vector) to a neighbor and compares it to the current relative position of that neighbor. The difference between these two vectors is the output of the behaviour. This generates a vector which always points towards the location in space at which the agent satisfies its chosen formation condition. The magnitude of this vector, and thus the weight it occupies in the agent's decision for movement, depends on the degree to which the agent is out of formation.

The *Go to Goal* behavior is extremely simple. It simply results in a vector, pointing towards the desired final destination of the formation. The magnitude of this vector is constant, reflecting a constant desire to move toward a goal (which is ultimately affected by the blending of the other two behaviours).

The *Avoid Obstacles* behaviour calculates a vector intended to direct the agent away from nearby obstacles, with a magnitude relative to the inverse square of the current distance to that obstacle. The only obstacles considered by this calculation are those that are visible, and within a minimum distance. This is similar to the technique used by [2] for obstacle avoidance, and common in many behaviour-based agents. The primary difference between this work and [2] is that we do not consider other agents separately from environmental obstacles. This is what causes a failed agent to be viewed as an obstacle and avoided (thus allowing an agent to separate itself from a dead neighbor and later re-join a formation).

One positive aspect of a behaviour-based approach is the ease of extending the capabilities of the agent. For example, if we were to adapt this technique to a team of searching robots, we may want to add a behaviour that would attract agents to any visible targets.

Similar to the techniques described in [2], the above behaviors each generate a vector, indicating the direction that this single behavior would cause us to go. These vectors are then scaled according to pre-defined parameters. For the purposes of this work, we used a scale factor of 2 for the *Go To Goal* behavior, 1 for the *Keep Formation* behavior and 3 for the *Avoid Obstacles* behaviour. These values were determined by trial and error, and are not necessarily optimal. The vector sum of these behaviors is then used to calculate the heading and speed of the agent.

Communication is handled outside of the behaviour-based architecture. In order to more accurately simulate real robots, direct inter-agent communication will be used. Both types of communication will have a limited range. Yoshida et al. [11] demonstrated the feasibility of using local communication in a formation control domain. They also referenced several results, suggesting that beyond approximately 10 agents, global communication would no longer be feasible.

Because Player/Stage is a simulated environment, it would be easy to have communication go across an unrealistic range. To ensure greater correspondence with the physical world, all messages are passed through a communication server. This server tracks the absolute positions of agents and uses these to filter messages, only delivering those to agents in range. It will also allow us to experiment in future with greater communication faults and difficulties. This server also serves as a convenient place to do tracking of statistics, as it has access to the absolute positions and formation conditions of all agents.

Five types of messages are implemented: *Formation Request, State, Capability, Neighbor,* and *Heartbeat* messages. All messages contain the ID of the sender, to allow for replies even if the recipient cannot identify the sender through perception.

The *Formation Request* message is sent when an agent encounters a new neighbor, and is intended to elicit the formation condition probabilities from that agent, as described in Section 3. If the ID of the neighbor is unknown, this message can not be sent. In response, the neighbor sends a *State Message*, containing the formation condition probabilities described in Section 3.

Upon encountering an agent whose sensing capabilities are unknown, an agent will send a *Capability* message. An agent receiving such a message responds with a description of its sensing capabilities.

If an agent is capable of sensing unique identifiers, it can share this information by sending out *Neighbor* messages. These messages are directed at a target agent, which lacks the ability to sense unique identifiers. A neighbor message contains the polar coordinates of a sensed agent, relative to the target agent.

A *Heartbeat* message is sent from an agent to its neighbor(s) at regular intervals. This is done to help agents to track if they have unsatisfied formation conditions. A lack of a heartbeat for an established period of time indicates that a neighbor is no longer present.

4.2   Evaluation

The approach as described above was implemented using the Player/Stage [4] simulation package. Agents were modeled as Pioneer 2DX robots as described above. Player/Stage's *fakelocalize* package was used to give absolute coordinates of the agents for tracking purposes. We ran a series of trials to examine the performance of the approach in general, and to examine the effect of local communication and the number of agents that could perceive the IDs of others. For a basis of comparison, we considered two measurements of error, Error measurements were taken by a human observer, at the first point in time when every agent in the system, was at the correct relative position and angle to satisfy any formation condition. We define a local error to be an agent following a formation condition that has zero probability, given the actual condition of its neighbor. We define a global error to be a measure of difference from the ideal formation of n agents. This can be determined by finding and counting the largest group of agents that are in positions consistent with the ideal formation, and subtracting this number from the total size of the formation.

Initial test runs showed that if the goal-seeking behavior was not strong enough, agents could deadlock, by each following one another. Increasing the weight of the goal-seeking behavior corrects this, by moving one agent out of the field of view of the other. A better solution for a future implementation would be some sort of negotiation when an agent chooses a neighbor. This could be skipped in a case where the ID of the neighbor is unknown.

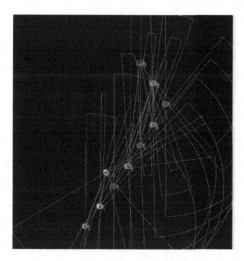

Fig. 2. An example of a mostly correct V formation achieved by our system

It was also noticed, as one might expect, that when the desired formation distance was close to the obstacle avoidance distance, formation members would drift out of formation. This often led to changes in the nearest neighbor over time.

Our first trial consisted of a line formation with five members. Varying the number of agents able to sense IDs had no noticeable impact on the time needed to establish the line formation. Times ranged between 58 and 64 seconds, with no discernable pattern. This makes sense, as the only piece of information communicated by agents is their current formation condition, and there is only one condition in this formation. It also goes without saying that there were no errors, as the line formation has only a single condition.

Next, we examined ten agents in a V formation. This formation is more interesting, as it has three formation conditions. each condition describing the arms of the V is valid if the neighbor shares that condition (as per Fig. 1). In the ideal V formation, there is a single agent in the central position that is a neighbor to two agents. Qualitatively, agents who can sense IDs do generate more straight line formations, where the group is reduced to one half of the V. This is likely due to the fact that agents are more likely to communicate with agents other than the one in front. The results suggest that the number of local errors increases as the number of agents who can sense IDs decreases. This makes sense, as agents who cannor sense IDs have limited communication abilities. The results also suggest that there is a critical number of agents who cannot sense IDs, beyond which the number of relative errors is roughly constant. The results of these trials are displayed in Tables 1 and 2.

We should also note that the formations established with no ID sensing were more prone to sudden change, since agents can't uniquely identify one-another without the help of an observer. This inability to identify one-another makes them more-likely to re-calculate their desired position in the formation. These re-calculations tend to result in formations where members shift around. The ability to uniquely identify others reduces this occurrence in the other two sets of trials.

Table 1. Number of global errors in the formation with ten agents

Number of agents who can sense Ids	Trial 1	Trial 2	Trial 3	Trial 4
10	1	4	4	4
5	4	4	2	2
0	3	4	4	3

Table 2. Number of local errors in the formation with ten agents

Number of agents who can sense Ids	Trial 1	Trial 2	Trial 3	Trial 4
10	0	0	0	0
5	2	0	1	2
0	2	1	1	2

5 Discussion

We are currently working on an evaluation of this approach using a larger team of physical robots. In order to support a large team in a small area, we are using twenty Citizen Micro-robots (Fig. 3, left), each approximately an inch in size. These operate in a mixed reality environment: the robots form their own physical reality, along with any other objects that are introduced, and a virtual reality is provided by running the robots on a large horizontally-mounted monitor (Fig. 3, right), allowing a global vision system to perceive both physical and virtual elements, and precisely track the movements of robots. In prior work [12], we have shown that this approach allows better control and repeatability of experiments while allowing large numbers of small robots to operate in a physical environment. Here, this approach will allow us to substitute human judgement with computer-vision based tracking to examine the accuracy of the formations and their adaptivity over time, as well as generate random obstacles and automatically track collisions between these and robots moving in formation. The micro-robots contain no laser scanners and currently have no local vision. Limited local vision, differing between types of agents, can be provided by restricting the viewpoint of the agents in our global vision system.

One limitation of our current approach is that we consider only formation conditions where the neighbor's position is a fixed value. An interesting extension would be to allow the neighbor's position to be based on a function instead. For example, we could vary the desired angle as distance between agents changes, and create curved formations.

Some positions in a formation are more important than others: in particular, there are situations where agents may be required to share a common neighbor. One potential mechanism to deal with this is the use of *mandatory* formation conditions. In a mandatory formation condition, specific neighbors are tagged as mandatory, and if there no neighbor(s) satisfy the condition, an agent will send a message to all neighbors it can identify, requesting that these conditions be filled. Upon receiving this message, an agent can decide to fill the request, or pass it along to its neighbors. An agent will only choose to fill the request if it cannot be passed along further.

Fig. 3. Left, Citizen Micro-robots; Right, a Mixed Reality application

Another possible addition would be to make an agent's state depend not only on the state of its nearest neighbor, but also an any known agents already following it. This would potentially resolve some of the issues associated with branching formations. For example, if the leading agent in a V formation has a left follower, but no right follower, it should be considered to be in a different state than if it has a left follower, but no right follower. This extension, combined with the mandatory conditions described above could help to overcome some of the difficulties that this system has in reducing the number of global errors.

In this paper, we have described an approach to heuristic formation control in groups of agents with different types of sensors, and have described an implementation using ten agents. Since the approach does not rely on each agent knowing the identity of all others, the approach is robust to agent failure and adapts to adding new agents as well. Though our current evaluation has not examined scalability, we argue that this approach should scale very well, because there are no communications bottlenecks, and the per-agent processing is not strongly related to the total number of agents in the system. In fact, we expect that results with very large numbers of agents will yield better results, as a larger number of agents increases the likelihood that all of the different conditions of the formation will be met. The adaptability and scalability should make this approach one that is useful for large numbers of agents, and environments such as search-and-rescue, where failure and agent replacement is not only possible, but expected.

References

1. Dimock, G.A., Selig, M.S.: The aerodynamic benefits of self-organization in bird flocks. In: Proceedings of the 41st AIAA Aerospace Sciences Meeting, Reno, NV (January 2003)
2. Balch, T., Arkin, R.: Behavior-based formation control for multirobot teams. IEEE Transactions on Robotics and Automation 14, 926–939 (1998)
3. Sugawara, K., Tanigawa, H., Kosuge, K., Hayakawa, Y., Mizuguchi, T., Sano, M.: Collective motion and formation in simple interacting robots. In: Proceedings of the 2006 IEEE International Conference on Robots and Systems (IROS), Beijing, China, October 2006, pp. 1062–1067 (2006)

4. Gerkey, B., Vaughan, R., Stoy, K., Howard, A., Sukhatme, G., Mataric, M.: Most valuable player: a robot device server for distributed control. In: Proceedings of the 2001 IEEE/RSJ International Conference on Intelligent Robots and Systems, November 2001, vol. 3, pp. 1226–1231 (2001)
5. Yamaguchi, H.: Adaptive formation control for distributed autonomous mobile robot groups. In: Proceedings of the 1997 IEEE International Conference on Robotics and Automation, April 1997, vol. 3, pp. 2300–2305 (1997)
6. Fredslund, J., Mataric, M.: A general algorithm for robot formations using local sensing and minimal communication. IEEE Transactions on Robotics and Automation 18(5), 837–846 (2002)
7. Howard, A., Parker, L.E., Sukhatme, G.S.: Experiments with a large heterogeneous mobile robot team: Exploration, mapping, deployment and detection. The International Journal of Robotics Research 1(5-6), 431–447 (2006)
8. Hattenberger, G., Lacroix, S., Alami, R.: Formation flight: Evaluation of autonomous configuration control algorithms. In: Proceedings of the 2007 IEEE/RSJ International Conference on Intelligent Robots and Systems (IROS), San Jose, USA, October 2007, pp. 2628–2633 (2007)
9. Spears, W., Gordon, D.: Using artificial physics to control agents. In: Proceedings of the 1999 International Conference on Information Intelligence and Systems, Bethesda, MD, October 1999, pp. 281–288 (1999)
10. Lee, G., Chong, N.Y.: Adaptive self-configurable robot swarms based on local interactions. In: Proceedings of the 2007 IEEE/RSJ International Conference on Intelligent Robots and Systems (IROS), San Jose, USA, October 2007, pp. 4182–4187 (2007)
11. Yoshida, E., Arai, T., Ota, J., Miki, T.: Effect of grouping in local communication system of multiple mobilerobots. In: Proceedings of the 1994 IEEE/RSJ/GI International Conference on Intelligent Robots and Systems (IROS), Munich, Germany, September 1994, vol. 2, pp. 805–815 (1994)
12. Anderson, J., Baltes, J., Tu, K.Y.: Improving robotics competitions for real-world evaluation of AI. In: Proceedings of the AAAI Spring Symposium on Experimental Design for Real-World Systems. AAAI Spring Symposium Series, Stanford, CA (March 2009)

Cooperative Multi-robot Map Merging Using Fast-SLAM

N. Ergin Özkucur and H. Levent Akın

Boğaziçi University, Department of Computer Engineering,
Artificial Intelligence Laboratory, 34342 Istanbul, Turkey
{nezih.ozkucur,akin}@boun.edu.tr

Abstract. Multi-robot map merging is an essential task for cooperative robot navigation. In the realistic case, the robots do not know the initial positions of the others and this adds extra challenges to the problem. Some approaches search transformation parameters using the local maps and some approaches assume the robots will observe each other and use robot to robot observations. This work extends a previous work which is based on EKF-SLAM to the Fast-SLAM algorithm. The robots can observe each other and non-unique landmarks using visual sensors and merge maps by propagating uncertainty. Another contribution is the calibration of noise parameters with supervised data using the Evolutionary Strategies method. The developed algorithms are tested in both simulated and real robot experiments and the improvements and applicability of the developed methods are shown with the results.

1 Introduction

One of the problems in robot navigation is the simultaneous localization and mapping problem (SLAM). The problem addresses a robot generating a map of an unknown environment and localizing itself in this map. The RoboCup @Home and Rescue [1] leagues already require exploration and map building for the high level planning tasks. On the other hand, localization in the soccer leagues like Standard Platform and Middle Size gets harder due to decreasing number of unique landmarks, so the problem moves towards the SLAM problem. Since the environment is unknown, landmark observations do not include identity information. This ambiguity adds an extra challenge and is called the *data association problem*. In multi-robot systems, the cooperative map building task introduces additional challenges. The robots should transform and merge their own maps and resolve ambiguities. The problem attracts researchers because the solution provides more autonomy to robots and allows them to operate in more realistic application domains.

The single-robot SLAM problem is a more or less solved problem. In the EKF-SLAM algorithm [2,3], the landmark positions and the robot pose (position and orientation) form an augmented state where the belief state is represented as a Gaussian distribution and is updated using Extended Kalman Filter (EKF) [2]. Another well-known solution method called Fast-SLAM [4,5] uses the fact that

J. Baltes et al. (Eds.): RoboCup 2009, LNAI 5949, pp. 449–460, 2010.

given the robot pose, the landmark positions are independent. The belief state in this case is a set of particles where each particle is a robot pose and the associated map hypothesis. In both algorithms, the uncertainty in the odometry and the observations is assumed to have a Gaussian distribution [6]. The performance of the methods highly depends on the parameters of the actual noise distributions, which are generally unknown. In [7], the problem is solved with the least square approach and in [8], the optimal parameters are found with Expectation Maximization. In this paper we propose employing Evolutionary Strategies to search for the optimal parameter set, using the robot's ground truth position information.

In multi-robot systems, current solution methods can be categorized with respect to their assumptions. The most simplistic assumption is that the robots know their initial positions. In this case, the single-robot solutions can easily be extended for the cooperative case [9]. In more realistic cases, the robots do not know or observe the positions of others. In [10,11,12,13], the transformation parameters are searched with different heuristics using only the local maps of the robots. In [14,15], the idea is about localizing the robots within the maps of the other robots and finding the transformation hypothesis. In another assumption category, the robots can observe each other and are expected to meet at some point. In this case, the problem reduces to finding suitable transformation parameters using the robot to robot observations. In [9], the robots record their observation history and provide it to the other robots when they meet, so that the other robots can use the history to build the map of unexplored areas using Fast-SLAM algorithm. In [16], similar to the previous method, the transformation parameters are found with robot to robot observations, and instead of using observation history, current map estimations are merged.

This work is an extension of the map merging algorithm based on EKF-SLAM in [16] to the Fast-SLAM algorithm. The map merging is performed when robots meet using the robot to robot observations. However, uncertainty propagation in Fast-SLAM differs from the EKF-SLAM case because each particle has its own map estimation instead of single map estimation. In [9], this problem is addressed differently by recording the observation history. In this paper, we exploit the Markovian assumption of state representation and merge the most recent map estimations.

The rest of this paper is organized as follows. In Section 2, our methodology and assumptions are detailed. In Section 3, the experiment setups in both simulation and real world are explained and results are discussed. Finally Section 4 summarizes and concludes our work and points out some possible future work.

2   Proposed Approach

2.1   Map-Merging

For the simplicity of illustration of the methods, in the multi-robot case, we assume that there are two robots, exploring some part of the environment and eventually meet at some point where they can observe each other. When they meet, they inform each other and share knowledge to merge the map of the other robot with their own-map. In the map-merging case, we only focus on the instant where a robot receives the estimation of the other robot and since our method is distributed, we will consider the situation from the point of view of only one robot.

The message from the other robot includes:

– M_{other}: the map estimation of the other robot. Each entity $m_{other,i}$ contains the position $p_{other,i}$ and a 2x2 covariance matrix $\Sigma_{other,i}$ of the ith landmark.
– p_{other}: the pose of the other robot. Note that p_{other} is the other robots pose and $p_{other,i}$ is the position of the ith landmark of other robot.
– $z_{other,self} = \{l_{other,self}, \theta_{other,self}\}$ is the observation of the other robot to self robot.

In the map estimation M_{other}, the correlation between the landmarks are omitted except the 2x2 covariances of landmark positions. Extracting this information from the EKF-SLAM is trivial, however in the Fast-SLAM, we take the weighted mean of the map estimation of particles with the importance weight.

When the robot receives the message from the other robot, it first calculates the transformation matrix between the coordinate frames:

$$T = \begin{bmatrix} \cos(tr_\theta) & -\sin(tr_\theta) & tr_x \\ \sin(tr_\theta) & \cos(tr_\theta) & tr_y \\ 0 & 0 & 1 \end{bmatrix} \tag{1}$$

which has the translation parameters $\{tr_x, tr_y\}$ and the rotation parameter tr_θ. Figure 1 illustrates the geometric configuration of coordinate frames and parameters in the information sharing step. Calculating the transformation parameters is an analytic geometry problem so it is skipped for simplicity. With the transformation matrix, each entity in the incoming map is transformed with the following equations:

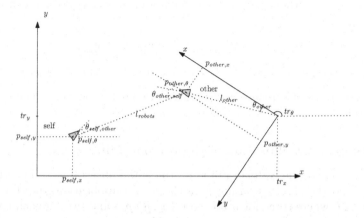

Fig. 1. The configuration and parameters when robots observe each other

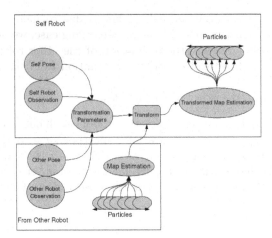

Fig. 2. Diagram of multi-robot map merging algorithm with the Fast-SLAM method

$$\begin{bmatrix} p_{other,x,i}' \\ p_{other,y,i}' \\ 1 \end{bmatrix} = T \begin{bmatrix} p_{other,x,i} \\ p_{other,y,i} \\ 1 \end{bmatrix} \tag{2}$$

$$\Sigma_{other,i}' = T^T \Sigma_{other,i} T \tag{3}$$

After transforming, the incoming map is merged into the map of each particle. Figure 2 gives the summary of the transformation and the merging operation in the Fast-SLAM algorithm. If N is the number of particles and M is the number of landmarks, the complexity of the merging step is $O(NM^2)$ with a naive implementation.

Nearest neighbor method is used to find duplicate landmarks when merging the map of the other robot with a single particle's map. If a landmark is a new landmark, it simply is added to the map. If the landmark is also known by itself, the other robot's estimation is considered as an evidence and the resulting state is calculated as:

$$\Sigma_{merged} = \Sigma_{self} - \Sigma_{self}\left[\Sigma_{self} + \Sigma_{other}'\right]^{-1}\Sigma_{self} \tag{4}$$

$$p_{merged} = p_{self} + \Sigma_{self}\left[\Sigma_{self} + \Sigma_{other}'\right]^{-1}(p_{other}' - p_{self}) \tag{5}$$

The merging of two Gaussians is displayed in Figure 3. Note that the uncertainty decreases in the resulting Gaussian.

2.2 Parameter Calibration of the Kalman Filter

The parameters of the Kalman Filter are the initial uncertainty, odometry reading noise and the observation noise. In the SLAM application, the state vector size is $3 + 2L$ where three for robot pose $\{p_x, p_y, p_\theta\}$ and two for each landmark $m_i = \{p_x^i, p_y^i\}$ in the map M with size L. The initial uncertainty has a covariance matrix P with size $(3 + 2L) \times (3 + 2L)$. An observation is represented as

(a) Two Gaussian   (b) Merged Gaussian

Fig. 3. Merging two Gaussian

the distance and orientation $z_i = \{l_i, \theta_i\}$ and therefore the observation noise covariance matrix Q has size 2×2. The odometry reading is the displacement of the robot in two dimensions and the change in the orientation $u = \{\Delta_x, \Delta_y, \Delta_\theta\}$ so the process noise covariance matrix R has size 3×3. These three covariance matrices form a very large parameter set, however we can reduce the number of parameters using some basic assumptions. The noise in all dimensions are assumed to be independent, so the correlation values become all zero. In addition, the noise on the x and y axes are assumed to be same. Finally, when we add the process noise or the landmark location, the parameter set becomes:

$$Q = \begin{pmatrix} \omega_1 & 0 \\ 0 & \omega_2 \end{pmatrix}, R = \begin{pmatrix} \omega_3 & 0 & 0 & 0 & 0 & \dots \\ 0 & \omega_3 & 0 & 0 & 0 & \dots \\ 0 & 0 & \omega_4 & 0 & 0 & \dots \\ 0 & 0 & 0 & \omega_5 & 0 & \dots \\ 0 & 0 & 0 & 0 & \omega_5 & \dots \\ . & . & . & . & . & . \\ . & . & . & . & . & . \end{pmatrix}, P = \begin{pmatrix} \omega_6 & 0 & 0 & 0 & \dots \\ 0 & \omega_6 & 0 & 0 & \dots \\ 0 & 0 & \omega_7 & 0 & \dots \\ 0 & 0 & 0 & \omega_8 & \dots \\ . & . & . & . & . \\ . & . & . & . & . \end{pmatrix} \quad (6)$$

where the vector $\omega = \{\omega_1, \omega_2, ..., \omega_8\}$ forms our actual parameter set to be estimated using Evolutionary Strategies [17]. A chromosome represents a possible combination of these values. An episode is executed with the parameter set of an individual to calculate the fitness value, which is the mean distance between the robot pose estimation and the actual (ground truth) pose. If the error grows very large, the episode is terminated to avoid unnecessary continuation and a small fitness value is returned with respect to the episode. As new generations are formed, they result with smaller errors on the robot position estimation. Note that only the ground truth position information of the robot is used as the supervised data.

In the Fast-SLAM method, the particles are sampled with a Gaussian distribution using the odometry readings. The importance weight calculations with the observations are also made with Gaussian likelihood function. For these reasons, the parameter set we estimated can be explicitly used for the Fast-SLAM algorithm. The only extra parameter is the number of particles which affects

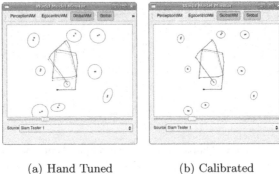

(a) Hand Tuned   (b) Calibrated

Fig. 4. Path comparison of the hand tuned and calibrated parameters in a single episode

the quality of the posterior distribution. In our method to select the number of particles, we perform successive experiments with increasing number of particles until performance convergences.

3   Experiments and Results

We used the Festo Robotino robot [18] as the hardware platform. It has omni-directional locomotion ability and a webcam which provides RGB images with dimensions 320x240 and 50 degrees field of view. We also equipped the robot with a URG laser range finder device [19] to generate occupancy grid map and to avoid obstacles. The laser range finder has 5 meters range and 270 degrees field of view, however physical placement of the device allows 140 degrees of field of view. We implemented the software with the *Player/Stage* framework [20], which includes a 2D simulator and provides the ability to test our algorithms on both simulation environment and real-world by only changing the configurations. In the simulation environment, ground truth positions are directly accessed. In the real world, we used an overhead camera system [21] to measure the global position of the robots.

3.1   Parameter Calibration Results

The first experiment is designed to compare the hand-tuned and the calibrated parameters in the simulation environment. The simulator provides perfect knowledge, however we inject Gausian noise to both observations and the odometry. The search for the optimal set is initiated from the hand tuned parameters. Figure 4 shows the result of the path and map estimations of both parameter sets. The straight line is the exact path, the dots are the estimated path, the cross marks are the exact location of the landmarks and the ellipses are the landmark estimations. For the simulation scenarios, the robots followed a predefined path.

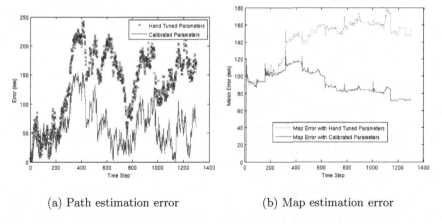

(a) Path estimation error   (b) Map estimation error

Fig. 5. Performance comparison of the hand tuned and calibrated parameters in a single episode

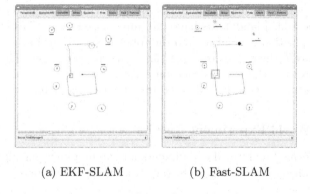

(a) EKF-SLAM   (b) Fast-SLAM

Fig. 6. Path and map estimation results in the simulated map-merging experiment

In Figure 5, the errors in landmark and path estimations are given. The path error plots show that the calibrated parameters bounded the odometric error more accurately. The map estimation error is the mean error of known landmark estimations. Another interesting result is that the accuracy of the map increases even if we discard the orientation and landmark errors in the fitness function. This shows that an accurate map estimation requires an accurate path estimation.

3.2   Simulated Experiment Results

In the first experiment of map merging, we demonstrated the applied methods for both the EKF-SLAM approach and the Fast-SLAM approach in the simulation environment. In Figure 6, the path and map estimations are given for one robot. The squared region in the path is the meeting point where map-merging

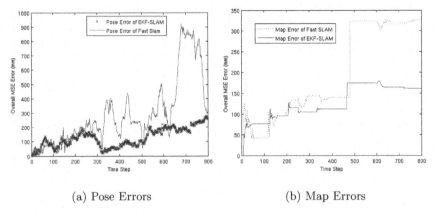

(a) Pose Errors   (b) Map Errors

Fig. 7. Pose and map errors for EKF-SLAM and Fast-SLAM methods in the simulated map-merging experiment

occurred. The overlined landmarks also exists in the incoming message, so they are merged, and underlined landmarks are added from the incoming message. Note that the added landmarks have the same biased error. This is the effect of the errors in position estimation and the robot observations in the map-merging step. In Figure 7, the path and map estimation errors are given. The biased error can be observed in the map estimation error. The error increases considerably when the landmarks with biased error are added to the map.

3.3   Real World Experiment Results

We also performed real world experiments for the map-merging algorithm using Fast-SLAM algorithm. In Figure 8, the experiment setup for the real-world is

Fig. 8. Experiment setup of the real-world map-merging experiment

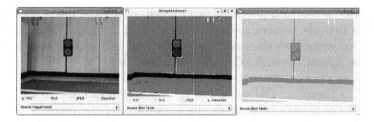

Fig. 9. Marker observation process. From left to right: raw image, color segmented image, and formed blobs.

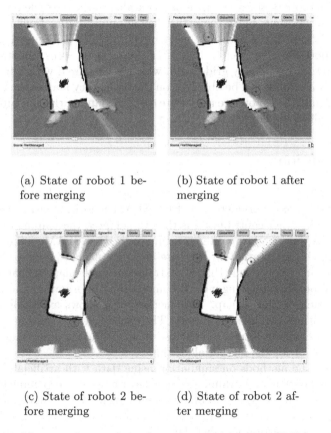

(a) State of robot 1 before merging

(b) State of robot 1 after merging

(c) State of robot 2 before merging

(d) State of robot 2 after merging

Fig. 10. Map-merging result in the real world with Fast-SLAM method

given. The robots can observe colored landmarks around the world and start facing opposite directions. They are also marked with colored paper in order to enable robot to robot observations.

Landmark Detection. We used two-colored markers as landmarks. The images taken from the camera are in RGB format. To recognize the regions in the markers, we segmented the color space into four colors using the Generalized Regression Neural Network (GRNN) [22]. After segmenting the pixels, markers are observed as regions of interested colors. The snapshots of these steps are given in Figure 9. The robot has monocular camera and to calculate the distance of the marker, we exploit the known size of the markers. We simply calibrated a non-linear function of region size which gives the distance of the marker.

The robots detect obstacles with the laser range finder in the direction of movement. The detected obstacles are represented as a line formed by laser readings. The robots always turn right as parallel to the obstacle line when they encounter an obstacle. Since they start facing opposite directions, they explore separate portions of the area and meet in the middle. In Figure 10, the states of both robots are given before and after the merging steps. As stated before, we also generate occupancy grid-world, which is visualized with intensity of the grids. The other robot can be observed on the occupancy map of the both robots. Before merging, the robots are not aware of all the landmarks, but after merging, they know all of the landmarks.

4   Conclusions

In this paper, we have two notable contributions to the literature. First, we calibrated the noise parameters of EKF-SLAM algorithm using Evolutionary Strategies. The other contribution is that we adapted the map merging method for EKF-SLAM method from the literature to the Fast-SLAM method. The main difficulty in this adaptation is the representation of multiple maps in the Fast-SLAM method. To overcome this, we extracted a single map estimation from the Fast-SLAM belief state along with the uncertainty information and merged this map with each particle's map. This method also allows execution of different SLAM algorithms on different robots. The exchanged map estimation between robots has a common format, so the estimator of the incoming map estimation is not important for the merger algorithms.

We tested the methods on simulation using data with artificially introduced noise for evaluation and training. The algorithms are also implemented and tested on autonomous robots in indoor environment. We showed that the multi-robot map-merging method can also be applied to the Fast-SLAM algorithm without loss of uncertainty knowledge.

The major obstacle between this work and a real life scalable multi-robot SLAM application is the assumption of unique markers in the robots. With non-unique robot markers, correspondence problem for robot observations makes the problem ore complicated and should be addressed.

Acknowledgments

This project is supported by Boğaziçi University Research Fund Project 06HA102 and Scientific and Technological Research Council of Turkey (TÜBİTAK) under grant 106E172 and 2228.

References

1. Robocup official site maintained by the RoboCup Federation (2009), http://www.robocup.org/
2. Welch, G., Bishop, G.: An introduction to the kalman filter. Technical report, Chapel Hill, NC, USA (1995)
3. Smith, R., Self, M., Cheeseman, P.: Estimating uncertain spatial relationships in robotics. In: Autonomous robot vehicles, pp. 167–193 (1990)
4. Montemerlo, M.: FastSLAM: A factored solution to the simultaneous localization and mapping problem with unknown data association. In: CMU Robotics Institute (2003)
5. Montemerlo, M., Thrun, S., Koller, D., Wegbreit, B.: FastSLAM 2.0: An improved particle filtering algorithm for simultaneous localization and mapping that provably converges. In: Gottlob, G., Walsh, T. (eds.) IJCIA, pp. 1151–1156. Morgan Kaufmann, San Francisco (2003)
6. Thrun, S., Burgard, W., Fox, D.: Probabilistic Robotics. MIT Press, Cambridge (2005)
7. Carew, B., Belanger, P.: Identification of optimum filter steady-state gain for systems with unknown noise covariances. IEEE Transactions on Automatic Control 18(6), 582–587 (1973)
8. Ghahramani, Z., Hinton, G.E.: Parameter estimation for linear dynamical systems. Technical report (1996)
9. Howard, A.: Multi-robot simultaneous localization and mapping using particle filters. Int. J. Rob. Res. 25(12), 1243–1256 (2006)
10. Thrun, S., Liu, Y.: Multi-robot slam with sparse extended information filers. I. J. Robotic Res. 15, 254–266 (2005)
11. Birk, A., Carpin, S.: Merging occupancy grid maps from multiple robots. Proceedings of the IEEE 94(7), 1384–1397 (2006)
12. Carpin, S.: Fast and accurate map merging for multi-robot systems. Auton. Robots 25(3), 305–316 (2008)
13. Huang, W.H., Beevers, K.R.: Topological map merging. Int. J. Rob. Res. 24(8), 601–613 (2005)
14. Konolige, K., Fox, D., Limketkai, B., Ko, J., Stewart, B.: Map merging for distributed robot navigation. In: Proceedings of 2003 IEEE/RSJ International Conference on Intelligent Robots and Systems, 2003 (IROS 2003), October 2003, vol. 1, pp. 212–217 (2003)
15. Fox, D., Konolige, K., Limketkai, B., Ko, J., Schulz, D., Stewart, B.: Distributed multi-robot exploration and mapping. In: Proceedings of the 2nd Canadian Conference on Computer and Robot Vision, 2005, pp. XV–XV (May 2005)
16. Zhou, X.S., Roumeliotis, S.I.: Multi-robot slam with unknown initial correspondence: The robot rendezvous case. In: 2006 IEEE/RSJ International Conference on Intelligent Robots and Systems, October 2006, pp. 1785–1792 (2006)

17. Beyer, H.-G.: The Theory of Evolution Strategies. Springer-Verlag, Heidelberg (2001)
18. Festo robotino,
    http://www.festo-didactic.com/int-en/learning-systems/
    education-and-research-robots-robotino/
19. Urg laser range finder,
    http://www.hokuyo-aut.jp/02sensor/07scanner/urg.html
20. Gerkey, B., Vaughan, R., Howard, A.: The player/stage project: Tools for multi-robot and distributed sensor systems. In: 11th International Conference on Advanced Robotics (ICAR 2003), Coimbra, Portugal (June 2003)
21. Kavaklıoğlu, C.: Developing a probabilistic post perception module for mobile robotics. Master's thesis, Boğaziçi University, Turkey (2009)
22. Alpaydın, E.: Machine Learning. MIT Press, Cambridge (2004)

Author Index